T0325867

PRINCIPLES OF SCATTERING AND TRANSPORT OF LIGHT

Light scattering is one of the most well-studied phenomena in nature. It occupies a central place in optical physics and plays a key role in multiple fields of science and engineering. This volume presents a comprehensive introduction to the subject. For the first time, the authors bring together in a self-contained and systematic manner the physical concepts and mathematical tools that are used in the modern theory of light scattering and transport, presenting them in a clear, accessible way. The power of these tools is demonstrated by a framework that links various aspects of the subject: scattering theory to radiative transport, radiative transport to diffusion, and field correlations to the statistics of speckle patterns. For graduate students and researchers in optical physics and optical engineering, this book is an invaluable resource on the interaction of light with complex media and the theory of light scattering in disordered and complex systems.

RÉMI CARMINATI is Professor of Physics at ESPCI Paris - PSL, before which he held a faculty position at École Centrale Paris. He was awarded the Fabry-de Gramont prize of the French Optical Society and is a Fellow of the Optical Society of America.

JOHN C. SCHOTLAND is Professor of Mathematics at Yale University. He has held faculty positions at the University of Pennsylvania and the University of Michigan, where he was the founding director of the Michigan Center for Applied and Interdisciplinary Mathematics.

PRINCIPLES OF SCATTERING AND TRANSPORT OF LIGHT

RÉMI CARMINATI
ESPCI Paris - PSL University

JOHN C. SCHOTLAND
Yale University

CAMBRIDGE
UNIVERSITY PRESS

University Printing House, Cambridge CB2 8BS, United Kingdom

One Liberty Plaza, 20th Floor, New York, NY 10006, USA

477 Williamstown Road, Port Melbourne, VIC 3207, Australia

314–321, 3rd Floor, Plot 3, Splendor Forum, Jasola District Centre, New Delhi – 110025, India

79 Anson Road, #06–04/06, Singapore 079906

Cambridge University Press is part of the University of Cambridge.

It furthers the University's mission by disseminating knowledge in the pursuit of education, learning, and research at the highest international levels of excellence.

www.cambridge.org
Information on this title: www.cambridge.org/9781107146938
DOI: 10.1017/9781316544693

First published 2021

A catalogue record for this publication is available from the British Library.

Library of Congress Cataloging-in-Publication Data
Names: Carminati, Rémi, author. | Schotland, John C., author.
Title: Principles of scattering and transport of light / Rémi Carminati, ESPCI Paris, John C. Schotland, Yale University.
Description: New York : Cambridge University Press, 2021. | Includes bibliographical references and index.
Identifiers: LCCN 2020053665 | ISBN 9781107146938 (hardback) | ISBN 9781316544693 (ebook)
Subjects: LCSH: Light–Scattering. | Radiative transfer.
Classification: LCC QC427.4 .C37 2021 | DDC 535/.43–dc23
LC record available at https://lccn.loc.gov/2020053665

ISBN 978-1-107-14693-8 Hardback

To our families

Emmanuelle, Lena and Thibaut

Helena, Sam, Marilyn and Nina

and our parents

Monique and René

Marilyn and Donald

Contents

Part V Speckle and Interference Phenomena

Foreword

Light has been studied and utilized since time immemorial. The effects of light scattering are seen in many places in nature. Until recently, much of the theory of light scattering was limited to either single scattering or to transport theory. In the 1980s, the discovery of universal conductance fluctuations signaled a revolution in condensed-matter physics. I remember that my colleagues working in optics were, at that moment, not very impressed by the discovery. "Condensed-matter physicists have rediscovered speckle," I heard regularly around me. It soon became clear those physicists working in optics were very wrong. The "speckles" discovered in condensed-matter physics were of a totally different nature than ordinary optical speckles. From that time on, the fields of optics and condensed-matter physics have been inspiring and stimulating each other. Many new phenomena were found or predicted, explained and applied. The field of light scattering had entered a new era.

The present book, *Principles of Scattering and Transport of Light*, fills a long-standing gap. The authors both won their spurs in the field of scattering and transport of light. I am very happy with its publication and note that many modern developments are covered in ample depth. The book is self-contained and very useful for independent study, but it could certainly also serve as a text for courses given to graduate students.

Why did it take so long for a book of this nature to appear? The most universal experiment in physics is a scattering experiment, but in fields such as high-energy physics and nuclear physics, it is almost the only type of experiment. Developing a quantum-mechanical scattering theory was therefore essential for the fields of nuclear and high-energy physics, resulting in quite a number of mature treatises on the subject. For a long while, it seemed that there was nothing more to be said about scattering theory, including the scattering of light. Notwithstanding the mathematical similarities between, for instance, elastic electron scattering and elastic light scattering, the differences have become so large that books on scattering theory not explicitly dealing with light scattering have become of limited value for workers studying light scattering alone. In addition, traditional books handle single scattering only. The interaction between matter and light is so strong that researchers working with light often are confronted with multiple scattering. In earlier days, the occurrence of multiple light scattering was seen as a nuisance, but nowadays it is looked upon as an area of opportunity, with high potential for applications. Finally, the field of (multiple) light scattering has its own book – and rightly so.

When students in my laboratory become interested in modern light scattering, they usually ask me to supply them with the title of a book that could be their guide to the field. To my shame, I always had to answer that such a monograph did not yet exist. Now I can point them to this book.

Ad Lagendijk
Professor of Physics, University of Twente
Emeritus Distinguished University Professor, University of Amsterdam

Preface

The best way to become acquainted with a subject is to write a book about it.

Benjamin Disraeli

In this book, we present a comprehensive introduction to the topic of scattering and transport of light. This is a classical subject that is still evolving, while occupying a central place in modern optical physics. At the same time, it plays a key role in multiple disciplines of applied science. Our approach is to begin with the basic theory of wave propagation and construct a coherent framework that connects various aspects of the subject: scattering theory to radiative transport, radiative transport to diffusion, and field correlations to the statistics of speckle patterns. We note that the selection of topics is strongly influenced by our views on the foundations of the field, which questions are fundamental and, quite evidently, our taste and interests. As a consequence, many important topics are omitted. These include Anderson localization, applications of random matrix theory to multiple scattering, non-Gaussian speckle correlations, inverse problems and imaging and effects due to nonclassical states of light, among others.

The book is an outgrowth of our teaching and is geared toward graduate students and researchers who wish to enter the field. Our goal is primarily pedagogical. We aim to develop physical ideas from first principles, in sufficient detail such that calculations can be easily followed and reproduced. Indeed, the book consists of short chapters, each of which may be envisioned as a lecture. The chapters are grouped into parts, emphasizing the logical structure of the subject matter. Each part concludes with a set of exercises at various levels of difficulty. Only basic references are provided, which take the form of an annotated bibliography. We apologize to those scientists whose work we may have failed to cite properly.

Scientific knowledge results from a long journey, dotted with personalities who directly or indirectly contributed to the emergence of this book. R.C. would like to thank Jean-Jacques Greffet, who introduced him to light scattering and many other aspects of modern optics, and Manuel Nieto-Vesperinas, Claude Boccara and Mathias Fink, who reinforced his interest in the field of waves in complex media. He is also grateful to many colleagues, collaborators, students, postdocs and friends, in particular Florian Bigourdan, Matthieu Boffety, Emmanuel Bossy, Etienne Castanié, Alexandre Cazé, Aristide Dogariu, Rachid Elaloufi, Nikos Fayard, Luis Froufe, Antonio García Martín, Sylvain Gigan, Arthur Goetschy, William Guérin, Carsten Henkel, Karl Joulain, Robin Kaiser, Valentina Krachmalnicoff, Ad Lagendijk, Olivier Leseur, Philippe Réfrégier, Jorge Ripoll, Julien de Rosny, Juan José Sáenz, Riccardo Sapienza, Frank Scheffold, Patrick Sebbah, Anne Sentenac, Boris Shapiro, Sergey Skipetrov, Bart van Tiggelen, Arnaud Tourin, Kevin Vynck and Yannick De Wilde. His colleague and former student Romain Pierrat deserves a specific acknowledgment for his enthusiasm for wave physics and scientific computing, which makes him a key collaborator, and for designing the images on the book cover. Finally, he is indebted to Institut Langevin, ESPCI Paris and CNRS for constant support and for providing a stimulating scientific environment in which to work.

J.C.S. would like to thank his teachers Britton Chance and Jack Leigh, who introduced him to the field of light scattering, for their long-standing support. He is also grateful to his collaborators Simon Arridge, Guillaume Bal, Liliana Borcea, Scott Carney, Alexandre Cazé, Francis Chung, Charlie Epstein, Lucia Florescu, Anna Gilbert, Alex Govyadinov, John Haselgrove, Arnold Kim, Soren Konecky, Howard Levinson, Vladimir Lukic, Manabu Machida, Vadim Markel, Imran Mirza, Shari Moskow, Ted Norris, George Panasyuk, Hala Shehadeh, Jin Sun, Bruce Tromberg, Emil Wolf and Zhengmin Wang, from whom he learned much about the subject of this book. He would like to thank his graduate students Jeremy Hoskins, Joe Kraisler and Wei Li, for tirelessly reading early drafts of the book, and for their feedback and suggestions. He is obliged to his colleagues Anna Gilbert and Ted Norris, who steadfastly attended the lectures on which some of this book was based, for sharing their invaluable insights into the subjects treated herein. Finally, he is indebted to the NSF Division of Mathematical Sciences and Arje Nachman's program at AFOSR for their financial support of his research over the years.

The authors gratefully acknowledge the hospitality and support of Institut Langevin and ESPCI Paris, University of Michigan and Donostia International Physics Center, where much of this book was written. We are especially thankful to Ad Lagendijk for contributing the foreword. Frank Natterer's advice to "write what you know, not what you want to know" was instrumental in finishing the book. We would also like to express our gratitude to Simon Capelin and the staff of Cambridge University Press for their support, enthusiasm and patience throughout the duration of this project.

Finally, completing this book has meant many sacrifices by our families. We are most grateful to our wives, Emmanuelle and Helena, for their encouragement and understanding.

1

Introduction

A child of five could understand this. Fetch me a child of five.
Groucho Marx

The problem of light scattering and transport in disordered systems can be addressed from several points of view. Throughout this book, by *disordered* we mean nonperiodic systems, such as clouds, milk, paint and biological tissue. In this setting, by beginning from microscopic theory based on wave physics, one can construct a macroscopic theory for the propagation of light at large scales. This bottom-up approach is commonplace in mesoscopic physics of electronic transport. Alternatively, one can employ a macroscopic theory, based on a phenomenological diffusion picture, as a model for light transport. This top-down approach has been widely applied, especially in biomedical optics. Both viewpoints are treated by several excellent monographs or textbooks, which, most of the time, favor one approach or the other. The goal of this book is to connect the microscopic theory of light scattering to macroscopic transport theory, while explicating the study of intensity fluctuations (speckle). The emphasis is on basic principles and the foundations of the subject, without necessarily covering all of its most advanced aspects. In a nutshell, the central idea of the book is to treat light propagation in disordered systems as the study of waves in random media. Averaging of randomness at the microscopic level leads to deterministic behavior at the macroscopic level.

Disorder, Randomness and Averaging

A typical experiment in light scattering from a disordered medium is shown in Fig. 1.1. Here a laser beam propagates through an aqueous suspension of scattering particles (milk). In the single-scattering regime, the density of particles is small enough to ensure that light is not scattered more than once on average before exiting the medium. One observes a ballistic beam propagating along the incident direction, which is weakly attenuated by scattering. Upon increasing the density of

Fig. 1.1 Propagation of a laser beam through a glass of water containing a few drops of milk. The beam enters the medium on the left. The milk concentration increases from the left to the right images. Left: single-scattering regime. Middle: multiple-scattering regime. Right: diffusion regime. Courtesy of Nina Schotland.

particles, the ballistic beam disappears, and the intensity distribution is dominated by multiply-scattered diffuse light.

In this experiment, the particles undergo Brownian motion, and the resulting image can be understood as a time average over the motions of the particles. Since the positions of the particles are unknown, it will prove to be advantageous to regard them as random. The recorded image then corresponds to the average intensity $\langle I \rangle$. Here $\langle \cdots \rangle$ denotes the statistical average over realizations of the random medium. We have also made a crucial assumption regarding ergodicity, and have equated a time average with an ensemble average. The derivation of transport and diffusion equations for the average intensity is a major achievement of multiple-scattering theory; it is the main topic of Parts III and IV of the book.

It is important to note that treating a scattering medium as random is purely a matter of convenience. As in statistical physics, this allows one to predict observables that depend only on a small number of parameters, defined by suitable averages. Consider, for instance, a stack of several sheets of white paper. Each sheet has the same reflectance, although the microscopic structure of one sheet differs from another. That is, although every sheet has its own microstructure, the global reflectance, averaged over a large area, depends on a few statistical parameters. Radiative transport theory provides a method to compute this global reflectance, without knowledge of the microscopic structure of a given sheet of paper.

Returning to the experiment in Fig. 1.1, it is instructive to consider the complex amplitude of the optical field U scattered by one realization of the disordered medium as the sum of an average value and a fluctuation:

$$U = \langle U \rangle + \delta U \, , \tag{1.1}$$

where $\langle \delta U \rangle = 0$. Given that the intensity $I = |U|^2$, we have that

$$\langle I \rangle = |\langle U \rangle|^2 + \left\langle |\delta U|^2 \right\rangle . \tag{1.2}$$

Evidently, the intensity consists of two contributions. The first term on the right-hand side of (1.2) is the intensity of the average field. It coincides with the ballistic intensity described above. The second term corresponds to the intensity of the fluctuating part of the field. The image in the left panel in Fig. 1.1 is dominated by $|\langle U \rangle|^2$, while the image in the right panel is primarily $\langle |\delta U|^2 \rangle$. Describing these contributions to the average intensity is one of the main goals of radiative transport theory, as explained in Part III.

Intensity Fluctuations in Speckle Patterns

The intensity image produced by a specific realization of a disordered sample (a single sheet of paper) forms a speckle pattern , as shown in the left panel in Fig. 1.2. The intensity can be written as

$$I = \langle I \rangle + \delta I \,. \tag{1.3}$$

Upon averaging over many realizations of the disordered medium, the fluctuations δI average to zero, producing the diffuse light image shown in the right panel. Although transport theory describes the average intensity $\langle I \rangle$, the statistical characterization of the speckle pattern requires theoretical developments beyond transport theory. In particular, intensity fluctuations result from wave interferences, and a speckle pattern is the result of such interferences. Here we emphasize a crucial point. Despite the deterministic nature of a given speckle pattern, predicting its

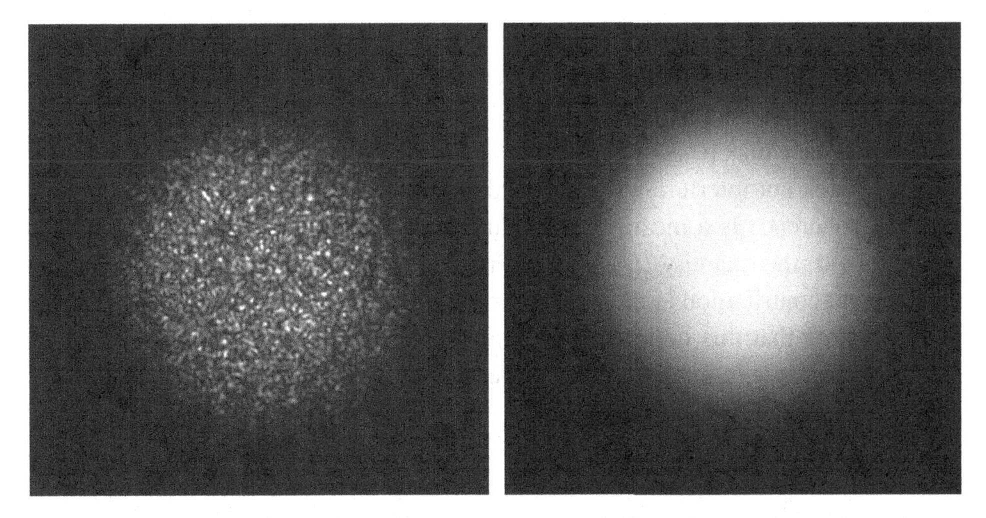

Fig. 1.2 Left: Speckle pattern produced by transmission of a laser beam through a static disordered medium. Right: Intensity distribution resulting from averaging over a large ensemble of speckle patterns. Courtesy of Ignacio Izeddin and Yannick De Wilde.

structure is not possible, since the scattering medium is generally unknown. To handle this problem, the statistical properties of an ensemble of speckle patterns (originating from different realizations of the disordered medium) are analyzed. The correlation functions $\langle \delta I \delta I' \rangle$, where δI and $\delta I'$ are intensity fluctuations at two different points or two different times, are an example of a statistical quantity of great practical interest. The study of statistical fluctuations in speckle patterns is an important part of the subject of light scattering in disordered media, the basics of which are treated in Part V.

Averaging and Coherence

Optical coherence is concerned with the characterization of statistical fluctuations of fields. Coherence theory was developed to describe the statistical fluctuations produced by randomness in primary sources. Indeed, even the light emitted by a laser is subject to random fluctuations due to spontaneous emission. In the theory of multiple scattering, it is often assumed that the incident light is fully coherent. Moreover, when only elastic scattering processes are considered, there is no loss of coherence of the scattered field. As a consequence, the scattered light comprising a speckle pattern is fully coherent. That is, there is no loss of coherence resulting from multiple scattering; rather, decoherence is due to averaging over disorder.

Let U_0 denote the incident field, which is taken to be deterministic, and introduce the scattered field $U_s = U - U_0$. The average intensity is then given by

$$\langle I \rangle = |U_0 + \langle U_s \rangle|^2 + \langle |\delta U_s|^2 \rangle , \tag{1.4}$$

where we made use of the fact that $\langle \delta U_s \rangle = 0$. The first term on the right-hand side describes the interference of the average scattered field with the incident field. This destructive interference is responsible for the extinction of the ballistic intensity, namely the fact that $|U_0 + \langle U_s \rangle| \ll |U_0|$. In addition, the fluctuating scattered field δU_s does not interfere with the incident field after averaging. In view of the fact that coherence is a measure of the interference of the average field with the incident field, the quantity $\langle U \rangle$ is sometimes referred to as the coherent field, or the coherent contribution to the field. Likewise, δU is referred to as the incoherent field. We note that this notion of coherence differs from the conventional use of the term coherence in optics, and we prefer the terms *average field* and *fluctuating field* to avoid confusion.

Scalar versus Vector Waves

Many aspects of multiple scattering are captured by a theory of scalar waves. Nevertheless, there are features of electromagnetic scattering in which the vector nature of the electromagnetic field must be taken into account. Depolarization of polarized light by multiple scattering and the influence of quasi-static near-fields on speckle patterns observed at subwavelength distances are two examples. Moreover, applications to imaging and communications, among others, make use of polarization degrees of freedom. The essential features of the scattering of vector waves, including radiative transport and speckle correlations are presented in Part VI.

Structure of This Book

The book is composed of six parts, each consisting of short chapters. Part I (*Wave Optics*) is an introduction to physical optics, including the scalar theory of the optical field, reflection and transmission, Green's functions, plane-wave expansions, diffraction, and a brief presentation of coherence theory. Part II (*Scattering of Waves*) is a general presentation of scattering theory, that includes a discussion of integral equations, Born series and multiple scattering, optical theorem and principle of reciprocity, and the analysis of simple scattering systems. Part III (*Wave Transport*) is concerned with multiple scattering of waves in random media, and is the cornerstone of the book. It includes the introduction of the basic equations governing the propagation of the average field and intensity, and the derivation of the radiative transport equation (RTE). The role of spatial correlations is discussed, as is the concept of energy velocity. Part IV (*Radiative Transport and Diffusion*) is dedicated to the study of the RTE. It includes an analysis of the mathematical structure of the RTE and some of its solutions, the derivation of the diffusion approximation, and the treatment of diffusive transport in simple geometries. Diffuse optics, including the scattering of diffuse waves is also examined in some detail. We note that this part of the book can be read independently by those mainly interested in the applications of the RTE, for example to biomedical optics. Part V (*Speckle and Interference Phenomena*) addresses the problem of intensity fluctuations and correlations in speckle patterns. Emphasis is given to the study of speckles in the framework of Gaussian statistics. This part also includes a description of other interference phenomena, such as coherent backscattering and temporal fluctuations in dynamic speckles. Finally, Part VI (*Electromagnetic Waves and Near-Field Scattering*) introduces the essential features of multiple scattering of vector electromagnetic waves. This part is complementary to the content of Parts III, IV and V, and emphasizes those aspects of the subject in which the behavior of electromagnetic waves differs from the scalar case.

How to Use This Book

The book was designed to be read by students and working scientists who have a basic knowledge of electromagnetic theory and optical physics. In the

United States, this would correspond to second-year graduate students in physics, applied mathematics or engineering. In France and other European countries, it is appropriate for students in a master's degree program. In either case, a variety of courses can be taught from the book. An introductory course would begin with Part I, and cover Parts II, III and selected portions of IV. A more advanced course could begin with Part III, and cover Parts IV, V and VI. Other variations are also possible. A course for students whose primary interest lies in biomedical optics, could cover portions of Part III, treat Part IV in detail, and discuss some aspects of Part V. All three variants have been tested in the classroom.

As noted in the Preface, there is a broad landscape of works on the subject of scattering and transport of light that complement this book. The books by van de Hulst, and Bohren and Huffman are classic treatises on scattering from particles. The book by Ishimaru takes an engineering approach to scattering and transport, as does the series by Tsang, Kong and coworkers. The books by Apresyan and Kravtsov; Rytov, Kravtsov, and Tatarskii; and Mishchenko present the viewpoint of the Russian school, which has a long and distinguished history. The more recent volume by Akkermans and Montambaux is concerned with electronic and photonic transport from a mesoscopic physics point of view, as is the book by Sheng. Case and Zweifel's book presents a timeless account of radiative transport theory. Finally, the text by Ripoll is an authoritative treatment of biomedical applications.

References and Additional Reading

K.M. Case and P.F. Zweifel, *Linear Transport Theory* (Addison-Wesley, Reading, 1967).

H.C. van de Hulst, *Light Scattering by Small Particles* (Dover, New York, 1981).

C.F. Bohren and D.R. Human, *Absorption and Scattering of Light by Small Particles* (Wiley, New York, 1983).

S.M. Rytov, Y.A. Kravtsov and V.I. Tatarskii, *Principles of Statistical Radio-Physics* (Springer, Berlin, 1989), Vol. 4.

P. Sheng, *Introduction to Wave Scattering, Localization, and Mesoscopic Phenomena* (Academic Press, San Diego, 1995).

L.A. Apresyan and Y.A. Kravtsov, *Radiation Transfer. Statistical and Wave Aspects* (Gordon and Breach, 1996).

A. Ishimaru, *Wave Propagation and Scattering in Random Media* (IEEE Press, Piscataway, 1997).

L. Tsang, J.A. Kong and K.H. Ding, *Scattering of Electromagnetic Waves: Theories and Applications* (Wiley, New York, 2000); L. Tsang, J.A. Kong, K.H. Ding and C.O. Ao, *Scattering of Electromagnetic Waves: Numerical Simulations* (Wiley, New York, 2001); L. Tsang and J.A. Kong, *Scattering of Electromagnetic Waves: Advanced Topics* (Wiley, New York, 2001).

E. Akkermans and G. Montambaux, *Mesoscopic Physics of Electrons and Photons* (Cambridge University Press, Cambridge, 2007).

J. Ripoll Lorenzo, *Principles of Diffuse Light Propagation* (World Scientific, Singapore, 2012).

M.I. Mishchenko, *Electromagnetic Scattering by Particles and Particle Groups: An Introduction* (Cambridge University Press, Cambridge, 2014).

Part I

Wave Optics

2
Electromagnetic Waves

In this chapter, we recall some basic principles of electromagnetic theory. We derive the wave equation for the electric field from the Maxwell equations in material media. We also introduce the scalar model as a simplified description of the electromagnetic field. Although the scalar model is used in most of the book, vector electromagnetic waves are studied in Part VI.

2.1 Macroscopic Maxwell's Equations

Maxwell's equations govern the classical description of all electromagnetic phenomena. In SI units, the macroscopic Maxwell equations are of the form

$$\nabla \times \mathbf{E}(\mathbf{r}, t) + \frac{\partial \mathbf{B}}{\partial t}(\mathbf{r}, t) = 0 , \tag{2.1}$$

$$\nabla \cdot \mathbf{D}(\mathbf{r}, t) = \rho_{\text{ext}}(\mathbf{r}, t) , \tag{2.2}$$

$$\nabla \times \mathbf{H}(\mathbf{r}, t) - \frac{\partial \mathbf{D}}{\partial t}(\mathbf{r}, t) = \mathbf{j}_{\text{ext}}(\mathbf{r}, t) , \tag{2.3}$$

$$\nabla \cdot \mathbf{B}(\mathbf{r}, t) = 0 . \tag{2.4}$$

Here $\mathbf{E}$ and $\mathbf{B}$ are the electric and magnetic fields, and ρ_{ext} and $\mathbf{j}_{\text{ext}}$ are the external charge and current densities, which act as sources of the fields. In a material medium, the response to applied electric and magnetic fields is described by the $\mathbf{D}$ and $\mathbf{H}$ fields:

$$\mathbf{D}(\mathbf{r}, t) = \varepsilon_0 \mathbf{E}(\mathbf{r}, t) + \mathbf{P}(\mathbf{r}, t) , \tag{2.5}$$

$$\mathbf{B}(\mathbf{r}, t) = \mu_0 \mathbf{H}(\mathbf{r}, t) + \mathbf{M}(\mathbf{r}, t) , \tag{2.6}$$

where $\mathbf{P}$ and $\mathbf{M}$ are the polarization and magnetization. Note that the conservation of charge, in the form of the continuity equation

$$\nabla \cdot \mathbf{j}_{\text{ext}}(\mathbf{r}, t) + \frac{\partial \rho_{\text{ext}}}{\partial t}(\mathbf{r}, t) = 0 , \tag{2.7}$$

is guaranteed by the Maxwell equations.

We now restrict our attention to monochromatic fields at a frequency ω: $\mathbf{E}(\mathbf{r}, t) = \mathrm{Re}[\mathbf{E}(\mathbf{r}) \exp(-i\omega t)]$ and $\mathbf{B}(\mathbf{r}, t) = \mathrm{Re}[\mathbf{B}(\mathbf{r}) \exp(-i\omega t)]$, where $\mathbf{E}(\mathbf{r})$ and $\mathbf{B}(\mathbf{r})$ are the corresponding complex amplitudes of the fields. The Maxwell equations for the amplitudes and sources thus become

$$\nabla \times \mathbf{E}(\mathbf{r}) - i\omega\mathbf{B}(\mathbf{r}) = 0 \,, \tag{2.8}$$

$$\nabla \cdot \mathbf{D}(\mathbf{r}) = \rho_{\text{ext}}(\mathbf{r}) \,, \tag{2.9}$$

$$\nabla \times \mathbf{H}(\mathbf{r}) + i\omega\mathbf{D}(\mathbf{r}) = \mathbf{j}_{\text{ext}}(\mathbf{r}) \,, \tag{2.10}$$

$$\nabla \cdot \mathbf{B}(\mathbf{r}) = 0 \,. \tag{2.11}$$

In material media, within the linear response regime, the polarization and magnetization are related to the applied fields according to the constitutive relations

$$\mathbf{P}(\mathbf{r}) = \varepsilon_0\, \chi_e(\mathbf{r}, \omega)\mathbf{E}(\mathbf{r}) \,, \tag{2.12}$$

$$\mathbf{M}(\mathbf{r}) = \mu_0\, \chi_m(\mathbf{r}, \omega)\mathbf{H}(\mathbf{r}) \,, \tag{2.13}$$

where χ_e is the electric susceptibility and χ_m is the magnetic susceptibility. In these relations, we have assumed that the medium exhibits a local and isotropic response. We note that χ_e and χ_m are generally frequency-dependent, as a consequence of temporal dispersion. In inhomogeneous media, they also depend on position. The macroscopic fields are thus given by

$$\mathbf{D}(\mathbf{r}) = \varepsilon_0\, \varepsilon(\mathbf{r}, \omega)\mathbf{E}(\mathbf{r}) \,, \tag{2.14}$$

$$\mathbf{B}(\mathbf{r}) = \mu_0\, \mu(\mathbf{r}, \omega)\mathbf{H}(\mathbf{r}) \,, \tag{2.15}$$

where

$$\varepsilon(\mathbf{r}, \omega) = 1 + \chi_e(\mathbf{r}, \omega) \,, \tag{2.16}$$

$$\mu(\mathbf{r}, \omega) = 1 + \chi_m(\mathbf{r}, \omega) \,. \tag{2.17}$$

These relations define the dielectric permittivity (or dielectric function) ε and the magnetic permeability μ.

2.2 Wave Equations

We now derive the wave equations for the electromagnetic field. At optical frequencies, naturally occurring materials have a vanishingly small magnetic response, and we thus consider only nonmagnetic media with $\chi_m = 0$. Eqs. (2.8)–(2.11) for an inhomogeneous medium in the absence of external sources become

$$\nabla \times \mathbf{E} - i\omega\mathbf{B} = 0 \,, \tag{2.18}$$

$$\nabla \cdot (\varepsilon\mathbf{E}) = 0 \,, \tag{2.19}$$

$$\nabla \times \mathbf{B} + i\frac{\omega}{c^2}\varepsilon\mathbf{E} = 0 \,, \tag{2.20}$$

$$\nabla \cdot \mathbf{B} = 0 \,, \tag{2.21}$$

where $c = 1/\sqrt{\varepsilon_0\mu_0}$ is the speed of light in vacuum. The wave equation for $\mathbf{E}$ is obtained by taking the curl of (2.18) and eliminating $\mathbf{B}$ using (2.20). This leads to

$$\nabla \times \nabla \times \mathbf{E} - k_0^2\, \varepsilon\, \mathbf{E} = 0 \,, \tag{2.22}$$

where $k_0 = \omega/c = 2\pi/\lambda$ is the free-space wavenumber, with λ being wavelength in vacuum. Equation (2.22) is the vector Helmholtz equation. It can be put into a different form by making use of the identity

$$\nabla \times \nabla \times \mathbf{E} = -\nabla^2\mathbf{E} + \nabla(\nabla \cdot \mathbf{E}) \tag{2.23}$$

and noting that $\nabla \cdot \mathbf{E} = -\nabla\varepsilon/\varepsilon \cdot \mathbf{E}$. We thereby obtain

$$\nabla^2\mathbf{E} + k_0^2\, \varepsilon\, \mathbf{E} + \nabla\left(\frac{\nabla\varepsilon}{\varepsilon} \cdot \mathbf{E}\right) = 0 \,. \tag{2.24}$$

The gradient term in the above equation couples the vector components of $\mathbf{E}$. If ε varies slowly on the scale of the wavelength, that is, if

$$\frac{|\nabla\varepsilon|}{\varepsilon} \ll k_0 \,, \tag{2.25}$$

then (2.24) becomes

$$\nabla^2\mathbf{E} + k_0^2\, \varepsilon\, \mathbf{E} = 0 \,. \tag{2.26}$$

Thus each component of $\mathbf{E}$ satisfies the scalar Helmholtz equation

$$\nabla^2 U + k_0^2\, \varepsilon\, U = 0 \,, \tag{2.27}$$

where U is a scalar monochromatic field. Evidently, the components of $\mathbf{E}$ are not independent, however, since $\nabla \cdot \mathbf{E} = 0$ within the accuracy of the approximation (2.25). If we restrict our attention to (2.27), we will say that we are working with the scalar model of the electromagnetic field. This book largely makes use of the scalar model, although Part VI is dedicated to certain aspects of the full vector theory.

2.3 Boundary Conditions

The boundary conditions satisfied by the fields $\mathbf{E}$, $\mathbf{B}$, $\mathbf{D}$ and $\mathbf{H}$ across an interface separating two homogeneous media 1 and 2, with dielectric constants ε_1 and ε_2, are given by

$$\mathbf{E}_1 \cdot \hat{\mathbf{t}} = \mathbf{E}_2 \cdot \hat{\mathbf{t}} \,, \tag{2.28}$$

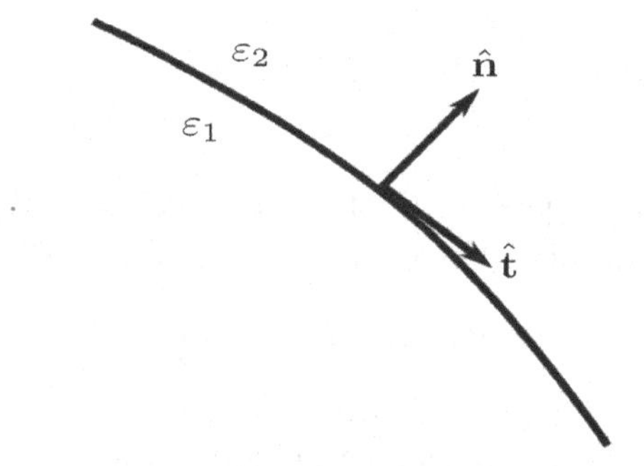

Fig. 2.1 Illustrating the boundary conditions at an interface separating two media with dielectric permittivities ε_1 and ε_2.

$$\mathbf{B}_1 \cdot \hat{\mathbf{n}} = \mathbf{B}_2 \cdot \hat{\mathbf{n}} \,, \tag{2.29}$$

$$\mathbf{D}_1 \cdot \hat{\mathbf{n}} = \mathbf{D}_2 \cdot \hat{\mathbf{n}} \,, \tag{2.30}$$

$$\mathbf{H}_1 \cdot \hat{\mathbf{t}} = \mathbf{H}_2 \cdot \hat{\mathbf{t}} \,. \tag{2.31}$$

See Fig. 2.1. Here the subscripts 1 and 2 are to be understood as the limiting values of the fields as the interface is approached from medium 1 and medium 2, respectively, $\hat{\mathbf{n}}$ denotes the unit normal to the interface pointing from medium 1 to medium 2, and $\hat{\mathbf{t}}$ denotes the unit vector tangent to the interface. These relations mean that the tangential components of **E** and **H** and the normal components of **B** and **D** are continuous.

The corresponding boundary conditions for the scalar field U are

$$U_1 = U_2 \,, \tag{2.32}$$

$$\nabla U_1 \cdot \hat{\mathbf{n}} = \nabla U_2 \cdot \hat{\mathbf{n}} \,. \tag{2.33}$$

Eqs. (2.32) and (2.33) express the continuity of the field and its normal derivative across the interface.

2.4 Energy Conservation

We will consider the conservation of energy for electromagnetic fields within the scalar model. In this setting, an energy current density replaces the Poynting vector. We will see that the current obeys a conservation law that governs the balance of electromagnetic energy in a fixed volume. That is, the flow of energy out of the volume must be precisely balanced by a decrease in energy within the volume.

We begin by rewriting the scalar wave equation (2.27) in the form

$$\nabla^2 U + k_0^2 U = -k_0^2(\varepsilon - 1)U \,. \tag{2.34}$$

The complex-conjugate U^* also obeys a wave equation of the form

$$\nabla^2 U^* + k_0^2 U^* = -k_0^2(\varepsilon^* - 1)U^* . \quad (2.35)$$

If we multiply (2.34) by U^* and (2.35) by U, and subtract the resulting expressions, we obtain the relation

$$U^*\nabla^2 U - U\nabla^2 U^* = -k_0^2 \left(\varepsilon - \varepsilon^*\right) |U|^2 . \quad (2.36)$$

Applying the identity

$$\nabla \cdot \left[U^*\nabla U - U\nabla U^*\right] = U^*\nabla^2 U - U\nabla^2 U^* , \quad (2.37)$$

(2.36) can be written in the form of the conservation law

$$\nabla \cdot \mathbf{J} + k_0 \,\mathrm{Im}\varepsilon |U|^2 = 0 . \quad (2.38)$$

Here the energy current $\mathbf{J}$ is defined by

$$\mathbf{J} = \frac{1}{2ik_0}\left(U^*\nabla U - U\nabla U^*\right) . \quad (2.39)$$

Note that $\mathbf{J}$ is normalized so that a unit amplitude plane wave has unit magnitude current. The current $\mathbf{J}$ plays the role of the Poynting vector in the scalar model. The term $|U|^2$ is proportional to the energy density of the field.

We now integrate both sides of (2.38) over a volume V that is bounded by a closed surface S with unit outward normal $\hat{\mathbf{n}}$. Upon applying the divergence theorem, we obtain

$$\int_S \mathbf{J} \cdot \hat{\mathbf{n}}\, d^2r + k_0 \mathrm{Im} \int_V \varepsilon |U|^2 \, d^3r \ = 0. \quad (2.40)$$

Equation (2.40) governs the conservation of energy for scalar electromagnetic fields. The first term on the left-hand side is the power passing through S. The second term corresponds to the power absorbed by the medium. That is, the absorbed power P_a is obtained by integrating the inward-going normal component of $\mathbf{J}$ over the closed surface S:

$$\begin{aligned} P_a &= -\int_S \mathbf{J} \cdot \hat{\mathbf{n}} d^2r \\ &= k_0 \mathrm{Im} \int_V \varepsilon |U|^2 d^3r . \end{aligned} \quad (2.41)$$

This result shows that absorption is associated with the imaginary part of the dielectric function. That is, if $\mathrm{Im}\varepsilon = 0$, the medium is non-absorbing. Finally, it is important to note that the current $\mathbf{J}$ has an unambiguous meaning only when it is integrated over a closed surface. Otherwise, the curl of an arbitrary vector field may be added to $\mathbf{J}$ without changing the conservation of energy as specified by (2.40).

References and Additional Reading

A comprehensive description of macroscopic electrodynamics and electromagnetic waves can be found in the following textbooks:

L.D. Landau and E.M. Lifshitz, *Classical Theory of Fields* (Butterworth-Heinemann, Amsterdam, 1980).

L.D. Landau, E.M. Lifshitz and L.P. Pitaevskii, *Electrodynamics of Continuous Media* (Pergamon Press, Oxford, 1984).

J.D. Jackson, *Classical Electrodynamics* (Wiley, New York, 1998).

In addition to a deep presentation of many aspects of electrodynamics, this book proposes an original and clear derivation of the macroscopic Maxwell's equations:

J. Schwinger, L.L. DeRaad Jr., K.A. Milton and W. Tsai, *Classical Electrodynamics* (Perseus, Reading, 1998).

These papers discuss the scalar model of electromagnetic fields:

H.S. Green and E. Wolf, Proc. Phys. Soc. A **66**, 1129 (1953).

E. Wolf, Proc. Phys. Soc. A **74**, 269 (1959).

P. Roman, Proc. Phys. Soc. A **74**, 281 (1959).

3

Geometrical Optics

Geometrical optics is concerned with phenomena that arise when the dielectric permittivity varies slowly on the scale of the wavelength. In this regime, we will see that the theory of wave propagation can be expressed in geometrical terms.

3.1 Plane Waves

Consider the wave equation (2.27) for a monochromatic scalar field in the absence of a source:

$$\nabla^2 U + k_0^2 n^2 U = 0 . \tag{3.1}$$

Here we have introduced the refractive index $n = \sqrt{\varepsilon}$, which we take to be real-valued, thus describing a non-absorbing medium. If the medium is infinite and homogeneous, then the simplest solution to (3.1) is a plane wave. A plane wave propagating in the direction of the unit vector $\hat{\mathbf{s}}$ has the functional form $U(\mathbf{r} \cdot \hat{\mathbf{s}})$ and is constant on the plane $\mathbf{r} \cdot \hat{\mathbf{s}} = \text{const}$. If we define the coordinate $\xi = \mathbf{r} \cdot \hat{\mathbf{s}}$, then it is easily seen that

$$\nabla^2 U = \frac{\partial^2 U}{\partial \xi^2} . \tag{3.2}$$

Equation (3.1) thus becomes

$$\frac{\partial^2 U}{\partial \xi^2} + k^2 U = 0 , \tag{3.3}$$

which has solutions $U = \exp(\pm ik\xi)$, with $k = nk_0$ the wavenumber in the medium. A plane wave of amplitude A propagating in the $\hat{\mathbf{s}}$ direction is therefore of the form

$$U(\mathbf{r}) = A \exp(ik\hat{\mathbf{s}} \cdot \mathbf{r}) . \tag{3.4}$$

Note that the vector $\hat{\mathbf{s}}$ is the direction in which energy flows, since $\mathbf{J} = n|A|^2\hat{\mathbf{s}}$, which follows from (2.39). Evidently, the wavelength in the medium is λ/n, λ being the wavelength in vacuum.

Making use of the linearity of the wave equation, we see that the general solution to (3.1) in a homogeneous medium is given by the superposition of plane waves

$$U(\mathbf{r}) = \int A(\hat{\mathbf{s}}) \exp(ik\hat{\mathbf{s}} \cdot \mathbf{r})\, d\hat{\mathbf{s}}, \tag{3.5}$$

where $d\hat{\mathbf{s}}$ indicates integration over the unit sphere. Such plane wave decompositions will be treated in detail in Chapter 6.

3.2 Eikonal Equation

We now consider an inhomogeneous medium whose refractive index varies slowly on the scale of the wavelength. Physically, we expect that the solution to (3.1) should behave nearly as a plane wave on scales large compared to the wavelength. It will thus prove useful to investigate the asymptotic behavior of the field in the high-frequency limit $k_0 \to \infty$. Accordingly, we expand the field in powers of $1/k_0$:

$$U(\mathbf{r}) = \exp(ik_0 S(\mathbf{r})) \left(U_0(\mathbf{r}) + \frac{1}{k_0} U_1(\mathbf{r}) + \cdots \right), \tag{3.6}$$

where S, U_0 and U_1 are to be determined. More formally, (3.6) can be viewed as an asymptotic expansion. We now substitute (3.6) into the wave equation (3.1). Upon collecting terms of order $O(1)$, we obtain

$$|\nabla S|^2 = n^2 . \tag{3.7}$$

Equation (3.7) is known as the eikonal equation, and S is called the eikonal. At order $O(1/k_0)$, we have

$$\nabla S \cdot \nabla U_0 + \frac{1}{2} U_0 \nabla^2 S = 0 , \tag{3.8}$$

which is referred to as the transport equation. The eikonal and transport equations are the fundamental equations governing the behavior of waves in geometrical optics.

The eikonal may be interpreted as defining a position-dependent wave vector for a plane-wave approximation to the field U. To see this, consider a point $\mathbf{r}_0$ and expand $S(\mathbf{r})$ around $\mathbf{r}_0$:

$$S(\mathbf{r}) = S(\mathbf{r}_0) + (\mathbf{r} - \mathbf{r}_0) \cdot \nabla S(\mathbf{r}_0) + \cdots . \tag{3.9}$$

We then define the local wave vector $\mathbf{k}(\mathbf{r}_0)$ as

$$\begin{aligned}\mathbf{k}(\mathbf{r}_0) &= k_0 \nabla S(\mathbf{r}_0) \\ &= k_0\, n(\mathbf{r}_0)\, \hat{\mathbf{k}}(\mathbf{r}_0) \,,\end{aligned} \tag{3.10}$$

where $\hat{\mathbf{k}}(\mathbf{r}_0)$ is a unit vector in the direction of $\nabla S(\mathbf{r}_0)$, and (3.7) has been used. Thus, near the point $\mathbf{r}_0$, the field $U(\mathbf{r})$ is locally approximated by the plane wave

$$U(\mathbf{r}) = \mathcal{A}(\mathbf{r}_0) \exp(i\mathbf{k}(\mathbf{r}_0) \cdot \mathbf{r}) \,. \tag{3.11}$$

Equation (3.11) corresponds to a plane wave with amplitude A, whose functional form is not given, and wave vector $\mathbf{k}$.

3.3 Ray Equation

The surfaces S = constant are known as wavefronts. Rays are the paths whose tangent vectors are normal to the wavefronts, as shown in Fig. 3.1. Consider a ray $\mathbf{r}(s)$ that is parametrized by its arc length s. Since $\hat{\mathbf{k}}$ is normal to the wavefront, we see that

$$\frac{d\mathbf{r}}{ds} = \hat{\mathbf{k}} \,. \tag{3.12}$$

We can use this result to obtain the differential equation obeyed by the rays. To proceed, we use the fact that $\nabla S = n\hat{\mathbf{k}}$ and find that Eq. (3.12) becomes

$$\frac{d}{ds}\left(n\frac{d\mathbf{r}}{ds}\right) = \frac{d}{ds}\nabla S \,. \tag{3.13}$$

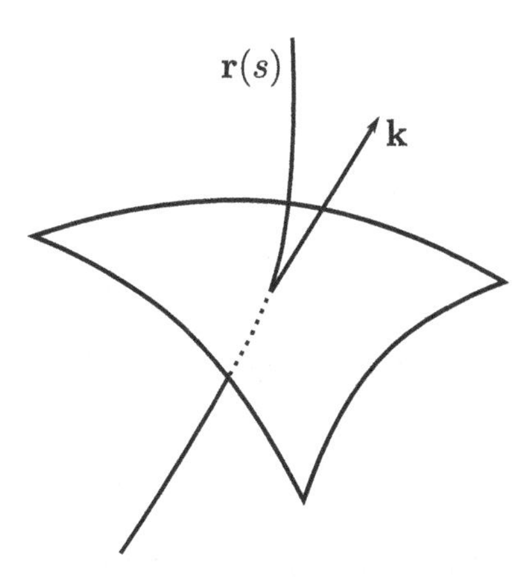

Fig. 3.1 Illustrating the ray geometry.

Now, it is easily seen that

$$\begin{aligned}\frac{dS}{ds} &= \nabla S \cdot \frac{d\mathbf{r}}{ds} \\ &= \hat{\mathbf{k}} \cdot \nabla S \\ &= n \,. \end{aligned} \tag{3.14}$$

Thus (3.13) becomes

$$\frac{d}{ds}\left(n\frac{d\mathbf{r}}{ds}\right) = \nabla n \,, \tag{3.15}$$

which is the equation obeyed by the rays.

In a homogeneous medium, with a constant index of refraction n, (3.15) has solutions that correspond to rays which are straight lines. In an inhomogeneous medium, (3.15) may be solved for the ray $\mathbf{r}(s)$ if the refractive index n is specified. Once the ray is known, it is possible to determine the eikonal S by integrating the differential equation (3.14):

$$S(\mathbf{r}(s)) = S(\mathbf{r}(s_0)) + \int_{s_0}^{s} n\,(\mathbf{r}(t))\,dt \,. \tag{3.16}$$

The field U_0 can be found by solving the transport equation (3.8). We note that

$$\frac{dU_0}{ds} = \nabla U_0 \cdot \frac{d\mathbf{r}}{ds} \tag{3.17}$$

$$= \frac{1}{n}\nabla U_0 \cdot \nabla S \,, \tag{3.18}$$

which follows from (3.12) and the relation $\nabla S = n\hat{\mathbf{k}}$. Using this result, (3.8) becomes

$$\frac{dU_0}{ds} = -\frac{1}{2n}U_0 \nabla^2 S \,, \tag{3.19}$$

which can be integrated to obtain

$$U_0(\mathbf{r}(s)) = U_0(\mathbf{r}(s_0)) \exp\left[-\frac{1}{2}\int_{s_0}^{s} \frac{1}{n(\mathbf{r}(t))}\nabla^2 S(\mathbf{r}(t))dt\right] \,. \tag{3.20}$$

We now consider the curvature κ of a ray $\mathbf{r}(s)$. We have

$$\frac{d^2\mathbf{r}}{ds^2} = \kappa \widehat{\mathbf{N}} \,, \tag{3.21}$$

where $\kappa = |d^2\mathbf{r}/ds^2|$ and $\widehat{\mathbf{N}}$ is the unit vector normal to the curve with $\hat{\mathbf{k}} \cdot \widehat{\mathbf{N}} = 0$. Using Eqs. (3.12) and (3.21), and carrying out the differentiation in (3.15), we

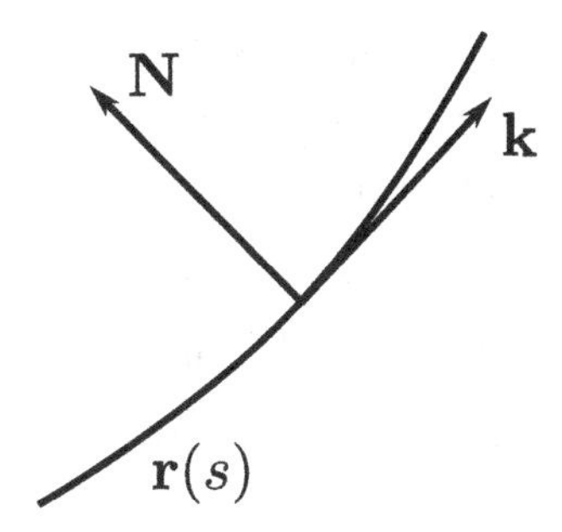

Fig. 3.2 Illustrating the curvature of the ray.

obtain

$$\frac{dn}{ds}\hat{\mathbf{k}} + n\,\kappa\,\widehat{\mathbf{N}} = \nabla n\,. \tag{3.22}$$

Thus the gradient of the refractive index lies in the plane defined by the tangent and normal vectors to the ray. Equation (3.22) implies that the curvature is given by the expression

$$\kappa = \widehat{\mathbf{N}} \cdot \nabla \ln n\,. \tag{3.23}$$

Evidently, since κ is nonnegative, rays are bent in the direction of increasing n along the ray normal as shown in Fig. 3.2.

3.4 Transport of Intensity

The energy current carried by the field is given by the expression

$$\mathbf{J} = |U_0|^2 \nabla S + O\left(\frac{1}{k_0}\right) \tag{3.24}$$

$$= n|U_0|^2 \hat{\mathbf{k}} + O\left(\frac{1}{k_0}\right), \tag{3.25}$$

where we have used (2.39) and (3.6). In a non-absorbing medium, the conservation law (2.38) can be restated as $\nabla \cdot \mathbf{J} = 0$. From (3.24), we see that as $k_0 \to \infty$ the conservation law becomes

$$\nabla \cdot nI\hat{\mathbf{k}} = 0\,, \tag{3.26}$$

where $I = |U|^2$ is the wave intensity. Thus the intensity obeys a transport equation of the form

$$\hat{\mathbf{k}} \cdot \nabla I + \mu_s I = 0\,, \tag{3.27}$$

where $\mu_s = \nabla \cdot (n\hat{\mathbf{k}})/n$ is the scattering coefficient. This transport equation holds in the geometrical optics limit. Transport equations for the intensity in the multiple scattering regime are studied in Part III.

References and Additional Reading

The foundations of geometrical optics, based on wave theory, are presented in:

M. Kline and I.W. Ray, *Electromagnetic Theory and Geometrical Optics* (Wiley, New York, 1965).

E. Hecht, *Optics*, 4th edition (Addison Wesley, San Francisco, 2002), chap. 4 and 5.

M. Born and E. Wolf, *Principles of Optics*, 7th edition (Cambridge University Press, Cambridge, 2005), chap. 3.

4

Waves at Interfaces

At an interface between media with different refractive indices, bending of rays, known as refraction, is observed. We first discuss refraction within the framework of geometrical optics. We then consider the wave theory of refraction, and derive the Fresnel formulas for reflection and transmission at a planar interface.

4.1 Geometrical Theory of Refraction

According to geometrical optics, the eikonal obeys the equation $\nabla S = n\hat{\mathbf{k}}$ derived in Chapter 3. Evidently,

$$\nabla \times (n\hat{\mathbf{k}}) = 0\,, \tag{4.1}$$

and thus, upon applying Stokes theorem, we obtain

$$\int_C n(\mathbf{r})\,\hat{\mathbf{k}}(\mathbf{r}) \cdot d\mathbf{r} = 0\,, \tag{4.2}$$

where C is an arbitrary closed curve.

Consider an interface separating two homogeneous media denoted 1 and 2, with dielectric constants ε_1 and ε_2, and refractive indices $n_1 = \sqrt{\varepsilon_1}$ and $n_1 = \sqrt{\varepsilon_2}$, which are assumed to be real-valued. Let the curve C enclose the interface as shown in Fig. 4.1. By shrinking C in the direction normal to the interface, we find that the tangential component of $n(\mathbf{r})\hat{\mathbf{k}}(\mathbf{r})$ is continuous across the interface. Thus we obtain the condition

$$\left(n_2\hat{\mathbf{k}}_2 - n_1\hat{\mathbf{k}}_1\right) \times \hat{\mathbf{n}} = 0\,, \tag{4.3}$$

where the subscripts 1 and 2 are to be understood as the limiting values of the corresponding quantities as the interface is approached from medium 1 and medium 2, respectively, and $\hat{\mathbf{n}}$ denotes the normal to the interface pointing from medium 1 to medium 2.

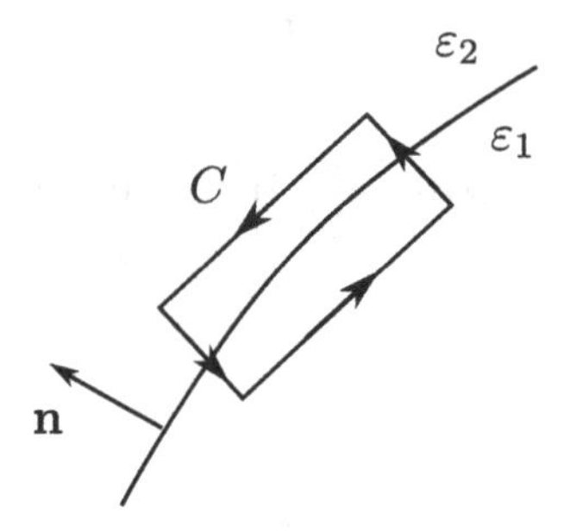

Fig. 4.1 Derivation of the interface boundary conditions.

We can now derive the law of refraction. Consider a ray that is incident on the interface from medium 1, and suppose that $n_2 > n_1$. Then there is a transmitted or refracted ray in medium 2 that is bent towards the direction normal to the interface. Using (4.3) we find that

$$n_1|\hat{\mathbf{k}}_1 \times \hat{\mathbf{n}}| = n_2|\hat{\mathbf{k}}_2 \times \hat{\mathbf{n}}| , \tag{4.4}$$

which becomes

$$n_1 \sin\theta_1 = n_2 \sin\theta_2 . \tag{4.5}$$

Here θ_1, θ_2 are the angles between the wave vectors $\hat{\mathbf{k}}_1, \hat{\mathbf{k}}_2$ and the normal $\hat{\mathbf{n}}$. The relation (4.5) is known as Snell's law. It follows that $\sin\theta_2 < \sin\theta_1$ when $n_2 > n_1$, which describes refraction of the incident field.

4.2 Wave Theory of Reflection and Transmission

The derivation of Snell's law given above may be criticized on the grounds that the discontinuity in the index of refraction across the interface violates the condition for the applicability of geometrical optics. In addition, geometrical optics cannot predict the relative intensities of the reflected and refracted rays. In this section, we address these difficulties within the framework of wave optics.

We consider a planar interface between two homogeneous media as shown in Fig. 4.2. Medium 1 consists of the half-space $z < 0$, and medium 2 the half-space $z \geq 0$. In either medium, the field satisfies the wave equation

$$\nabla^2 U + k^2 U = 0 , \tag{4.6}$$

where k is the wavenumber. In medium 1, the field obeys (4.6) with $k = n_1 k_0$. Suppose that a unit-amplitude plane wave is incident from medium 1. The field in medium 1 is the superposition of the incident and reflected waves:

$$U_1(\mathbf{r}) = \exp(i\mathbf{k}_1 \cdot \mathbf{r}) + R\exp(i\mathbf{k}_1' \cdot \mathbf{r}) , \tag{4.7}$$

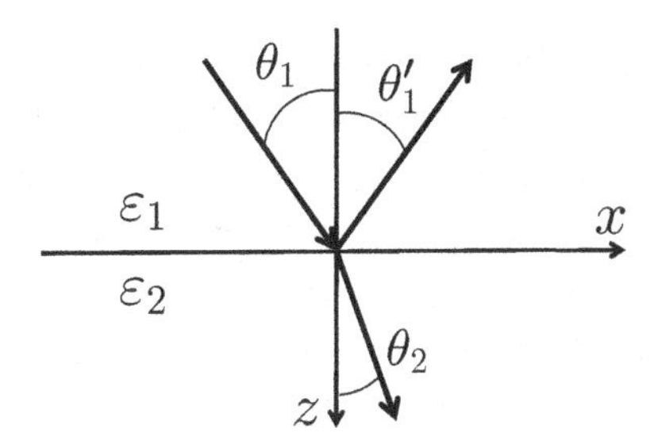

Fig. 4.2 Reflection and transmission at a planar interface between two homogeneous media with dielectric permittivities ε_1 and ε_2.

where $\mathbf{k}_1$ and $\mathbf{k}'_1$ are the wave vectors of the incident and reflected waves with $|\mathbf{k}_1| = |\mathbf{k}'_1| = n_1 k_0$, and R is the amplitude reflection coefficient. The field in medium 2

$$U_2(\mathbf{r}) = T \exp(i\mathbf{k}_2 \cdot \mathbf{r}) \tag{4.8}$$

obeys (4.6) with $k = n_2 k_0$ and corresponds to the transmitted wave. Here T is the amplitude transmission coefficient and $|\mathbf{k}_2| = n_2 k_0$.

According to (2.32) and (2.33), the field and its normal derivative must be continuous across the interface $z = 0$:

$$U_1|_{z=0} = U_2|_{z=0} \, , \tag{4.9}$$

$$\left.\frac{\partial U_1}{\partial z}\right|_{z=0} = \left.\frac{\partial U_2}{\partial z}\right|_{z=0} . \tag{4.10}$$

The boundary condition (4.9) implies that the phases of the incident, reflected, and transmitted plane waves must coincide on the plane $z = 0$. That is,

$$\mathbf{k}_1 \cdot \mathbf{r}|_{z=0} = \mathbf{k}'_1 \cdot \mathbf{r}|_{z=0} = \mathbf{k}_2 \cdot \mathbf{r}|_{z=0} \, . \tag{4.11}$$

We thus see that the wave vectors $\mathbf{k}_1$, $\mathbf{k}'_1$ and $\mathbf{k}_2$ all lie in the same plane, which we take to be the $x - z$ plane. Putting $y = 0$ in Eq. (4.11), we obtain

$$k_{1x} = k'_{1x} = k_{2x} \, . \tag{4.12}$$

Making use of the relations

$$k_{1x} = n_1 k_0 \sin\theta_1 \, , \tag{4.13}$$

$$k'_{1x} = n_1 k_0 \sin\theta'_1 \, , \tag{4.14}$$

$$k_{2x} = n_2 k_0 \sin\theta_2 \, , \tag{4.15}$$

we find that

$$\theta_1 = \theta'_1 \tag{4.16}$$

and

$$n_1 \sin\theta_1 = n_2 \sin\theta_2 \, . \tag{4.17}$$

Here $\theta_1, \theta_1', \theta_2$ are the angles between the wave vectors $\mathbf{k}_1, \mathbf{k}_1', \mathbf{k}_2$ and the normal to the interface, which are referred to as the angles of incidence, reflection and transmission, respectively. (4.16), which states that the angles of incidence and reflection are identical, is known as the law of reflection. (4.17) is Snell's law once again.

Applying the boundary conditions (4.9) and (4.10) to the field expressions (4.7) and (4.8), and taking into account the orientations of the wave vectors in both media, we find that the Fresnel coefficients R and T are given by the expressions

$$R = \frac{k_{1z} - k_{2z}}{k_{1z} + k_{2z}}, \tag{4.18}$$

$$T = \frac{2k_{1z}}{k_{1z} + k_{2z}}, \tag{4.19}$$

where $k_{1z} = (n_1^2 k_0^2 - k_{1x}^2)^{1/2}$ and $k_{2z} = (n_2^2 k_0^2 - k_{1x}^2)^{1/2}$ are the z-components of the wave vector in medium 1 and 2, respectively (the proper branch of the square root must be taken). Note that (4.7), (4.8), (4.18) and (4.19) determine the field everywhere.

Using Snell's law and the law of reflection, we can express R and T in terms of the angle of incidence θ_1:

$$R = \frac{n_1 \cos\theta_1 - \sqrt{n_2^2 - n_1^2 \sin^2\theta_1}}{n_1 \cos\theta_1 + \sqrt{n_2^2 - n_1^2 \sin^2\theta_1}}, \tag{4.20}$$

$$T = \frac{2n_1 \cos\theta_1}{n_1 \cos\theta_1 + \sqrt{n_2^2 - n_1^2 \sin^2\theta_1}}. \tag{4.21}$$

(4.20) and (4.21) are known as the Fresnel formulas; they allow us to connect the complex amplitudes of the reflected and transmitted waves to that of the incident wave.

We define the intensity reflection coefficient $\mathcal{R} = |R|^2$. In Fig. 4.3 we show the dependence of $\mathcal{R}$ on the angle of incidence θ_1 when $n_1 < n_2$.

4.3 Total Internal Reflection

Suppose that an incident plane wave illuminates the medium with higher refractive index, so that $n_1 > n_2$. Then, applying Snell's law, we see that $\theta_2 > \theta_1$, meaning that the refracted ray bends away from the normal. In this case, there is an angle θ_c, referred to as the critical angle, for which the reflection coefficient

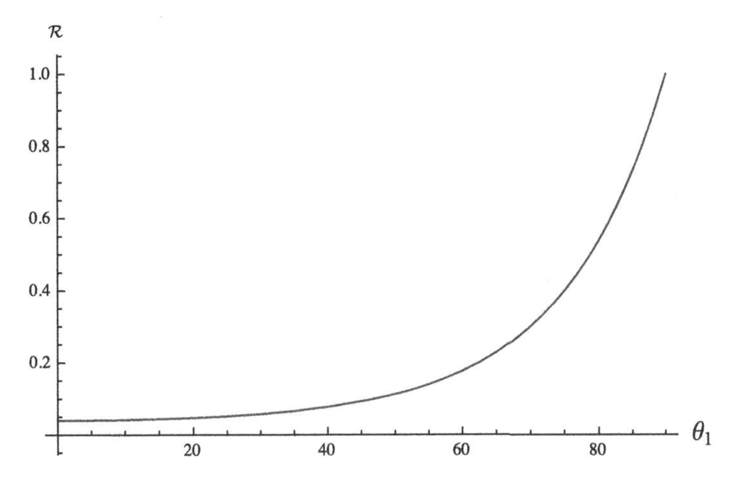

Fig. 4.3 Angular dependence of the intensity reflection coefficient $\mathcal{R}$ for $n_1 = 1$ and $n_2 = 1.5$.

$\mathcal{R} = 1$. According to the Fresnel formula (4.20) for the reflection coefficient, θ_c is determined from the relation

$$\sin\theta_c = \frac{n_2}{n_1} . \tag{4.22}$$

If $\theta_1 = \theta_c$, then the incident wave is completely reflected and the refracted wave travels along the interface with $\theta_2 = \pi/2$. This phenomenon is known as total internal reflection. If $\theta_1 > \theta_c$ then $\sin\theta_1 > n_2/n_1$ and, from Snell's law, we see that $\sin\theta_2 > 1$. This condition cannot be satisfied for any real θ_2. Indeed we find that the z-component of the transmitted wavevector

$$k_{2z} = in_2k_0\sqrt{\sin^2\theta_2 - 1} \tag{4.23}$$

is purely imaginary. From (4.18) we see that the reflection coefficient R has unit modulus, and thus the incident wave is totally reflected, albeit with a phase shift. Despite the total reflection of the incident wave, the transmitted field penetrates into the medium of lower refractive index, and its complex amplitude is

$$\begin{aligned} U_2(\mathbf{r}) &= T\exp\left(in_2k_0\sin\theta_2\, x + in_2k_0\cos\theta_2\, z\right) \\ &= T\exp\left[in_2k_0\frac{\sin\theta_1}{\sin\theta_c}x - n_2k_0\left(\frac{\sin^2\theta_1}{\sin^2\theta_c} - 1\right)^{1/2} z\right] . \end{aligned} \tag{4.24}$$

The above result shows that if $\theta_1 = \theta_c$, then the refracted wave propagates along the interface. If $\theta_1 > \theta_c$, the wave propagates parallel to the interface while decaying

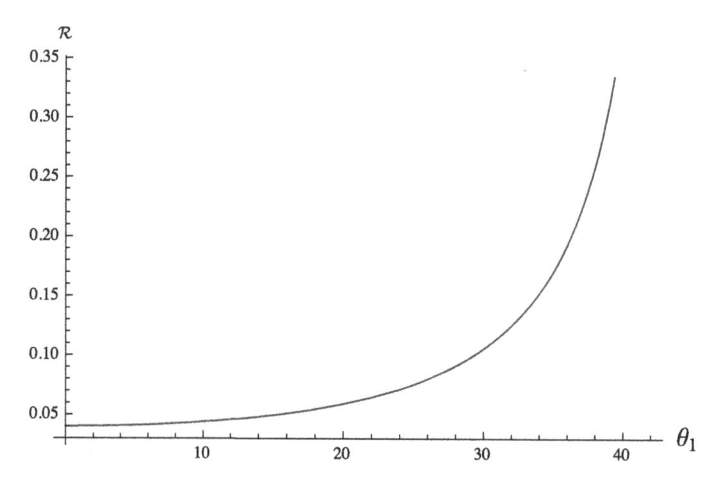

Fig. 4.4 Angular dependence of the intensity reflection coefficient $\mathcal{R}$ for total internal reflection with $n_1 = 1.5$ and $n_2 = 1.0$. The critical angle is $\theta_c \simeq 42$ degrees.

exponentially in the normal direction. This wave is known as an evanescent wave. In Fig. 4.4, we show the angular dependence of the intensity reflection coefficient $\mathcal{R}$ for total internal reflection.

The normal component of the energy current that is carried by the evanescent wave is given by

$$\mathbf{J} \cdot \hat{\mathbf{n}} = \frac{1}{k_0} |T|^2 \mathrm{Re}(k_{2z}) , \tag{4.25}$$

where we have used (2.39) and (4.8). Since k_{2z} is pure imaginary, $\mathbf{J} \cdot \hat{\mathbf{n}}$ vanishes. The absence of energy flow through the interface is consistent with the fact that the incident power is reflected.

References and Additional Reading

The reflection and transmission of optical waves at flat interfaces are treated in many textbooks. See for example:

J.D. Jackson, *Classical Electrodynamics* (Wiley, New York, 1962), chap. 7.

E. Hecht, *Theory and Problems in Optics* (McGraw Hill, New York, 1975), chap. 3.

L.D. Landau, E.M. Lifshitz and L.P. Pitaevskii, *Electrodynamics of Continuous Media* (Pergamon Press, Oxford, 1984), chap. X.

M. Born and E. Wolf, *Principles of Optics* (Cambridge University Press, Cambridge, 1999), chap. I.

The reflection and transmission of scalar waves is treated, for example, in:

J.A. DeSanto, *Scalar Wave Theory* (Springer-Verlag, Berlin, 1992), chap. 3.

5

Green's Functions and Integral Representations

In this chapter, we introduce Green's functions and related integral representations of solutions to the wave equation. We confine our discussion to homogeneous media. Inhomogeneous media will be taken up in Chapter 10 on scattering theory.

5.1 Kirchhoff Integral Formula

We consider the wave equation for a monochromatic scalar field U in a homogeneous medium in the presence of a source S:

$$\nabla^2 U + k^2 U = -S \, . \tag{5.1}$$

The Green's function $G(\mathbf{r}, \mathbf{r}')$ is defined as the solution of the equation

$$\nabla^2 G(\mathbf{r}, \mathbf{r}') + k^2 G(\mathbf{r}, \mathbf{r}') = -\delta(\mathbf{r} - \mathbf{r}') \, . \tag{5.2}$$

The Green's function may be interpreted as the solution to the wave equation (5.1) when S corresponds to a point source.

We now derive an integral representation for the solution to the wave equation (5.1). We use this result to specify the boundary conditions obeyed by the Green's function and the field. To proceed, we recall Green's identity

$$\int_V \left[f(\mathbf{r}')\nabla^2 g(\mathbf{r}') - g(\mathbf{r}')\nabla^2 f(\mathbf{r}') \right] d^3r' = \int_{\partial V} \left[f(\mathbf{r}')\nabla g(\mathbf{r}') - g(\mathbf{r}')\nabla f(\mathbf{r}') \right] \cdot \hat{\mathbf{n}} \, d^2r' \, , \tag{5.3}$$

which holds for functions f and g defined in a volume V bounded by a surface ∂V with unit outward normal $\hat{\mathbf{n}}$. Applying (5.3) with $f(\mathbf{r}') = G(\mathbf{r}, \mathbf{r}')$ and $g(\mathbf{r}') = U(\mathbf{r}')$, and using (5.1) and (5.2), we obtain the identity

$$\int_V \delta(\mathbf{r}-\mathbf{r}')U(\mathbf{r}')\,d^3r' = \int_V G(\mathbf{r},\mathbf{r}')S(\mathbf{r}')\,d^3r' + \int_{\partial V}\left[G(\mathbf{r},\mathbf{r}')\nabla U(\mathbf{r}') - U(\mathbf{r}')\nabla_{\mathbf{r}'}G(\mathbf{r},\mathbf{r}')\right]\cdot\hat{\mathbf{n}}\,d^2r' . \tag{5.4}$$

If $\mathbf{r} \in V$, (5.4) becomes

$$U(\mathbf{r}) = \int_V G(\mathbf{r},\mathbf{r}')S(\mathbf{r}')\,d^3r' + \int_{\partial V}\left[G(\mathbf{r},\mathbf{r}')\nabla U(\mathbf{r}') - U(\mathbf{r}')\nabla_{\mathbf{r}'}G(\mathbf{r},\mathbf{r}')\right]\cdot\hat{\mathbf{n}}\,d^2r' , \tag{5.5}$$

while if $\mathbf{r} \notin V$, we obtain

$$0 = \int_V G(\mathbf{r},\mathbf{r}')S(\mathbf{r}')\,d^3r' + \int_{\partial V}\left[G(\mathbf{r},\mathbf{r}')\nabla U(\mathbf{r}') - U(\mathbf{r}')\nabla_{\mathbf{r}'}G(\mathbf{r},\mathbf{r}')\right]\cdot\hat{\mathbf{n}}\,d^2r' . \tag{5.6}$$

(5.5) is known as the Kirchhoff integral formula. When $S = 0$, (5.5) forms the basis of the scalar theory of diffraction, as presented in Chapter 7. (5.6) gives rise to the so-called extinction theorem of scattering theory.

The Kirchhoff integral formula does not directly define a solution to the wave equation. Instead, (5.5) provides an integral representation for solutions to the wave equation. If, however, G is taken to obey the same boundary conditions as U on the surface ∂V, then (5.5) does indeed provide a solution to the wave equation. For example, suppose we impose that $U = 0$ on ∂V, which corresponds to the Dirichlet boundary condition. Then the surface integral in (5.5) vanishes and we have

$$U(\mathbf{r}) = \int_V G(\mathbf{r},\mathbf{r}')S(\mathbf{r}')\,d^3r' . \tag{5.7}$$

In an infinite medium, we will employ the boundary condition

$$\lim_{r\to\infty} r\left(\frac{\partial U}{\partial r} - ikU\right) = 0 , \tag{5.8}$$

which is known as the Sommerfeld radiation condition. This condition means that U behaves as an outgoing spherical wave. Taking ∂V to be a sphere, and applying (5.8) to both U and G, we easily see that the surface integral in (5.5) vanishes. As a result, the solution to (5.1) is given by (5.7). Note that (5.7) is the mathematical embodiment of Huygen's principle, which states that a general source radiates as a superposition of point sources.

An important property of G is the reciprocity relation $G(\mathbf{r},\mathbf{r}') = G(\mathbf{r}',\mathbf{r})$, which asserts that sources and detectors are interchangeable. Reciprocity in wave scattering is studied in detail in Chapter 14, which includes a derivation of the reciprocity relation for inhomogeneous media.

5.2 The Green's Function in an Infinite Medium

In an infinite medium, the Green's function depends only on the difference of its arguments, consistent with translational invariance. Accordingly, (5.2) can be solved by Fourier transformation. We define the Fourier transform of $G(\mathbf{r}, \mathbf{r}')$ by

$$\tilde{G}(\mathbf{q}, \mathbf{q}') = \int d^3r d^3r' \exp\left(i\mathbf{q}\cdot\mathbf{r} - i\mathbf{q}'\cdot\mathbf{r}'\right) G(\mathbf{r}, \mathbf{r}') \tag{5.9}$$

$$= (2\pi)^3 G(\mathbf{q})\delta(\mathbf{q} - \mathbf{q}') , \tag{5.10}$$

which defines $G(\mathbf{q})$. It follows from (5.2) that $G(\mathbf{q})$ obeys

$$(q^2 - k^2)G(\mathbf{q}) = 1. \tag{5.11}$$

The solution to the above equation can be expressed in the sense of distributions as

$$G(\mathbf{q}) = P\left(\frac{1}{q^2 - k^2}\right) + A\,\delta(q - k) + B\,\delta(q + k). \tag{5.12}$$

where P denotes the principal value, and A and B are constants. To determine $G(\mathbf{r}, \mathbf{r}')$, we compute the inverse Fourier transform

$$G(\mathbf{r}, \mathbf{r}') = \int \frac{d^3q}{(2\pi)^3} \exp\left(-i\mathbf{q}\cdot(\mathbf{r} - \mathbf{r}')\right) G(\mathbf{q}) . \tag{5.13}$$

Since $G(\mathbf{q})$ depends only on $|\mathbf{q}|$, we can express (5.13) as the one-dimensional integral

$$G(\mathbf{r}, \mathbf{r}') = \frac{1}{2\pi^2|\mathbf{r} - \mathbf{r}'|} \operatorname{Im} \int_0^\infty q\, G(\mathbf{q}) \exp(iq|\mathbf{r} - \mathbf{r}'|)\, dq . \tag{5.14}$$

After inserting (5.12) into (5.14), we are left with three integrals to perform. The integral corresponding to the first term in (5.12) can be calculated as a principal value by contour integration in the complex plane. The integral, corresponding to the second term, involves a delta function and is readily evaluated. Finally, the integral corresponding to the third term vanishes for $k > 0$. We thereby arrive at the result

$$\begin{aligned} G(\mathbf{r}, \mathbf{r}') = &\left(\frac{1}{8\pi|\mathbf{r} - \mathbf{r}'|} + \frac{Ak}{4i\pi^2|\mathbf{r} - \mathbf{r}'|}\right) \exp(ik|\mathbf{r} - \mathbf{r}'|) \\ &+ \left(\frac{1}{8\pi|\mathbf{r} - \mathbf{r}'|} - \frac{Ak}{4i\pi^2|\mathbf{r} - \mathbf{r}'|}\right) \exp(-ik|\mathbf{r} - \mathbf{r}'|) . \end{aligned} \tag{5.15}$$

Since G must satisfy the radiation condition, the second term on the right-hand side in (5.15) must vanish. This condition gives $A = i\pi/(2k)$. We thus obtain

$$G(\mathbf{r}, \mathbf{r}') = \frac{\exp(ik|\mathbf{r} - \mathbf{r}'|)}{4\pi|\mathbf{r} - \mathbf{r}'|} , \tag{5.16}$$

which is the outgoing Green's function for an infinite medium. The above Green's function is also referred to as the *retarded Green's function*, since in the time domain it describes radiation with a retarded time dependence. The Green's function G^*, which corresponds to the solution of (5.2) that obeys an incoming wave condition, is called the *advanced Green's function.*

The calculation of the Green's function can also be performed using another approach. To see this, we note that the solution to (5.11) is given by

$$G(\mathbf{q}) = \frac{1}{q^2 - k^2 - i\epsilon}, \tag{5.17}$$

where the limit $\epsilon \to 0^+$ is implied. The small imaginary part in the denominator regularizes the Fourier integral

$$G(\mathbf{r}, \mathbf{r}') = \int \frac{d^3q}{(2\pi)^3} \frac{1}{q^2 - k^2 - i\epsilon} \exp[i\mathbf{q} \cdot (\mathbf{r} - \mathbf{r}')] , \tag{5.18}$$

which leads directly to the expression (5.16) for the retarded Green's function. Note that using

$$G(\mathbf{q}) = \frac{1}{q^2 - k^2 + i\epsilon} \tag{5.19}$$

rather than (5.17) yields the advanced Green's function.

The equivalence between the two approaches can be made explicit by making use of the identity

$$\frac{1}{x \pm i\epsilon} = P\left(\frac{1}{x}\right) \mp i\pi\delta(x) . \tag{5.20}$$

Applying this identity to (5.17) or (5.19) leads directly to

$$\frac{1}{q^2 - k^2 \pm i\epsilon} = P\left(\frac{1}{q^2 - k^2}\right) \mp i\pi\delta(q^2 - k^2) . \tag{5.21}$$

Recalling the identity

$$\delta(q^2 - k^2) = \frac{1}{2|k|}\left[\delta(q - k) + \delta(q + k)\right] , \tag{5.22}$$

we see that (5.21) plays the role of (5.12). The key difference is that in (5.21), the choice of the retarded or advanced Green's function is determined by the choice of sign of $i\epsilon$.

5.3 Far-Field Radiation Pattern

We now apply the above results to obtain the far-field radiation pattern produced by a localized source. The field is assumed to propagate in free space and is taken to obey (5.1) with $k = k_0$. Using (5.7) and (5.16), we find that

$$U(\mathbf{r}) = \int \frac{\exp(ik_0|\mathbf{r} - \mathbf{r}'|)}{4\pi|\mathbf{r} - \mathbf{r}'|} S(\mathbf{r}')\, d^3r' . \tag{5.23}$$

When $r \gg r'$, the following expansions hold:

$$|\mathbf{r} - \mathbf{r}'| = r - \hat{\mathbf{r}} \cdot \mathbf{r}' + O\left(\frac{1}{r}\right) , \tag{5.24}$$

$$\frac{1}{|\mathbf{r} - \mathbf{r}'|} = \frac{1}{r} + \frac{\hat{\mathbf{r}} \cdot \mathbf{r}'}{r^2} + O\left(\frac{1}{r^3}\right) . \tag{5.25}$$

The far-field asymptotic form of the Green's function is thus given by

$$\frac{\exp(ik_0|\mathbf{r} - \mathbf{r}'|)}{4\pi|\mathbf{r} - \mathbf{r}'|} \sim \frac{\exp(ik_0 r)}{4\pi r} \exp(-ik_0\, \hat{\mathbf{r}} \cdot \mathbf{r}') . \tag{5.26}$$

Note that in the phase term above a second-order term $k_0 r'^2/r$ has been neglected, which requires the condition $k_0 r'^2/r \ll 2\pi$ to be satisfied. Therefore, for a source of size L, the far-field expansion is valid under the conditions $r \gg L$ and $r \gg L^2/\lambda$.

If we define $\mathbf{k} = k_0\hat{\mathbf{r}}$ to be the wave vector in the outgoing direction, then in the far-field

$$U(\mathbf{r}) \sim \frac{\exp(ik_0 r)}{4\pi r} \int \exp(-i\mathbf{k} \cdot \mathbf{r}')\, S(\mathbf{r}')\, d^3r' \tag{5.27}$$

$$= \frac{\exp(ik_0 r)}{4\pi r} \tilde{S}(\mathbf{k}) , \tag{5.28}$$

where $\tilde{S}$ denotes the Fourier transform of S. Thus we see that in the far-zone of the source, the radiated field behaves as an outgoing spherical wave with amplitude $\tilde{S}$. Using this result, it is easily seen that the energy current $\mathbf{J}$ carried by the field, defined in (2.39), is given by

$$\mathbf{J} = \frac{|\tilde{S}|^2}{(4\pi)^2 r^2} \hat{\mathbf{r}} . \tag{5.29}$$

The power passing through an element of area $R^2 d\Omega$ on a spherical surface of radius R is of the form

$$dP = (\mathbf{J} \cdot \hat{\mathbf{r}})|_{r=R}\, R^2 d\Omega . \tag{5.30}$$

We thus define the angular distribution of radiated power by

$$\frac{dP}{d\Omega} = \lim_{R\to\infty} (\mathbf{J}\cdot\hat{\mathbf{r}})\,|_{r=R}\, R^2 . \tag{5.31}$$

The quantity $dP/d\Omega$ is of interest since it is independent of how far the observer is from the source. Using (5.29), the angular distribution of the radiated power is seen to be given by

$$\frac{dP}{d\Omega} = \frac{|\tilde{S}|^2}{(4\pi)^2} . \tag{5.32}$$

The finite energy flow at large distances is a consequence of the $1/r$ decay of the field. Note that the radiated power is isotropic for a point source. This result is the analog of the dipole radiation pattern for scalar fields. It differs from the anisotropic electric-dipole radiation pattern, which is a consequence of the transversality of the vector electromagnetic fields.

References and Additional Reading

For an introduction to Green's functions, see for example:

P.M. Morse and H. Feshbach, *Methods of Theoretical Physics* (McGraw Hill, New York, 1953), chap. 7 and 11.

F.W. Byron Jr. and R.W. Fuller, *Mathematics of Classical and Quantum Physics* (Dover, New York, 1992), chap. 7.

E.M. Economou, *Green's Functions in Quantum Physics*, 3rd edition (Springer, Berlin, 2006).

Green's function, integral theorems and the extinction theorem in optics are presented in the following textbooks:

E. Wolf, "A Generalized Extinction Theorem and Its Role in Scattering Theory," in *Coherence and Quantum Optics*, L. Mandel and E. Wolf, eds (Plenum Press, New York, 1973), p. 339.

J.A. DeSanto, *Scalar Wave Theory* (Springer-Verlag, Berlin, 1992).

M. Born and E. Wolf, *Principles of Optics*, 7th edition (Cambridge University Press, Cambridge, 2005), chap. 8.

M. Nieto-Vesperinas, *Scattering and Diffraction in Physical Optics*, 2nd edition (World Scientific, Singapore, 2006), chap. 1 and 6.

6
Plane-Wave Expansions

In this chapter, we introduce plane-wave decompositions for the scalar field and Green's function in homogeneous media. Such expansions are important mathematical tools for the treatment of many problems in scattering theory and physical optics.

6.1 Plane-Wave Modes

We begin by considering the propagation of a monochromatic scalar field in a source-free region. The field satisfies the wave equation

$$\nabla^2 U + k_0^2 U = 0, \tag{6.1}$$

where k_0 is the wavenumber in free space. The field can be written as the two-dimensional Fourier integral

$$U(\mathbf{r}) = \int \tilde{U}(\mathbf{q}, z) \exp(i\mathbf{q} \cdot \boldsymbol{\rho}) \frac{d^2 q}{(2\pi)^2}, \tag{6.2}$$

where $\mathbf{r} = (\boldsymbol{\rho}, z)$ and the z-axis has been chosen arbitrarily. In order to determine the Fourier components $\tilde{U}(\mathbf{q}, z)$, we insert the above expression into (6.1), which leads to

$$\int \left[\frac{\partial^2 \tilde{U}}{\partial z^2} + (k_0^2 - q^2)\tilde{U} \right] \exp(i\mathbf{q} \cdot \boldsymbol{\rho}) \frac{d^2 q}{(2\pi)^2} = 0. \tag{6.3}$$

Since (6.3) holds for each $\boldsymbol{\rho}$, $\tilde{U}(\mathbf{q}, z)$ obeys the differential equation

$$\frac{\partial^2 \tilde{U}}{\partial z^2} + k_z^2(q)\tilde{U} = 0, \tag{6.4}$$

where

$$k_z(q) = \sqrt{k_0^2 - q^2} \text{ when } q \le k_0 \tag{6.5}$$

$$k_z(q) = i\sqrt{q^2 - k_0^2} \text{ when } q > k_0 \,. \tag{6.6}$$

The general solution of (6.4) is

$$\tilde{U}(\mathbf{q}, z) = A(\mathbf{q}) \exp(ik_z(q)z) + B(\mathbf{q}) \exp(-ik_z(q)z) \,, \tag{6.7}$$

where $A(\mathbf{q})$ and $B(\mathbf{q})$ are arbitrary complex coefficients. Substituting this result in (6.2), we find that $U(\mathbf{r})$ can be expressed as a superposition of plane waves, or plane-wave expansion:

$$U(\mathbf{r}) = \int \left[A(\mathbf{q}) \exp\left(i\mathbf{q} \cdot \boldsymbol{\rho} + ik_z(q)z\right) + B(\mathbf{q}) \exp\left(i\mathbf{q} \cdot \boldsymbol{\rho} - ik_z(q)z\right) \right] \frac{d^2 q}{(2\pi)^2} \,. \tag{6.8}$$

The exponential terms on the right-hand side in (6.8) are known as plane-wave modes. The first term corresponds to plane waves propagating towards $z > 0$, while the second term corresponds to plane waves propagating towards $z < 0$. The modes may be classified according to whether $|\mathbf{q}| \le k_0$ or $|\mathbf{q}| > k_0$. In the former case, the modes correspond to propagating waves since $k_z(q)$ is real-valued. In the latter case, $k_z(q)$ is purely imaginary and the modes are evanescent. Evanescent waves are exponentially small on scales that are large compared to the wavelength. They are of primary importance in near-field optics, where the subwavelength structure of wave fields is of interest.

6.2 Weyl Formula

We now show that the free-space Green's function (5.16) can be expanded into plane-wave modes. To proceed, we expand the spherical wave amplitude in the form

$$\frac{\exp(ik_0 r)}{r} = \int A(\mathbf{q}) \exp\left(i\mathbf{q} \cdot \boldsymbol{\rho} + ik_z(q)z\right) \frac{d^2 q}{(2\pi)^2} \,, \tag{6.9}$$

where $z \ge 0$. To find $A(\mathbf{q})$, we put $z = 0$ and invert the two-dimensional Fourier transform:

$$\begin{aligned} A(\mathbf{q}) &= \int \frac{\exp(ik_0\rho)}{\rho} \exp(-i\mathbf{q} \cdot \boldsymbol{\rho})\, d^2\rho \\ &= 2\pi \int_0^\infty \exp(ik_0\rho)\, J_0(q\rho)\, d\rho \\ &= 2\pi i \frac{1}{\sqrt{k_0^2 - q^2}} \,. \end{aligned} \tag{6.10}$$

Hence, for $z \geq 0$ we obtain

$$\frac{\exp(ik_0 r)}{r} = \frac{i}{2\pi} \int \frac{1}{k_z(q)} \exp\left(i\mathbf{q} \cdot \boldsymbol{\rho} + ik_z(q)z\right) d^2q \,. \tag{6.11}$$

When $z \leq 0$, it can be seen that

$$\frac{\exp(ik_0 r)}{r} = \frac{i}{2\pi} \int \frac{1}{k_z(q)} \exp\left(i\mathbf{q} \cdot \boldsymbol{\rho} - ik_z(q)z\right) d^2q \,. \tag{6.12}$$

Using (6.11) and (6.12), we can write the free-space scalar Green's function G_0 in the form

$$\begin{aligned} G_0(\mathbf{r}, \mathbf{r}') &= \frac{\exp(ik_0|\mathbf{r} - \mathbf{r}'|)}{4\pi\,|\mathbf{r} - \mathbf{r}'|} \\ &= \frac{i}{8\pi^2} \int \frac{1}{k_z(q)} \exp\left(i\mathbf{q} \cdot (\boldsymbol{\rho} - \boldsymbol{\rho}') + ik_z(q)|z - z'|\right) d^2q \,. \end{aligned} \tag{6.13}$$

The above result, which expresses the free-space Green's function as an expansion into propagating and evanescent plane-wave modes, is known as the Weyl formula.

6.3 Beam-Like Fields

Beams are wave fields that propagate with a small angular deviation from a fixed direction, which defines the axis of the beam. Consider a field propagating into the half-space $z > 0$ represented as the plane-wave expansion

$$U(\mathbf{r}) = \int A(\mathbf{q}) \exp\left(i\mathbf{q} \cdot \boldsymbol{\rho} + ik_z(q)z\right) \frac{d^2q}{(2\pi)^2} \,. \tag{6.14}$$

If $A(\mathbf{q}) \neq 0$ only for $|\mathbf{q}| \ll k_0$, then (6.14) corresponds to a beam whose axis is along the z-direction and that propagates towards $z > 0$. In this situation, we make the paraxial approximation in which $k_z(q)$ is replaced by its second-order expansion in terms of $|\mathbf{q}|/k_0$:

$$k_z(q) = k_0 - \frac{1}{2k_0} q^2 + O(q^4) \,. \tag{6.15}$$

Thus (6.14) becomes

$$U(\mathbf{r}) = \exp(ik_0 z) \int_{|\mathbf{q}| \ll k_0} A(\mathbf{q}) \exp\left(i\mathbf{q} \cdot \boldsymbol{\rho} - \frac{iz}{2k_0} q^2\right) \frac{d^2q}{(2\pi)^2} \,. \tag{6.16}$$

Equation (6.14) is the general form of a paraxial wave field. It defines a beam propagating in the positive z-direction.

If U is specified on the plane $z = 0$, which is known as the waist plane of the beam, then $A(\mathbf{q})$ can be determined from the two-dimensional Fourier transform

$$A(\mathbf{q}) = \int U(\rho, 0) \exp(-i\mathbf{q} \cdot \rho)\, d^2\rho \,. \tag{6.17}$$

Using this result, we see that the beam propagates outward from the waist plane according to

$$U(\mathbf{r}) = \int K(\rho, \rho', z) U(\rho', 0)\, d^2\rho' \tag{6.18}$$

where K is given by the expression

$$K(\rho, \rho', z) = \exp(ik_0 z) \int_{|\mathbf{q}| \ll k_0} \exp\left[i\mathbf{q} \cdot (\rho - \rho') - \frac{iz}{2k_0} q^2 \right] \frac{d^2q}{(2\pi)^2} \,. \tag{6.19}$$

The integral in (6.19) can be evaluated by making use of the identity

$$\int_{-\infty}^{\infty} \exp(-ax^2 + bx) dx = \sqrt{\frac{\pi}{a}} \exp\left(\frac{b^2}{4a}\right) \,. \tag{6.20}$$

We thus obtain

$$K(\rho, \rho', z) = \frac{ik_0}{2\pi z} \exp\left[\frac{ik_0}{2z} (\rho - \rho')^2 + ik_0 z \right] , \tag{6.21}$$

after extending the integration in (6.19) to all of space. The function K is known as the beam propagator. It may be interpreted as the field produced by a secondary point source located in the waist plane of the beam. We note the quadratic phase in the transverse direction, which is typical of beam-like fields.

Evidently, the field U calculated in the paraxial approximation is not an exact solution to the wave equation. To understand the nature of this approximation, we set

$$U(\mathbf{r}) = \exp(ik_0 z)\, u(\mathbf{r}) \,, \tag{6.22}$$

where u is defined by (6.16). Then it is readily seen that u satisfies the equation

$$\Delta u + 2ik_0 \frac{\partial u}{\partial z} = 0 \,, \tag{6.23}$$

where Δ denotes the two-dimensional Laplacian with respect to ρ. Equation (6.23) is referred to as the paraxial wave equation. It has the form of a two-dimensional Schrodinger equation where the z coordinate plays the role of time.

The paraxial equation may also be obtained by regarding (6.22) as an ansatz that defines u. Then, inserting (6.22) into the wave equation

$$\nabla^2 U + k_0^2 U = 0 \,, \tag{6.24}$$

we obtain

$$\Delta u + \frac{\partial^2 u}{\partial z^2} + 2ik_0 \frac{\partial u}{\partial z} = 0 \,. \tag{6.25}$$

Comparing (6.25) and (6.23), we see that the paraxial approximation holds when the condition

$$\left| \frac{\partial^2 u}{\partial z^2} \right| \ll k_0 \left| \frac{\partial u}{\partial z} \right| \,, \tag{6.26}$$

that is, when u varies slowly on the scale of the wavelength.

References and Additional Reading

For a discussion of plane-wave expansions, the Weyl formula, and properties of the beam propagator, see:

L. Mandel and E. Wolf, *Optical Coherence and Quantum Optics* (Cambridge University Press, Cambridge, 1995), chap. 3.

M. Nieto-Vesperinas, *Scattering and Diffraction in Physical Optics*, 2nd edition (World Scientific, Singapore, 2006), chap. 5 and 6.

Plane-wave expansions are the basic ingredient of Fourier optics, as described in:

J.W. Goodman, *Introduction to Fourier Optics*, 2nd edition (McGraw Hill, New York, 1996), chap. 3.

Plane-wave expansions with evanescent modes are useful in the description of near-field optics:

J.-J. Greffet and R. Carminati, “Image formation in near-field optics", Prog. Surf. Sci. **56**, 133 (1997).

L. Novotny and B. Hect, *Principles of Nano-Optics* (Cambridge University Press, Cambridge, 2006), chap. 2.

M. Nieto-Vesperinas, *Scattering and Diffraction in Physical Optics*, 2nd edition (World Scientific, Singapore, 2006), chap. 11.

7
Diffraction

The term *diffraction* refers to a form of scattering in which the scatterer is characterized by boundary conditions rather than a scattering potential. Diffraction occurs when the linear dimensions of the scattering object are of the order of the wavelength. In this circumstance, deviations from the predictions of geometrical optics occur. Well-known examples are the diffraction of light by an aperture or an opaque obstacle.

7.1 Rayleigh–Sommerfeld Formulas

We begin by considering the propagation of light through an aperture in an opaque surface, here referred to as a *screen*. According to geometrical optics, there should be a shadow behind the screen that is clearly distinct from the region that is illuminated. Instead, as we will see, the effects of diffraction lead to the formation of an interference pattern.

We first develop the theory of diffraction for an aperture in a planar screen coinciding with the plane $z = 0$. We consider a monochromatic scalar field whose source is taken to lie to the left of the screen in the $z < 0$ half-space, and we wish to determine the field to the right of the screen in the $z > 0$ half-space. In this half-space, the field obeys the equation

$$\nabla^2 U + k_0^2 U = 0 \tag{7.1}$$

with k_0 the free-space wavenumber. Following the development in Chapter 6, we introduce the plane-wave expansion of U:

$$U(\mathbf{r}) = \int A(\mathbf{q}) \exp\left(i\mathbf{q} \cdot \boldsymbol{\rho} + ik_z(q)z\right) \frac{d^2q}{(2\pi)^2}, \tag{7.2}$$

where $k_z(q)$ is given by (6.5) and (6.6). If U is prescribed on the plane $z = 0$, the so-called Dirichlet problem, then we may determine $A(\mathbf{q})$ from the Fourier integral

$$A(\mathbf{q}) = \int U(\boldsymbol{\rho}, 0) \exp(-i\mathbf{q} \cdot \boldsymbol{\rho})\, d^2\rho \,, \tag{7.3}$$

where $\boldsymbol{\rho}$ is the coordinate in the $z = 0$ plane. Using this expression, (7.2) becomes

$$U(\mathbf{r}) = \int K_D(\boldsymbol{\rho}, \boldsymbol{\rho}', z)\, U(\boldsymbol{\rho}', 0)\, d^2\rho' \,, \tag{7.4}$$

where the Dirichlet propagator K_D,

$$K_D(\boldsymbol{\rho}, \boldsymbol{\rho}', z) = \int \exp\left(i\mathbf{q} \cdot (\boldsymbol{\rho} - \boldsymbol{\rho}') + ik_z(q)z\right) \frac{d^2q}{(2\pi)^2} \,. \tag{7.5}$$

Equation (7.4) can be used to propagate the field from the $z = 0$ plane to any point $\mathbf{r} = (\boldsymbol{\rho}, z)$ in the half-space $z > 0$.

Note that in the far-field, the evanescent plane-wave modes in (7.5) are exponentially small and only low spatial frequencies with wave vectors $|\mathbf{q}| \leq k_0$ survive propagation. Such wave vectors correspond to features at the scale of the wavelength $\lambda = 2\pi/k_0$ or larger. That is, upon propagation to the far-zone, all features of the field varying at scales smaller than the wavelength λ are filtered out. This may be interpreted to mean that the far-field resolution of optical systems is of order λ, which is often referred to as the diffraction limit.

We can now derive the Rayleigh–Sommerfeld formulas. To proceed, we express K_D in terms of the free-space Green's function G_0 and make use of the plane-wave expansion (6.13) to obtain

$$\frac{\partial}{\partial z} G_0(\mathbf{r}, \mathbf{r}')\big|_{z'=0} = -\frac{1}{8\pi^2} \int \exp\left(i\mathbf{q} \cdot (\boldsymbol{\rho} - \boldsymbol{\rho}') + ik_z(q)z\right) d^2q \,. \tag{7.6}$$

By comparing the above result with (7.5), we find that

$$K_D(\boldsymbol{\rho}, \boldsymbol{\rho}', z) = -2\frac{\partial}{\partial z} G_0(\mathbf{r}, \mathbf{r}')\big|_{z'=0} \,. \tag{7.7}$$

Inserting (7.7) into (7.4), we obtain

$$U(\mathbf{r}) = -2 \int_{z'=0} \frac{\partial}{\partial z} G_0(\mathbf{r}, \mathbf{r}') U(\mathbf{r}')\, d^2r' \,, \tag{7.8}$$

where the integral is over the plane containing the aperture. Equation (7.8) is known as the first Rayleigh–Sommerfeld formula.

In the Neumann problem, the derivative $\partial U/\partial z$ is specified on the $z = 0$ plane. Following reasoning similar to the above, we obtain a result analogous to (7.4):

$$U(\mathbf{r}) = \int K_N(\boldsymbol{\rho}, \boldsymbol{\rho}', z) \frac{\partial U}{\partial z}(\boldsymbol{\rho}', 0)\, d^2\rho', \tag{7.9}$$

where the Neumann propagator K_N is given by

$$K_N(\boldsymbol{\rho}, \boldsymbol{\rho}', z) = \int \frac{i}{k_z(q)} \exp\left(i\mathbf{q}\cdot(\boldsymbol{\rho}-\boldsymbol{\rho}') + ik_z(q)z\right) \frac{d^2q}{(2\pi)^2} . \tag{7.10}$$

Making use once again of the plane-wave expansion (6.13), we find that

$$K_N(\boldsymbol{\rho}, \boldsymbol{\rho}', z) = 2G_0(\mathbf{r}, \mathbf{r}')\big|_{z'=0} . \tag{7.11}$$

Thus we obtain

$$U(\mathbf{r}) = 2\int_{z'=0} G_0(\mathbf{r}, \mathbf{r}') \frac{\partial}{\partial z'} U(\mathbf{r}')\, d^2r' , \tag{7.12}$$

which is referred to as the second Rayleigh–Sommerfeld formula.

The Rayleigh–Sommerfeld formulas determine the diffracted field self-consistently. An extremely useful simplification is possible if the linear dimensions of the aperture are large compared to the wavelength. In this situation, the field on the right-hand side in (7.8) and (7.12) may be taken to be approximately equal to the incident field U_0 within the aperture, and to vanish everywhere else in the plane of the screen. Under these conditions, which define the so-called Kirchhoff approximation, (7.8) and (7.12) assume the form

$$U(\mathbf{r}) = -2\int_{\mathcal{A}} \frac{\partial}{\partial z} G_0(\mathbf{r}, \mathbf{r}') U_0(\mathbf{r}')\, d^2r' , \tag{7.13}$$

$$U(\mathbf{r}) = 2\int_{\mathcal{A}} G_0(\mathbf{r}, \mathbf{r}') \frac{\partial}{\partial z'} U_0(\mathbf{r}')\, d^2r' , \tag{7.14}$$

where the incident field U_0 denotes the field that would exist in the absence of the screen, and $\mathcal{A}$ denotes the aperture. We note that the above result provides an accurate approximation to the field, except at points which are close to the edge of the aperture.

In principle, either (7.13) or (7.14) can be used to compute U. Alternatively, they can be combined into the formula

$$U(\mathbf{r}) = \int_{\mathcal{A}} \left[G(\mathbf{r}, \mathbf{r}') \frac{\partial}{\partial z'} U_0(\mathbf{r}') - \frac{\partial}{\partial z} G(\mathbf{r}, \mathbf{r}') U_0(\mathbf{r}') \right] d^2r' . \tag{7.15}$$

This result, which is known as the Kirchhoff diffraction integral, can also be derived from the Kirchhoff integral formula (5.5). There are well-known mathematical difficulties with this derivation, since the problem of simultaneously specifying both Dirichlet and Neumann data for the wave equation (7.1) is inconsistent.

7.2 Fresnel and Fraunhofer Diffraction

We consider two approximations that can be used to simplify calculations within diffraction theory. For simplicity, we restrict our attention to the case of the first Rayleigh–Sommerfeld formula.

We begin with the Fresnel approximation. In this case, we assume that the field propagates with a small angular deviation from the optical axis and make the paraxial approximation

$$k_z(q) = k_0 - \frac{1}{2k_0}q^2 \tag{7.16}$$

that was introduced in Chapter 6. The Dirichlet propagator simplifies into

$$K_D(\boldsymbol{\rho}, \boldsymbol{\rho}'; z) = \frac{ik_0}{2\pi z} \exp\left(\frac{ik_0}{2z}\left(\boldsymbol{\rho} - \boldsymbol{\rho}'\right)^2 + ik_0 z\right) . \tag{7.17}$$

If we insert this result into (7.13), we find that the field propagates to the right of the aperture according to

$$U(\mathbf{r}) = \frac{ik_0}{2\pi}\frac{\exp(ik_0 z)}{z} \int_A \exp\left(\frac{ik_0}{2z}\left(\boldsymbol{\rho} - \boldsymbol{\rho}'\right)^2\right) U_0(\boldsymbol{\rho}', 0)\, d^2\rho' . \tag{7.18}$$

The above result is called the Fresnel diffraction formula. We note that Fresnel diffraction is closely related to the theory of beam-like fields as discussed in Chapter 6. In fact, (6.18) and (6.21) imply that (7.18) may be interpreted as defining the propagation of a beam which coincides with the field U_0 in its waist plane.

We now turn to the Fraunhofer approximation. Here we assume that the point of observation is in the far-zone of the aperture. We may thus make the far-field approximation to the free-space Green's function

$$G_0(\mathbf{r}, \mathbf{r}') \sim \frac{\exp(ik_0 r)}{4\pi r} \exp(-ik_0 \hat{\mathbf{r}} \cdot \mathbf{r}') , \tag{7.19}$$

which is valid when $r \gg r'$ and $r \gg r'^2/\lambda$. According to the first Rayleigh–Sommerfeld formula, we require the derivative

$$\frac{\partial}{\partial z'} G_0(\mathbf{r}, \mathbf{r}') \sim -ik_0 \frac{\exp(ik_0 r)}{4\pi r} \exp(-ik_0 \hat{\mathbf{r}} \cdot \mathbf{r}') \cos\theta , \tag{7.20}$$

where θ is the angle between the direction of observation $\hat{\mathbf{r}}$ and the z-axis. Using this result, (7.13) becomes

$$U(\mathbf{r}) \sim \frac{ik_0}{2\pi}\frac{\exp(ik_0 r)}{r} \cos\theta \int_A \exp(-ik_0 \hat{\mathbf{r}} \cdot \mathbf{r}') U_0(\mathbf{r}')\, d^2 r' , \tag{7.21}$$

which is known as the Fraunhofer diffraction formula.

Evidently, within the accuracy of the Fraunhofer approximation, the diffracted field behaves as an outgoing spherical wave of the form

$$U(\mathbf{r}) = A\,\frac{\exp(ik_0 r)}{4\pi r}\,, \tag{7.22}$$

where the amplitude A is defined by

$$A = 2ik_0 \cos\theta \int_A \exp(-ik_0\hat{\mathbf{r}}\cdot\mathbf{r}')\, U_0(\mathbf{r}')\, d^2r'\,. \tag{7.23}$$

From (5.28) and (5.32), it can be seen that the angular distribution of diffracted intensity is given by

$$\frac{dP}{d\Omega} = \frac{|A|^2}{(4\pi)^2}\,. \tag{7.24}$$

7.3 Circular Aperture

As a simple example, we compute the Fraunhofer diffraction pattern for a circular aperture of radius a. We assume that the incident field is a unit-amplitude plane wave at normal incidence:

$$U_0(\mathbf{r}) = \exp(ik_0 z)\,. \tag{7.25}$$

The integral defining the amplitude A can then be performed:

$$A = 2ik_0\cos\theta \int_0^{2\pi}\int_0^a r'\exp\left[-ik_0 r'\sin\theta\cos(\phi-\phi')\right] dr'd\phi' \tag{7.26}$$

$$= 4\pi i k_0 \cos\theta \int_0^a r' J_0(k_0 r'\sin\theta) dr' \tag{7.27}$$

$$= 4\pi i k_0 a^2 \cos\theta\, \frac{J_1(k_0 a\sin\theta)}{k_0 a \sin\theta}\,. \tag{7.28}$$

Here ϕ is the azimuthal angle in spherical polar coordinates, ϕ' is the polar angle in the plane of the aperture, and we have used the identity

$$\int_0^x t J_0(t) dt = x J_1(x)\,. \tag{7.29}$$

From (7.24), we see that the angular distribution of radiated power is given by

$$\frac{dP}{d\Omega} = (k_0 a^2)^2 \cos^2\theta \left[\frac{J_1(k_0 a\sin\theta)}{k_0 a\sin\theta}\right]^2\,. \tag{7.30}$$

If $k_0 a \gg 1$, meaning that the radius of the aperture is large compared to the wavelength, then $J_1(k_0 a\sin\theta)/(k_0 a\sin\theta)$ and $dP/d\Omega$ are sharply peaked around $\theta = 0$ with a small angular width $\delta\theta \sim 1/(k_0 a)$. As a result, most of the incident wave passes through the aperture in the forward direction, as may be expected from

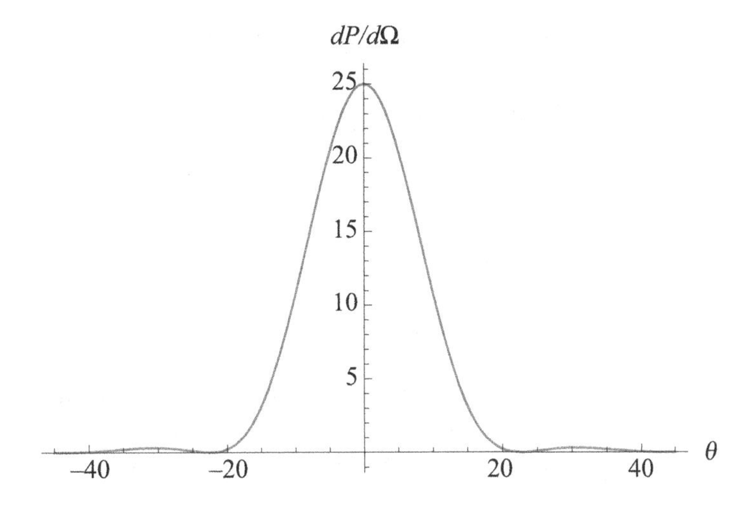

Fig. 7.1 Diffraction pattern produced from a circular aperture with $k_0 a = 10$.

geometrical optics. We note that the diffraction pattern has azimuthal symmetry about the optical axis and has minima that correspond to the zeros of $J_1(k_0 a \sin\theta)$, as shown in Fig. 7.1. The first zero of $J_1(x)$ is at $x = 3.83$ and thus the first minimum in the diffraction pattern occurs when $\sin\theta = 0.61\lambda/a$. This result appears in the usual formulas for the resolution limit of optical instruments using circular apertures, such as telescope or microscopes.

If $k_0 a \sim 1$, then interference effects are present and the predictions of geometrical optics must be modified. This point may be appreciated by examining the behavior of the scattering cross section σ_s, defined by

$$\sigma_s = \int \frac{d\sigma}{d\Omega} d\Omega \, . \tag{7.31}$$

Here the differential cross section $d\sigma_s/d\Omega$ is the radiated power normalized by the intensity of the incident field. We note that if the incident field is a plane wave of unit amplitude, then $d\sigma_s/d\Omega$ and $dP/d\Omega$ coincide. When $k_0 a \gg 1$, we expect that $\sigma_s \simeq \pi a^2$, which is the area of the aperture. To see this, we consider the cross section for scattering into the forward hemisphere, which is given by the expression

$$\sigma_s = 2\pi a^2 \int_0^{\pi/2} \cos^2\theta \frac{J_1^2(k_0 a \sin\theta)}{\sin\theta} d\theta \, . \tag{7.32}$$

We may evaluate the integral by making the small-angle approximation $\sin\theta \simeq \theta$ and $\cos\theta \simeq 1$, with the result

$$\sigma_s = 2\pi a^2 \int_0^{k_0 a\pi/2} \frac{J_1^2(x)}{x} dx \tag{7.33}$$

$$= \pi a^2 \left[1 - J_0^2(k_0 a\pi/2) - J_1^2(k_0 a\pi/2)\right] \, . \tag{7.34}$$

We may now use the asymptotic form of the Bessel functions

$$J_0(x) \sim \sqrt{\frac{2}{\pi x}} \cos(x - \pi/4) , \tag{7.35}$$

$$J_1(x) \sim \sqrt{\frac{2}{\pi x}} \sin(x - \pi/4) , \tag{7.36}$$

for $x \gg 1$. This leads to the result

$$\sigma_s = \pi a^2 \left(1 - \frac{4}{\pi^2} \frac{1}{k_0 a}\right) , \tag{7.37}$$

which exhibits the corrections to the geometrical optics cross section due to diffraction. In Fig. 7.2 we show the frequency-dependence of σ_s, obtained by numerical evaluation of the integral in (7.32). Also shown is the small-angle approximation (7.34), which is seen to be quite accurate at high frequencies.

If $k_0 a \ll 1$, diffraction theory is not expected to be valid (due to the failure of the Kirchhoff approximation). Nevertheless, it is instructive to examine the behavior of σ_s in this limit. We require the asymptotic form of the Bessel function $J_1(x)$ for $x \ll 1$:

$$J_1(x) = \frac{1}{2}x + O(x^3) . \tag{7.38}$$

Equation (7.32) thus becomes

$$\sigma_s = \frac{\pi}{2} a^2 (k_0 a)^2 \int_0^{\pi/2} \cos^2\theta \sin\theta d\theta \tag{7.39}$$

$$= \frac{\pi}{6} a^2 (k_0 a)^2 . \tag{7.40}$$

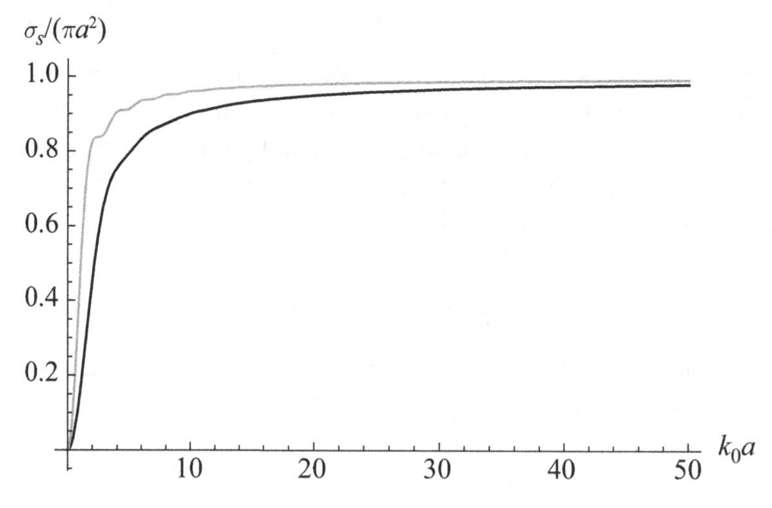

Fig. 7.2 Frequency-dependence of the normalized cross section σ_s for diffraction from a circular aperture. Top curve: small-angle approximation. Bottom curve: exact result.

We note that σ_s is small for small apertures and vanishes when the diameter of the aperture shrinks to zero.

References and Additional Reading

The following textbooks contain chapters dedicated to the theory of diffraction:

J.W. Goodman, *Introduction to Fourier Optics*, 2nd edition (McGraw Hill, New York, 1996), chap. 3 and 4.

M. Born and E. Wolf, *Principles of Optics*, 7th edition (Cambridge University Press, Cambridge, 2005), chap. 8.

M. Nieto-Vesperinas, *Scattering and Diffraction in Physical Optics*, 2nd edition (World Scientific, Singapore, 2006), chap. 6.

A more formal treatment of diffraction problems can also be found in:

P.M. Morse and H. Feshbach, *Methods of Theoretical Physics* (McGraw-Hill, New York, 1953), vol. 2, chap. 9 and 11.

8

Coherence Theory: Basic Concepts

In this chapter, we introduce the theory of optical coherence, which is concerned with characterizing the statistical fluctuations of optical fields. We will see that such fluctuations can be described in terms of temporal and spatial correlation functions, and that these correlation functions arise in the description of interferometric experiments.

8.1 Analytic Signal Representation

Until now, we have restricted our attention to deterministic optical fields. However, it is often useful to regard optical fields as random. For instance, the light emitted by a thermal source exhibits random fluctuations in time due to spontaneous emission from the atoms that comprise the source. Likewise, the output intensity emitted from a laser has some degree of randomness arising from spontaneous emission and vibrations in the laser cavity, among other causes. The presence of temporal fluctuations has an important physical consequence, namely that the field is no longer monochromatic.

We begin by considering the propagation of a scalar field $U(\mathbf{r}, t)$ in free space, which obeys the wave equation

$$\nabla^2 U = \frac{1}{c^2}\frac{\partial^2 U}{\partial t^2}, \tag{8.1}$$

where c is the speed of light in vacuum. The field U can be represented as the Fourier integral

$$U(\mathbf{r}, t) = \int_{-\infty}^{\infty} \widetilde{U}(\mathbf{r}, \omega) \exp(-i\omega t)\, \frac{d\omega}{2\pi}\,, \tag{8.2}$$

where $\widetilde{U}(\mathbf{r}, \omega)$ is the time-domain Fourier transform of $U(\mathbf{r}, t)$. Since U is real-valued, it follows that $\widetilde{U}(\mathbf{r}, -\omega) = \widetilde{U}^*(\mathbf{r}, \omega)$. Thus U is fully determined by its frequency components at positive frequencies. This leads us to introduce the complex-valued function

$$V(\mathbf{r}, t) = \int_0^\infty \widetilde{U}(\mathbf{r}, \omega) \exp(-i\omega t) \frac{d\omega}{2\pi} . \tag{8.3}$$

We see immediately that

$$U(\mathbf{r}, t) = V(\mathbf{r}, t) + V^*(\mathbf{r}, t) . \tag{8.4}$$

We note that V and V^* correspond to the positive and negative frequency parts of U, respectively. The function V is known as the analytic signal, and (8.4) is known as the analytic signal representation of U. The name stems from the fact that V is an analytic function in the lower-half t-plane. It follows that $\mathrm{Re}V$ and $\mathrm{Im}V$ form a Hilbert transform pair:

$$\mathrm{Re}V(\mathbf{r}, t) = -P \int_{-\infty}^{\infty} \frac{\mathrm{Im}V(\mathbf{r}, t')}{t - t'} \frac{dt'}{\pi} , \tag{8.5}$$

$$\mathrm{Im}V(\mathbf{r}, t) = P \int_{-\infty}^{\infty} \frac{\mathrm{Re}V(\mathbf{r}, t')}{t - t'} \frac{dt'}{\pi} , \tag{8.6}$$

where P stands for principal value.

The analytic signal V satisfies the wave equation (8.1). This follows from (8.3) and the fact that $\widetilde{U}$ obeys the frequency-domain wave equation

$$\nabla^2 \widetilde{U}(\mathbf{r}, \omega) + k_0^2 \widetilde{U}(\mathbf{r}, \omega) = 0 , \tag{8.7}$$

where $k_0 = \omega/c = 2\pi/\lambda$. We define the instantaneous intensity by

$$I(\mathbf{r}, t) = |V(\mathbf{r}, t)|^2 . \tag{8.8}$$

We note that I does not correspond to the intensity of the field U. However, if U is quasi-monochromatic, then the time average of I does correspond to the measurable intensity. To see this, suppose that the field U is of the form

$$U(\mathbf{r}, t) = A(\mathbf{r}, t) \cos\left(\omega t + \phi(\mathbf{r}, t)\right) , \tag{8.9}$$

where the amplitude A and phase ϕ are slowly varying on the time scale $1/\Delta\omega$ with $\Delta\omega/\omega \ll 1$. That is, the field has a central frequency ω and bandwidth $\Delta\omega$. The analytic signal representation of U is given by

$$V(\mathbf{r}, t) = \frac{1}{2} A(\mathbf{r}, t) \exp(i \left(\omega t + \phi(\mathbf{r}, t)\right)) . \tag{8.10}$$

The intensity recorded by a detector is proportional to the average

$$\overline{U^2}(\mathbf{r}, t) = \frac{1}{2T} \int_{t-T}^{t+T} U^2(\mathbf{r}, t') \, dt' , \tag{8.11}$$

where T is the integration time of the detector. Here $\omega T \gg 1$, so that the detector integrates over many periods of oscillation of the field at the frequency ω. Making use of (8.4) and (8.10), we find that $\overline{U^2}(\mathbf{r}, t) = |V^*(\mathbf{r}, t)|^2/2$, which follows from the fact that $|V|^2$ is slowly varying and contributes to the time average, while V^2 and V^{*2} are rapidly varying and do not contribute.

8.2 Random Fields and Coherence Functions

We now turn to the theory of random optical fields. We assume that the analytic signal V is a random process and introduce the correlation function

$$\Gamma(\mathbf{r}_1, t_1; \mathbf{r}_2, t_2) = \langle V^*(\mathbf{r}_1, t_1) V(\mathbf{r}_2, t_2) \rangle , \tag{8.12}$$

where $\langle \cdots \rangle$ denotes the statistical average. In coherence theory, Γ is known as the mutual coherence function. We will often assume that V is statistically stationary, which means that Γ is invariant under translations in time. We then write

$$\Gamma(\mathbf{r}_1, \mathbf{r}_2, \tau) = \langle V^*(\mathbf{r}_1, t) V(\mathbf{r}_2, t + \tau) \rangle , \tag{8.13}$$

where the dependence of the mutual coherence function on the time difference τ has been made explicit. We note that stationarity may be expected to hold when the sampling rate of the detector is much faster than the rate of variation of the optical field, so that on average, measurements of the field are independent of the origin of time. If the field is ergodic, then ensemble averages are equal to time averages, and Γ is also given by

$$\Gamma(\mathbf{r}_1, \mathbf{r}_2, \tau) = \lim_{T \to \infty} \frac{1}{2T} \int_{-T}^{T} V^*(\mathbf{r}_1, t) V(\mathbf{r}_2, t + \tau) \, dt . \tag{8.14}$$

The mutual coherence function Γ is the fundamental quantity in the theory of coherence. We will see that it is possible to determine Γ from interferometric measurements. We note that the average instantaneous intensity is given by

$$\langle I(\mathbf{r}, t) \rangle = \langle V^*(\mathbf{r}, t) V(\mathbf{r}, t) \rangle \tag{8.15}$$

$$= \Gamma(\mathbf{r}, \mathbf{r}, 0) . \tag{8.16}$$

It follows that for a statistically stationary random field, $\langle I \rangle$ is a time-independent quantity. We note the following useful properties of Γ:

$$\Gamma(\mathbf{r}_1, \mathbf{r}_2, \tau) = \Gamma(\mathbf{r}_2, \mathbf{r}_1, -\tau)^* , \tag{8.17}$$

$$\Gamma(\mathbf{r}, \mathbf{r}, \tau) \geq 0 , \tag{8.18}$$

$$|\Gamma(\mathbf{r}_1, \mathbf{r}_2, \tau)|^2 \leq \Gamma(\mathbf{r}_1, \mathbf{r}_1, \tau) \Gamma(\mathbf{r}_2, \mathbf{r}_2, \tau) . \tag{8.19}$$

Equation (8.19) follows from the Cauchy-Schwarz inequality applied to (8.14).

It will prove useful to normalize the mutual coherence function as follows:

$$\gamma(\mathbf{r}_1, \mathbf{r}_2, \tau) = \frac{\Gamma(\mathbf{r}_1, \mathbf{r}_2, \tau)}{\sqrt{\Gamma(\mathbf{r}_1, \mathbf{r}_1, 0)}\sqrt{\Gamma(\mathbf{r}_2, \mathbf{r}_2, 0)}}, \tag{8.20}$$

which defines the quantity γ known as the complex degree of coherence. It follows immediately from (8.19) that

$$0 \leq |\gamma(\mathbf{r}_1, \mathbf{r}_2, \tau)| \leq 1\,. \tag{8.21}$$

We now establish the relation between the mutual coherence function and the properties of the field in the spectral domain. To proceed, we introduce the time-domain Fourier transform of the mutual coherence function:

$$W(\mathbf{r}_1, \mathbf{r}_2, \omega) = \int \Gamma(\mathbf{r}_1, \mathbf{r}_2, \tau)\ \exp(i\omega\tau)\, d\tau\ , \tag{8.22}$$

which is known as the cross-spectral density. To understand its physical meaning, we consider the Fourier transform of the analytic signal V, which is defined by

$$\widetilde{V}(\mathbf{r}, \omega) = \int V(\mathbf{r}, t)\ \exp(i\omega t)\, dt\ , \tag{8.23}$$

where $\omega \geq 0$. Note that if V is statistically stationary, its Fourier transform must be viewed as a distribution. Proceeding formally, we find that the correlation function of $\widetilde{V}$ can be written in the form

$$\left\langle \widetilde{V}^*(\mathbf{r}_1, \omega_1) V(\mathbf{r}_2, \omega_2) \right\rangle = 2\pi\ W(\mathbf{r}_1, \mathbf{r}_2, \omega_1)\, \delta(\omega_1 - \omega_2)\,. \tag{8.24}$$

This relation shows that for a stationary field, two different frequencies are uncorrelated, and that the strength of the correlation at frequency ω_1 is $W(\mathbf{r}_1, \mathbf{r}_2, \omega_1)$. Inverting the Fourier integral (8.22) leads to

$$\Gamma(\mathbf{r}_1, \mathbf{r}_2, \tau) = \int_0^\infty W(\mathbf{r}_1, \mathbf{r}_2, \omega)\ \exp(-i\omega\tau)\, \frac{d\omega}{2\pi}\,. \tag{8.25}$$

It follows immediately that Γ is an analytic signal. The origin of the term cross-spectral density for $W(\mathbf{r}_1, \mathbf{r}_2, \omega)$ can be understood by setting $\mathbf{r}_1 = \mathbf{r}_2 = \mathbf{r}$ and $\tau = 0$ in Eq. (8.25), and making use of (8.16). We thus obtain

$$\langle I(\mathbf{r})\rangle = \int_0^\infty W(\mathbf{r}, \mathbf{r}, \omega)\, \frac{d\omega}{2\pi}\,. \tag{8.26}$$

This relation shows that the quantity $W(\mathbf{r}, \mathbf{r}, \omega)$ can be interpreted as a power spectral density. Rewriting (8.22) for $\mathbf{r}_1 = \mathbf{r}_2 = \mathbf{r}$ leads to

$$W(\mathbf{r}, \mathbf{r}, \omega) = \int \Gamma(\mathbf{r}, \mathbf{r}, \tau)\ \exp(i\omega\tau)\, d\tau\ , \tag{8.27}$$

which means that the power spectral density is the Fourier transform of the time correlation function of the field. This result is known as the Wiener-Khintchine theorem.

8.3 Interferometry

The mutual coherence function is observable in various interferometric experiments. We begin by considering the Michelson interferometer, in which a beam is split and made to interfere with itself by means of a pair of mirrors, as illustrated in Fig. 8.1.

If V_0 denotes the field that is incident on the beamsplitter, then the total field after passing through the interferometer is given by

$$V(\mathbf{r}, t) = V_0(\mathbf{r}, t - t_1) + V_0(\mathbf{r}, t - t_2) , \tag{8.28}$$

where $\mathbf{r}$ is the position of the detector. Here $t_1 = 2L_1/c$ and $t_2 = 2L_2/c$ denote the time delays of the beam, where L_1 and L_2 denote the corresponding path lengths in the interferometer. The average intensity measured by the detector is proportional to

$$\langle I(\mathbf{r})\rangle = \langle V^*(\mathbf{r}, t)V(\mathbf{r}, t)\rangle \tag{8.29}$$

$$= 2\Gamma(\mathbf{r}, \mathbf{r}, 0)\left[1 + \mathrm{Re}\Gamma(\mathbf{r}, \mathbf{r}, t_2 - t_1)\right] , \tag{8.30}$$

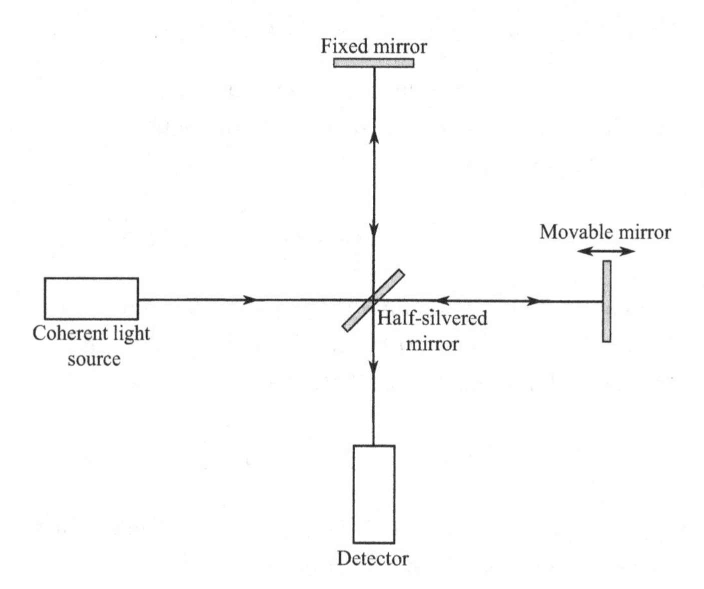

Fig. 8.1 The Michelson interferometer.

where $\Gamma(\mathbf{r}, \mathbf{r}', \tau) = \langle V_0^*(\mathbf{r}, t)V_0(\mathbf{r}', t + \tau)\rangle$ is the mutual coherence function of V_0, which is assumed to be statistically stationary. Upon introducing the complex degree of coherence γ, (8.30) becomes

$$\langle I(\mathbf{r})\rangle = 2\Gamma(\mathbf{r}, \mathbf{r}, 0)\left[1 + |\gamma(\mathbf{r}, \mathbf{r}, t_2 - t_1)| \cos\phi(\mathbf{r}, \mathbf{r}, t_2 - t_1)\right] , \tag{8.31}$$

where ϕ is the phase of γ. Equation (8.31) governs the interference of a statistically stationary optical field with a copy of itself shifted in time. We see that the intensity consists of contributions from each path, plus an interference term. If $\gamma = 0$, there is no interference and the field is said to be incoherent. If $|\gamma| = 1$, there is maximum interference and the field is coherent. If $|\gamma| < 1$, the field is called partially coherent. The one-point complex degree of coherence $\gamma(\mathbf{r}, \mathbf{r}, \tau)$ is a measure of the temporal coherence of the field. The width τ_c of $\gamma(\mathbf{r}, \mathbf{r}, \tau)$, considered as a function of τ, is the coherence time of the field. Using the Fourier transform relationship (8.25), we see that if $\Delta\omega$ is the bandwidth of the field, defined as the spectral width of W, we find that $\Delta\omega\tau_c \simeq 2\pi$. The relation

$$\tau_c = \frac{2\pi}{\Delta\omega} \tag{8.32}$$

illustrates the intimate connection between the coherence time and the bandwidth.

Next, we consider Young's two-slit experiment in which a thermal source is incident on a screen containing two slits, as shown in Fig. 8.2. The transmitted light is then detected at a point on a second screen. It follows from diffraction theory (within the Kirchhoff approximation) that if the slits are sufficiently large

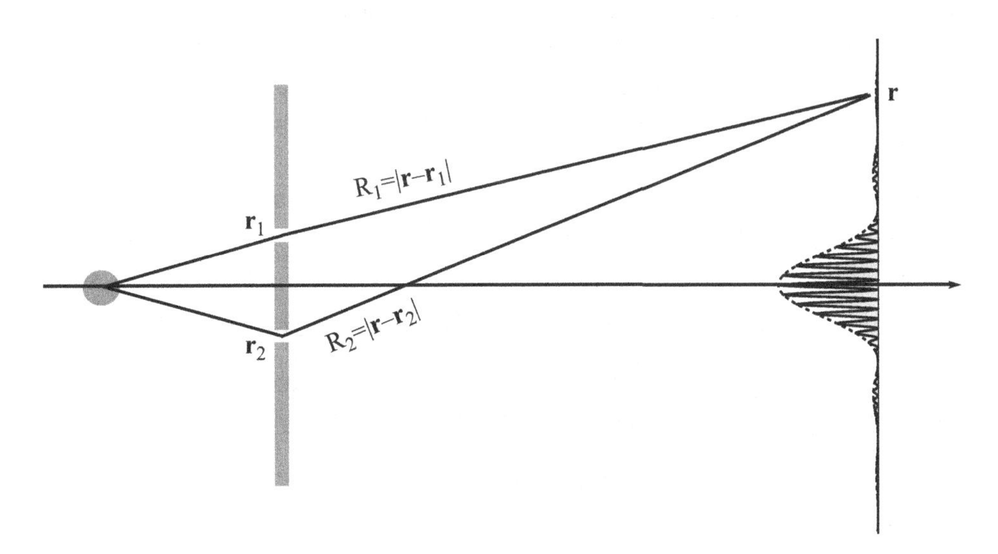

Fig. 8.2 Young's two-slit experiment.

and far from the second screen, then the total field is of the form

$$V(\mathbf{r}, t) = A_1 V(\mathbf{r}_1, t - t_1) + A_2 V(\mathbf{r}_2, t - t_2) , \tag{8.33}$$

where

$$t_1 = |\mathbf{r} - \mathbf{r}_1|/c , \quad t_2 = |\mathbf{r} - \mathbf{r}_2|/c , \tag{8.34}$$

are the travel times from the slits to the point of observation $\mathbf{r}$ on the second screen, and the amplitudes A_1, A_2 are purely imaginary. The average intensity is given by

$$\langle I(\mathbf{r}) \rangle = \left\langle V^*(\mathbf{r}, t) V(\mathbf{r}, t) \right\rangle . \tag{8.35}$$

Inserting (8.33) into (8.35) and making use of the definition of the mutual coherence function (8.13), we find that

$$\langle I(\mathbf{r}) \rangle = |A_1|^2 \langle I(\mathbf{r}_1) \rangle + |A_2|^2 \langle I(\mathbf{r}_2) \rangle + 2A_1^* A_2 \mathrm{Re}\Gamma(\mathbf{r}_1, \mathbf{r}_2, t_2 - t_1) . \tag{8.36}$$

Introducing the complex degree of coherence γ, the above result becomes

$$I(\mathbf{r}) = |A_1|^2 \langle I(\mathbf{r}_1) \rangle + |A_2|^2 \langle I(\mathbf{r}_2) \rangle + 2A_1^* A_2 \sqrt{I(\mathbf{r}_1) I(\mathbf{r}_2)} \mathrm{Re}\gamma(\mathbf{r}_1, \mathbf{r}_2, t_2 - t_1) . \tag{8.37}$$

(8.37) describes the interference between statistically stationary optical fields at two different points $\mathbf{r}_1$ and $\mathbf{r}_2$ and at two different times t_1 and t_2. The intensity consists of contributions from each slit, plus an interference term. As is the case for the Michelson interferometer, when $\gamma = 0$ the field is incoherent and there is no interference. If $|\gamma| = 1$, the field is coherent and there is maximum interference. Finally, $|\gamma| < 1$ corresponds to a partially coherent field. Similarly, the equal-time complex degree of coherence $\gamma(\mathbf{r}_1, \mathbf{r}_2, 0)$ describes the interference between two samples of the field taken at two different points at the same time. We say that $\gamma(\mathbf{r}_1, \mathbf{r}_2, 0)$ is a measure of the spatial coherence of the field at the points $\mathbf{r}_1$ and $\mathbf{r}_2$.

To further interpret (8.37), we make the simplifying assumptions that the slits are identical, the geometric factors $A_1 = A_2 = A$, and $\langle I(\mathbf{r}_1) \rangle = \langle I(\mathbf{r}_2) \rangle = I$. We thereby obtain

$$\langle I(\mathbf{r}) \rangle = 2|A|^2 I^2 \left[1 + |\gamma(\mathbf{r}_1, \mathbf{r}_2, t_2 - t_1)| \cos \phi(\mathbf{r}_1, \mathbf{r}_2, t_2 - t_1) \right] , \tag{8.38}$$

where ϕ is the phase of γ. If we vary the point of observation $\mathbf{r}$, then interference fringes are formed. The maximum and minimum intensities are given by

$$\langle I \rangle_{\mathrm{max}} = 2|A|^2 I^2 \left[1 + |\gamma| \right] , \quad \langle I \rangle_{\mathrm{min}} = 2|A|^2 I^2 \left[1 - |\gamma| \right] . \tag{8.39}$$

We define the visibility as

$$\mathcal{V} = \frac{\langle I \rangle_{\mathrm{max}} - \langle I \rangle_{\mathrm{min}}}{\langle I \rangle_{\mathrm{max}} + \langle I \rangle_{\mathrm{min}}} . \tag{8.40}$$

Note that $\mathcal{V} = |\gamma|$, so that the visibility of the interference pattern is a direct measure of the coherence of the field.

References and Additional Reading

For an introduction to optical coherence theory, see:

R. Loudon, *The Quantum Theory of Light* (Oxford University Press, Oxford, 1983), chap. 3.

J.W. Goodman, *Statistical Optics* (Wiley, New York, 1985), chap. 5.

L. Mandel and E. Wolf, *Optical Coherence and Quantum Optics* (Cambridge University Press, Cambridge, 1995), chap. 4.

The coherence of classical electromagnetic vector fields, and its connection to polarization, is presented in:

C. Brosseau, *Fundamentals of Polarized Light: A Statistical Optics Approach* (Wiley, New York, 1998).

E. Wolf, *Introduction to the Theory of Coherence and Polarization of Light* (Cambridge University Press, Cambridge, 2007).

The following textbooks cover the quantum theory of optical coherence:

R. Loudon, *The Quantum Theory of Light* (Oxford University Press, Oxford, 1983), chap. 6.

L. Mandel and E. Wolf, *Optical Coherence and Quantum Optics* (Cambridge University Press, Cambridge, 1995), chap. 12.

M.O. Scully and M.S. Zubairy, *Quantum Optics* (Cambridge University Press, Cambridge, 1997), chap. 3.

G. Grynberg, A. Aspect and C. Fabre, *Introduction to Quantum Optics: From the Semi-classical Approach to Quantized Light* (Cambridge University Press, Cambridge, 2000), chap. 5.

9

Coherence Theory: Propagation of Correlations

In this chapter, we explore a crucial unifying principle in coherence theory, namely that correlation functions obey wave equations and may thus be propagated according to the laws of physical optics. The van Cittert–Zernike theorem is an important consequence of this principle.

9.1 Wolf Equations

We consider a field $U(\mathbf{r}, t)$ propagating in free space and represented by its analytic signal $V(\mathbf{r}, t)$. We have seen in Chapter 8 that V obeys the wave equation

$$\nabla^2 V(\mathbf{r}_1, t_1) = \frac{1}{c^2} \frac{\partial^2 V(\mathbf{r}_1, t_1)}{\partial t_1^2} . \tag{9.1}$$

It is readily seen that the product $V^*(\mathbf{r}_1, t_1) V(\mathbf{r}_2, t_2)$ also obeys (9.1). Upon taking the ensemble average, it follows that the mutual coherence function

$$\Gamma(\mathbf{r}_1, t_1; \mathbf{r}_2, t_2) = \left\langle V^*(\mathbf{r}_1, t_1) V(\mathbf{r}_2, t_2) \right\rangle \tag{9.2}$$

obeys the wave equation

$$\nabla^2_{\mathbf{r}_1} \Gamma(\mathbf{r}_1, t_1; \mathbf{r}_2, t_2) = \frac{1}{c^2} \frac{\partial^2 \Gamma(\mathbf{r}_1, t_1; \mathbf{r}_2, t_2)}{\partial t_1^2} . \tag{9.3}$$

Likewise, Γ also obeys a wave equation with respect to its second set of arguments:

$$\nabla^2_{\mathbf{r}_2} \Gamma(\mathbf{r}_1, t_1; \mathbf{r}_2, t_2) = \frac{1}{c^2} \frac{\partial^2 \Gamma(\mathbf{r}_1, t_1; \mathbf{r}_2, t_2)}{\partial t_2^2} . \tag{9.4}$$

The wave equations (9.3) and (9.4) are known as the Wolf equations.

If V is statistically stationary, then $\Gamma(\mathbf{r}_1, t_1; \mathbf{r}_2, t_2)$ depends only on the difference $t_2 - t_1$. Accordingly, the cross-spectral density is given by the Fourier transform

$$W(\mathbf{r}_1, \mathbf{r}_2, \omega) = \int \Gamma(\mathbf{r}_1, t_1; \mathbf{r}_2, t_2) \exp(i\omega(t_2 - t_1))\, d(t_2 - t_1) . \tag{9.5}$$

It follows from (9.3), (9.4) and (9.5) that W obeys the pair of wave equations

$$\nabla^2_{\mathbf{r}_1} W(\mathbf{r}_1, \mathbf{r}_2, \omega) + k_0^2\, W(\mathbf{r}_1, \mathbf{r}_2, \omega) = 0 , \tag{9.6}$$

$$\nabla^2_{\mathbf{r}_2} W(\mathbf{r}_1, \mathbf{r}_2, \omega) + k_0^2\, W(\mathbf{r}_1, \mathbf{r}_2, \omega) = 0 , \tag{9.7}$$

where $k_0 = \omega/c = 2\pi/\lambda$.

We note the following properties of the cross-spectral density, which are analogous to (8.17)–(8.19):

$$W^*(\mathbf{r}_1, \mathbf{r}_2, \omega) = W(\mathbf{r}_2, \mathbf{r}_1, \omega) , \tag{9.8}$$

$$W(\mathbf{r}, \mathbf{r}, \omega) \geq 0 , \tag{9.9}$$

$$|W(\mathbf{r}_1, \mathbf{r}_2, \omega)|^2 \leq W(\mathbf{r}_1, \mathbf{r}_1, \omega) W(\mathbf{r}_2, \mathbf{r}_2, \omega) . \tag{9.10}$$

In addition, W is positive definite,

$$\int W(\mathbf{r}_1, \mathbf{r}_2, \omega) f^*(\mathbf{r}_1) f(\mathbf{r}_2)\, d^3r_1 d^3r_2 \geq 0, \tag{9.11}$$

for all complex-valued functions f. (9.11) follows from the inequality

$$\left\langle \left| \int \widetilde{V}(\mathbf{r}, \omega) f(\mathbf{r})\, d^3r \right|^2 \right\rangle \geq 0 , \tag{9.12}$$

where $\widetilde{V}$ is the Fourier transform of V, along with the relation

$$\langle \widetilde{V}^*(\mathbf{r}_1, \omega_1) V(\mathbf{r}_2, \omega_2) \rangle = 2\pi\, W(\mathbf{r}_1, \mathbf{r}_2, \omega_1)\delta(\omega_1 - \omega_2) , \tag{9.13}$$

which was derived in Chapter 8.

We will find it useful to introduce the normalized cross-spectral density,

$$\mu(\mathbf{r}_1, \mathbf{r}_2, \omega) = \frac{W(\mathbf{r}_1, \mathbf{r}_2, \omega)}{\sqrt{W(\mathbf{r}_1, \mathbf{r}_1, \omega)}\sqrt{W(\mathbf{r}_2, \mathbf{r}_2, \omega)}} , \tag{9.14}$$

which is known as the spectral degree of coherence. It follows immediately from (9.10) that $0 \leq |\mu| \leq 1$. The spectral degree of coherence is a measure of the coherence of an optical field at a fixed frequency. It can be determined from measurements carried out with narrow-band filters in a Young's two-slit experiment. As usual, if $\mu = 1$ the field is said to coherent and if $|\mu| = 0$ it is incoherent. If $|\mu| < 1$, the field is called partially coherent. The width of $\mu(\mathbf{r}_1, \mathbf{r}_2, \omega)$, considered as a function of $|\mathbf{r}_1 - \mathbf{r}_2|$, defines the spatial coherence length ℓ_{coh} of the field at frequency ω.

9.2 van Cittert–Zernike Theorem

We now consider the propagation of correlations of optical fields in free space. For simplicity, we study the case of the cross-spectral density, which is simpler than the mutual coherence function, since the former obeys the Helmholtz equation and the latter obeys the time-domain wave equation. We will make use of the propagation and diffraction laws described in Chapter 7.

Suppose that the cross-spectral density W is known on the planes $z_1 = z_2 = 0$, as shown in Fig. 9.1. We will show that W can be propagated into the half-spaces $z_1, z_2 > 0$. To proceed, we note that according to (9.13), $W(\mathbf{r}_1, \mathbf{r}_2, \omega)$ behaves as an outgoing wave in the $\mathbf{r}_2$ coordinates and as an incoming wave with respect to $\mathbf{r}_1$. It then follows from the first Rayleigh–Sommerfeld formula (7.8) that W can be propagated into the $z_2 > 0$ half-space according to

$$W(\mathbf{r}_1, \mathbf{r}_2, \omega) = \int_{z_2'=0} K_D(\mathbf{r}_2, \mathbf{r}_2') \, W(\mathbf{r}_1, \mathbf{r}_2', \omega) \, d^2 r_2' \,, \tag{9.15}$$

where

$$K_D(\mathbf{r}, \mathbf{r}') = -\frac{1}{2\pi} \frac{\partial}{\partial z} \frac{\exp(ik_0|\mathbf{r} - \mathbf{r}'|)}{|\mathbf{r} - \mathbf{r}'|}\Big|_{z'=0} \tag{9.16}$$

is the Dirichlet propagator. Similarly, it can be seen that

$$W(\mathbf{r}_1, \mathbf{r}_2, \omega) = \int_{z_1'=0} K_D^*(\mathbf{r}_1, \mathbf{r}_1') \, W(\mathbf{r}_1', \mathbf{r}_2, \omega) \, d^2 r_1' \,, \tag{9.17}$$

where the complex conjugate ensures the incoming sense of propagation with respect to $\mathbf{r}_1$. Making use of (9.17) and (9.15) yields

$$W(\mathbf{r}_1, \mathbf{r}_2, \omega) = \int_{z_1'=0} d^2 r_1' \int_{z_2'=0} d^2 r_2' \, K_D^*(\mathbf{r}_1, \mathbf{r}_1') \, K_D(\mathbf{r}_2, \mathbf{r}_2') \, W(\mathbf{r}_1', \mathbf{r}_2', \omega) \,, \tag{9.18}$$

which allows for the propagation of W.

As an application of (9.18), let us assume that the initial field is spatially incoherent, so that

$$W(\mathbf{r}_1, \mathbf{r}_2, \omega) = h(\mathbf{r}_2, \omega) \, \delta(\mathbf{r}_1 - \mathbf{r}_2) \tag{9.19}$$

on the plane $z_1 = z_2 = 0$. Here the function h describes the intensity distribution across the plane $z_1 = z_2 = 0$, hereafter referred to as the source plane. Equation (9.18) thus becomes

$$W(\mathbf{r}_1, \mathbf{r}_2, \omega) = \int K_D^*(\mathbf{r}_1, \mathbf{r}_2') \, K_D(\mathbf{r}_2, \mathbf{r}_2') \, h(\mathbf{r}_2', \omega) \, d^2 r_2' \,. \tag{9.20}$$

Note that if $h \neq 0$, then $W(\mathbf{r}_1, \mathbf{r}_2, \omega)$ is generally nonvanishing for $\mathbf{r}_1 \neq \mathbf{r}_2$. If follows that the field becomes spatially coherent upon propagation.

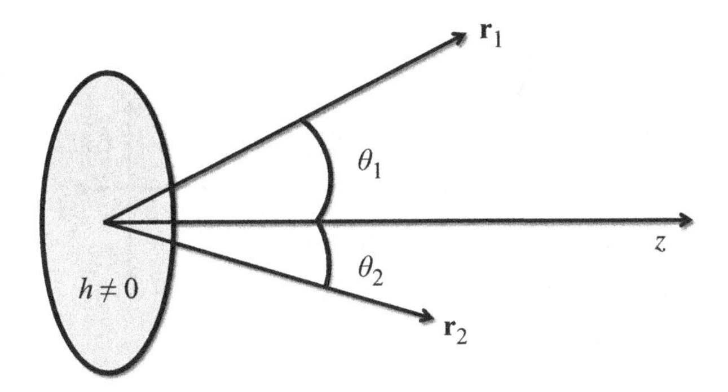

Fig. 9.1 Illustrating the geometry of the van Cittert–Zernike theorem.

Suppose that the points of observation are in the far-field of the source plane, so that $k_0|\mathbf{r}_1 - \mathbf{r}'_1| \gg 1$ and $k_0|\mathbf{r}_2 - \mathbf{r}'_2| \gg 1$. Making use of the asymptotic result

$$K_D(\mathbf{r}, \mathbf{r}') \sim -\frac{ik_0}{2\pi} \frac{\exp(ik_0|\mathbf{r} - \mathbf{r}'|)}{|\mathbf{r} - \mathbf{r}'|^2}(z - z') , \tag{9.21}$$

which holds for $k_0|\mathbf{r} - \mathbf{r}'| \gg 1$, we obtain

$$W(\mathbf{r}_1, \mathbf{r}_2, \omega) = \left(\frac{k_0}{2\pi}\right)^2 \int \frac{\exp(-ik_0|\mathbf{r}_1 - \mathbf{r}|)}{|\mathbf{r}_1 - \mathbf{r}|} \frac{\exp(ik_0|\mathbf{r}_2 - \mathbf{r}|)}{|\mathbf{r}_2 - \mathbf{r}|} \times (z_1 - z)(z_2 - z) h(\mathbf{r}, \omega)\, d^2r . \tag{9.22}$$

If we make the further assumption that $|\mathbf{r}_1| \gg |\mathbf{r}|$ and $|\mathbf{r}_2| \gg |\mathbf{r}|$, and define

$$\cos\theta_1 = \frac{z_1}{r_1} , \quad \cos\theta_2 = \frac{z_2}{r_2} \tag{9.23}$$

as shown in Fig. 9.2, we find that

$$W(\mathbf{r}_1, \mathbf{r}_2, \omega) = \left(\frac{k_0}{2\pi}\right)^2 \exp(-ik_0(r_1 - r_2)) \cos\theta_1 \cos\theta_2 \times \int \exp(ik_0(\hat{\mathbf{r}}_1 - \hat{\mathbf{r}}_2) \cdot \mathbf{r})\, h(\mathbf{r}, \omega)\, d^2r . \tag{9.24}$$

Equation (9.24) is known as the van Cittert–Zernike theorem. Mathematically, it says that the degree of coherence is determined by the Fourier transform of the intensity distribution in the source plane. In physical terms, it means that a spatially incoherent field becomes coherent upon propagation.

As an application of the van Cittert–Zernike theorem, we consider an incoherent source of linear size L and observe the field at a distance z in a plane perpendicular to the z-axis. We assume that the observation points $\mathbf{r}_1$ and $\mathbf{r}_2$ are symmetric around the z-axis, as shown in Fig. 9.2. We note that the integral in Eq. (9.24) is a two-dimensional Fourier transform of $h(\mathbf{r}, \omega)$. Since $h(\mathbf{r}, \omega)$ is non-vanishing over a surface of size L, the Fourier transform is non-negligible for wave vectors

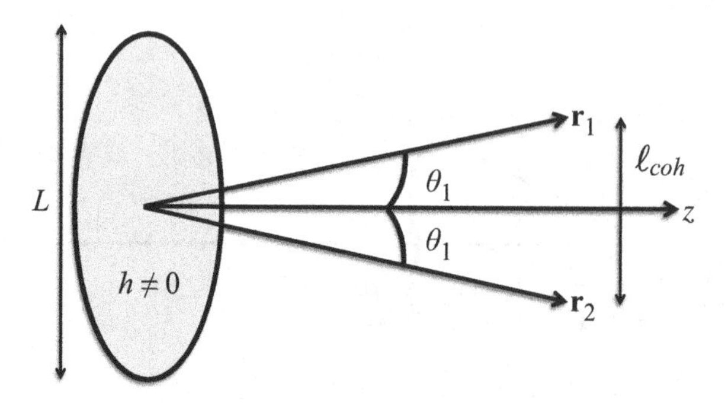

Fig. 9.2 Geometry used to compute the spatial coherence length of the field propagating from a planar incoherent source.

$k_0(\hat{\mathbf{r}}_1 - \hat{\mathbf{r}}_2)$ such that $k_0|\hat{\mathbf{r}}_1 - \hat{\mathbf{r}}_2|\, L \leq 2\pi$. In the geometry in Fig. 9.2, this leads to the condition $2k_0 \sin(\theta_1)_{\max} L = 2\pi$, where $(\theta_1)_{\max}$ is the value of θ_1 above which the integral vanishes. Assuming that $\theta_1 \ll 2\pi$, the spatial coherence length is $\ell_{coh} = 2\tan(\theta_1)_{\max}\, z \simeq 2\sin(\theta_1)_{\max}\, z$, which leads to

$$\ell_{coh} = \frac{\lambda\, z}{L} = \frac{\lambda}{\Delta\theta}\,, \tag{9.25}$$

with $\Delta\theta = L/z$ the angular size of the source as seen from the observation plane. We see that the spatial coherence length can be increased by reducing the size of the source or making observations at larger distances.

9.3 Coherent Mode Representation

It follows from (9.8) and (9.11) that $W(\mathbf{r}_1, \mathbf{r}_2, \omega)$ can be regarded as the kernel of a positive-definite Hermitian operator. Making use of the spectral theorem, we expand W as

$$W(\mathbf{r}_1, \mathbf{r}_2, \omega) = \sum_n \lambda_n f_n(\mathbf{r}_1) f_n^*(\mathbf{r}_2)\,. \tag{9.26}$$

Here f_n is an eigenfunction of W with eigenvalue λ_n. That is,

$$\int W(\mathbf{r}_1, \mathbf{r}_2, \omega) f_n(\mathbf{r}_2)\, d^3r_2 = \lambda_n f_n(\mathbf{r}_1)\,, \tag{9.27}$$

where λ_n is nonnegative. Moreover, the f_n are orthogonal and can be chosen to be orthonormal:

$$\int f_n(\mathbf{r}) f_m^*(\mathbf{r})\, d^3r = \delta_{nm}\,. \tag{9.28}$$

The result (9.26) is known as the coherent mode representation. The name is due to the fact that the modes

$$W_n(\mathbf{r}_1, \mathbf{r}_2, \omega) = f_n(\mathbf{r}_1) f_n^*(\mathbf{r}_2) \tag{9.29}$$

are spatially coherent in the sense that the spectral degree of coherence

$$\mu_n = \frac{W_n(\mathbf{r}_1, \mathbf{r}_2, \omega)}{\sqrt{f_n(\mathbf{r}_1) f_n^*(\mathbf{r}_1)}\sqrt{f_n(\mathbf{r}_2) f_n^*(\mathbf{r}_2)}} \tag{9.30}$$

has modulus $|\mu_n| = 1$. Therefore, the cross-spectral density $W(\mathbf{r}_1, \mathbf{r}_2, \omega)$ can be expanded in modes that are spatially coherent.

References and Additional Reading

The following paper introduced the description of optical coherence in terms of the propagation of correlation functions:

E. Wolf, Nuovo Cimento **12**, 884 (1954).

The propagation of correlations, and the van Cittert–Zernike theorem, are treated in classical textbooks on coherence theory:

J.W. Goodman, *Statistical Optics* (Wiley, New York, 1985), chap. 5.
L. Mandel and E. Wolf, *Optical Coherence and Quantum Optics* (Cambridge University Press, Cambridge, 1995), chap. 4.
E. Wolf, *Introduction to the Theory of Coherence and Polarization of Light* (Cambridge University Press, Cambridge, 2007), chap. 3.

Exercises

I.1 Following the approach of Chapter 3, develop the theory of geometrical optics in one dimension. Show that it is possible to solve the resulting eikonal and transport equations explicitly.

I.2 Consider a scalar monochromatic wave at normal incidence on a flat interface separating air from a homogeneous material with refractive index n. The wavelength is $\lambda = 633$ nm. Compute the intensity reflection factor for glass ($n = 1.5$), silicon ($n = 3.5$) and gold ($n = -9 + i$).

I.3 A scalar monochromatic wave, with wavelength $\lambda = 550$ nm, is incident on a flat glass-air interface from the glass side. The refractive index of glass is $n = 1.5$. Which condition on the incidence angle has to be satisfied to observe total internal reflection? What is the decay length of the transmitted evanescent wave for an angle of incidence $\theta_i = 45°$, and $\theta_i = 70°$? Total internal reflection can be used in optical microscopy to produce inhomogeneous illumination in the longitudinal direction.

I.4 In Chapter 5, we saw that (5.7) can be understood as the superposition of the radiation of elementary point sources forming the real source. Write a short computer program that sums the radiation from a small number of monochromatic point sources in an infinite homogeneous medium, and plot the intensity distribution in space. By positioning the point sources along a line, with a separation smaller than the wavelength, it is possible to simulate the radiation by a continuous line source. Show that the radiated intensity forms a diffraction pattern, and study the change of the pattern with the length of the line source.

I.5 A monochromatic plane wave with wavelength λ is transmitted through an aperture of linear size L in an opaque screen. The transmitted beam diffracts, with an angular aperture characterized by the angle θ. Using simple Fourier transform arguments, show that $\sin\theta \sim \lambda/L$.

I.6 A monochromatic plane wave with wavelength λ is normally incident on a square aperture of size $L \times L$ in an opaque screen. Using Kirchhoff's approximation, calculate the two-dimensional Fourier transform of the field amplitude in the plane of the screen. Find the plane-wave expansion of the transmitted field.

I.7 Consider a monochromatic field of wavelength λ that is incident on an object structured at a scale $d \ll \lambda$. In a plane close to the object, taken as the reference plane $z = 0$), the field amplitude $U(x, y, 0)$ also varies on the scale of d, as shown schematically in the figure below. Explain qualitatively why

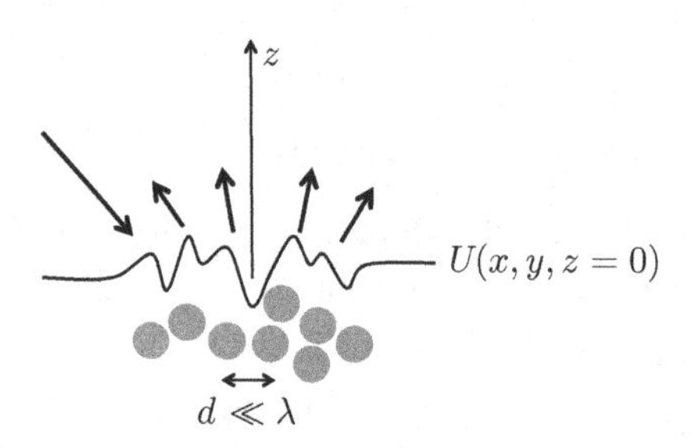

the plane-wave expansion of the field propagating in the half-space $z > 0$ contains evanescent waves that decay exponentially. Show that the decay length is $\delta \sim d/(2\pi)$. The filtering of the high spatial frequencies upon propagation is a key feature of near-field optics.

I.8 An extended planar source is assumed to be spatially incoherent and broadband. Draw schematically a setup that allows one to produce a beam with a higher degree of spatial and temporal coherence.
Hint: Use a pinhole, a lens and a filter.

I.9 A spatially incoherent source has the shape of a disk with diameter $D =$ 1 mm. It emits in a spectral bandwidth $\Delta\lambda = 1\ \mu$m around the wavelength $\lambda = 10\ \mu$m. What are the temporal and spatial coherence length of the emitted radiation, observed at a distance $z = 1$ m from the source plane?

I.10 Consider Young's two-slit experiment with an incoherent source of linear size L and emitting at frequency ω with a finite bandwidth $\Delta\omega \ll \omega$. Discuss qualitatively the influence of L and $\Delta\omega$ on the interference pattern.

I.11 From the van Cittert–Zernike theorem and the diffraction theory for a circular aperture in Section 7.3, compute the cross-spectral density of the far-field radiated from a planar circular incoherent source in free space. Deduce the spatial coherence length of the field at a fixed distance from the source.

Part II

Scattering of Waves

10

Scattering Theory

In this chapter, we study the propagation of waves in inhomogeneous media, and derive the integral equations of scattering theory.

10.1 Integral Equations

We consider the propagation of a monochromatic scalar wave field in an infinite inhomogeneous medium. The field satisfies the wave equation

$$\nabla^2 U(\mathbf{r}) + k_0^2 \varepsilon(\mathbf{r})\, U(\mathbf{r}) = -S(\mathbf{r}) \,, \tag{10.1}$$

where k_0 is the wavenumber in free space, ε is the dielectric function of the medium and S is the source. Physically, we regard the medium as consisting of a scatterer characterized by a dielectric function with finite range. Suppose that the source creates an incident field U_i, defined to be the field that would exist in the absence of the scatterer. The incident field obeys the equation

$$\nabla^2 U_i(\mathbf{r}) + k_0^2\, U_i(\mathbf{r}) = -S(\mathbf{r}) \,, \tag{10.2}$$

which has the solution

$$U_i(\mathbf{r}) = \int G_0(\mathbf{r}, \mathbf{r}') S(\mathbf{r}')\, d^3 r' \,. \tag{10.3}$$

Here G_0 is the free-space Green's function which is given by

$$G_0(\mathbf{r}, \mathbf{r}') = \frac{\exp(i k_0 |\mathbf{r} - \mathbf{r}'|)}{4\pi\, |\mathbf{r} - \mathbf{r}'|} \,. \tag{10.4}$$

We will find it useful to decompose the field into incident and scattered parts:

$$U = U_i + U_s \,, \tag{10.5}$$

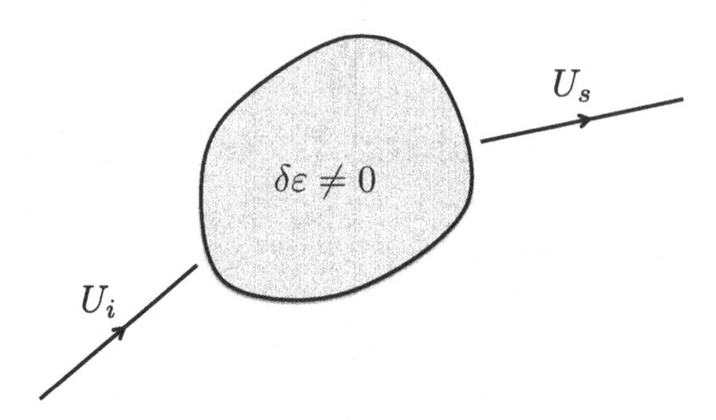

Fig. 10.1 Illustrating a scattering experiment.

where U_s is known as the scattered field. Evidently, the scattered field satisfies the equation

$$\nabla^2 U_s(\mathbf{r}) + k_0^2\, U_s(\mathbf{r}) = -k_0^2\, \delta\varepsilon(\mathbf{r}) U(\mathbf{r}), \tag{10.6}$$

where $\delta\varepsilon = \varepsilon - 1$. The solution to this equation may be expressed in the form

$$U_s(\mathbf{r}) = k_0^2 \int G_0(\mathbf{r}, \mathbf{r}')\delta\varepsilon(\mathbf{r}')U(\mathbf{r}')\, d^3r' \,. \tag{10.7}$$

Hence the total field obeys the integral equation

$$U(\mathbf{r}) = U_i(\mathbf{r}) + k_0^2 \int G_0(\mathbf{r}, \mathbf{r}')\delta\varepsilon(\mathbf{r}')U(\mathbf{r}')\, d^3r' \,. \tag{10.8}$$

This result, which is known as the Lippmann–Schwinger equation, gives the solution to the wave equation (10.1). It plays a central role in the theory of scattering. Physically, the scattered field is seen to be radiated by secondary sources of the form $k_0^2 \delta\varepsilon\, U$ that are present in the volume of the scatterer. A typical scattering experiment is illustrated in Fig. 10.1.

The Green's function corresponding to the wave equation (10.1) obeys

$$\nabla^2 G(\mathbf{r}, \mathbf{r}') + k_0^2 \varepsilon(\mathbf{r})\, G(\mathbf{r}, \mathbf{r}') = -\delta(\mathbf{r} - \mathbf{r}') \,. \tag{10.9}$$

Recalling that G_0 is the Green's function in the absence of scattering, and following the same procedure as above, we find that G satisfies the integral equation

$$G(\mathbf{r}, \mathbf{r}') = G_0(\mathbf{r}, \mathbf{r}') + k_0^2 \int G_0(\mathbf{r}, \mathbf{r}'')\delta\varepsilon(\mathbf{r}'')G(\mathbf{r}'', \mathbf{r}')\, d^3r'' \,, \tag{10.10}$$

which is known as the Dyson equation.

10.2 Born Series and Multiple Scattering

The Lippmann–Schwinger equation (10.8) provides only an implicit solution to the scattering problem, since the total field U appears on both the left- and right-hand sides of the equation. An explicit formula for U, involving only the incident field, can be obtained by iteration of (10.8) :

$$U(\mathbf{r}) = U_i(\mathbf{r}) + k_0^2 \int G_0(\mathbf{r}, \mathbf{r}')\delta\varepsilon(\mathbf{r}')U_i(\mathbf{r}')\, d^3r'$$

$$+ k_0^4 \int G_0(\mathbf{r}, \mathbf{r}')\delta\varepsilon(\mathbf{r}')G_0(\mathbf{r}', \mathbf{r}'')\delta\varepsilon(\mathbf{r}'')U_i(\mathbf{r}'')d^3r'd^3r'' + \cdots . \quad (10.11)$$

Equation (10.11) is known as the Born series. If the scattered field is much weaker than the incident field, then the Born series may be truncated at first order in $\delta\varepsilon$. The expression of the field thus becomes

$$U(\mathbf{r}) = U_i(\mathbf{r}) + k_0^2 \int G_0(\mathbf{r}, \mathbf{r}')U_i(\mathbf{r}')\delta\varepsilon(\mathbf{r}')\, d^3r' . \quad (10.12)$$

This result is referred to as the Born approximation. It is accurate for small, weak scatterers. Higher-order terms in the Born series correspond to multiple scattering of the incident field.

The Dyson equation (10.10) also leads to an infinite series for the Green's function in powers of $\delta\varepsilon$:

$$G(\mathbf{r}, \mathbf{r}') = G_0(\mathbf{r}, \mathbf{r}') + k_0^2 \int G_0(\mathbf{r}, \mathbf{r}'')\delta\varepsilon(\mathbf{r}'')G_0(\mathbf{r}'', \mathbf{r}')\, d^3r''$$

$$+ k_0^4 \int G_0(\mathbf{r}, \mathbf{r}'')\delta\varepsilon(\mathbf{r}'')G_0(\mathbf{r}'', \mathbf{r}''')\delta\varepsilon(\mathbf{r}''')G_0(\mathbf{r}''', \mathbf{r}')\, d^3r''d^3r''' + \cdots . \quad (10.13)$$

It is convenient to write this series in diagrammatic form:

$$G = \text{——} + \text{—•—} + \text{—•—•—} + \text{—•—•—•—} + \cdots . \quad (10.14)$$

The diagrams are to be understood as follows. The entry point $\mathbf{r}'$ is on the right, and the exit point $\mathbf{r}$ on the left in each diagram. A solid line corresponds to the free-space Green's function G_0, and the vertex • to a factor of $k_0^2\delta\varepsilon$. Diagrams through the third order of multiple scattering are shown in the previous expression. We will see that diagrammatic expansions turn out to be very useful for the computation of average fields and intensities in disordered media.

10.3 Scattering Amplitude and Cross Sections

In the far-field, the free-space Green's function assumes the asymptotic form

$$G_0(\mathbf{r}, \mathbf{r}') \sim \frac{\exp(ik_0 r)}{4\pi r} \exp(-ik_0\hat{\mathbf{r}} \cdot \mathbf{r}') , \quad (10.15)$$

which holds for $r \gg r'$ and $r \gg r'^2/\lambda$. The scattered field thus behaves as an outgoing spherical wave:

$$U_s(\mathbf{r}) \sim A\frac{\exp(ik_0 r)}{r} . \tag{10.16}$$

Here the scattering amplitude A is defined by the expression

$$A = \frac{k_0^2}{4\pi}\int \exp(-ik_0\hat{\mathbf{r}}\cdot\mathbf{r}')U(\mathbf{r}')\delta\varepsilon(\mathbf{r}')\, d^3r' . \tag{10.17}$$

The power radiated in the direction $\hat{\mathbf{r}}$, per unit solid angle, is defined by

$$\frac{dP_s}{d\Omega} = \lim_{R\to\infty} \left(\mathbf{J}_s\cdot\hat{\mathbf{r}}\right)|_{r=R}R^2. \tag{10.18}$$

In this expression, the energy current carried by the scattered field is given by

$$\begin{aligned}\mathbf{J}_s &= \frac{1}{2ik_0}\left(U_s^*\nabla U_s - U_s\nabla U_s^*\right)\\ &= \frac{|A|^2}{r^2}\hat{\mathbf{r}} ,\end{aligned} \tag{10.19}$$

where we have used (10.16) in the last line. Hence

$$\frac{dP_s}{d\Omega} = |A|^2 . \tag{10.20}$$

The differential scattering cross section, which is the effective area for scattering into a given element of solid angle, is defined to be

$$\frac{d\sigma_s}{d\Omega} = \frac{dP_s/d\Omega}{|\mathbf{J}_i|} , \tag{10.21}$$

where $\mathbf{J}_i$ is the current carried by the incident field. The scattering cross section is defined by

$$\sigma_s = \int \frac{d\sigma_s}{d\Omega}\, d\hat{\mathbf{r}} , \tag{10.22}$$

where the integral is taken over all directions on the unit sphere. We immediately see that the product $\sigma_s|\mathbf{J}_i|$ is the total scattered power P_s. Since $|\mathbf{J}_i|$ is the incident flux per unit area, the scattering cross section can be understood as the effective cross-sectional area of the scatterer.

Following the above approach, we define the absorption cross section σ_a so that $\sigma_a|\mathbf{J}_i| = P_a$, with P_a the power absorbed by the scatterer as defined by (2.41). The albedo ϖ, defined as

$$\varpi = \frac{\sigma_s}{\sigma_a + \sigma_s} , \tag{10.23}$$

is a measure of the "whiteness" of the scatterer. If the scatterer is nonabsorbing ($\sigma_a = 0$) then $\varpi = 1$.

Suppose that the incident field is a unit-amplitude plane wave of the form

$$U_i(\mathbf{r}) = \exp(i\mathbf{k} \cdot \mathbf{r}) , \tag{10.24}$$

where $\mathbf{k}$ is the incident wave vector and $|\mathbf{k}| = k_0$. The scattering amplitude then depends explicitly upon the incoming wave vector $\mathbf{k}$ and the outgoing wave vector $\mathbf{k}' = k_0\hat{\mathbf{r}}$, and is denoted $A(\mathbf{k}, \mathbf{k}')$. A plane wave of the form (10.24) has $\mathbf{J}_i = \hat{\mathbf{k}}$. The differential cross section is therefore given by

$$\frac{d\sigma_s}{d\Omega} = |A(\mathbf{k}, \mathbf{k}')|^2 . \tag{10.25}$$

Within the accuracy of the Born approximation, $A(\mathbf{k}, \mathbf{k}')$ is given by

$$A(\mathbf{k}, \mathbf{k}') = \frac{k_0^2}{4\pi} \int \exp[i(\mathbf{k} - \mathbf{k}') \cdot \mathbf{r}]\delta\varepsilon(\mathbf{r})\, d^3r . \tag{10.26}$$

Hence

$$\frac{d\sigma_s}{d\Omega} = \frac{k_0^4}{16\pi^2} |\widetilde{\delta\varepsilon}(\mathbf{k} - \mathbf{k}')|^2 , \tag{10.27}$$

where $\widetilde{\delta\varepsilon}$ denotes the Fourier transform of $\delta\varepsilon$.

As a simple example, we compute the differential cross section for a homogeneous spherical scatterer with small dielectric contrast $\delta\varepsilon_s \ll 1$ and radius $a \ll \lambda$. In this situation, the Born approximation is quite accurate. The Fourier transform of the dielectric function is easily seen to be

$$\begin{aligned} \widetilde{\delta\varepsilon}(\mathbf{k}) &= \frac{4\pi\,\delta\varepsilon_s}{k} \int_0^a r \sin(kr)dr \\ &\simeq \frac{4\pi}{3}\delta\varepsilon_s\, a^3 , \end{aligned} \tag{10.28}$$

when $k_0 a \ll 1$. Thus, the differential cross section is given by

$$\frac{d\sigma_s}{d\Omega} = k_0^4 \left(\delta\varepsilon_s \frac{a^3}{3}\right)^2 . \tag{10.29}$$

We note that $d\sigma_s/d\Omega$ is not angularly dependent, thus describing isotropic scattering. The scattering cross section becomes

$$\sigma_s = 4\pi k_0^4 \left(\delta\varepsilon_s \frac{a^3}{3}\right)^2 . \tag{10.30}$$

We further note that higher frequencies are scattered most strongly. The k_0^4 frequency dependence of σ_s is characteristic of scattering by non-resonant small particles, a regime known as Rayleigh scattering.

10.4 *T*-matrix

Until now, we have defined Green's functions as solutions to partial differential equations. More abstractly, they can be viewed as inverses of differential operators. From this standpoint, the Dyson equation (10.10) can be written in operator notation in the form

$$G = G_0 + G_0 V G \,, \tag{10.31}$$

where V is regarded as an operator with the kernel

$$V(\mathbf{r}, \mathbf{r}') = k_0^2 \delta\varepsilon(\mathbf{r}) \delta(\mathbf{r} - \mathbf{r}') \,. \tag{10.32}$$

Upon iterating (10.31), we obtain

$$G = G_0 + G_0 V G_0 + G_0 V G_0 V G_0 + \cdots \,, \tag{10.33}$$

which corresponds to (10.13). We can now rewrite the above series as

$$G = G_0 + G_0 T G_0 \,, \tag{10.34}$$

where the T-matrix is defined by

$$T = V + V G_0 V + V G_0 V G_0 V + \cdots \,. \tag{10.35}$$

By summing the geometric series, we obtain

$$T = V (1 - G_0 V)^{-1} \,. \tag{10.36}$$

A similar calculation, starting from (10.35), shows that

$$T = V + V G_0 T \,, \tag{10.37}$$

from which we see that, within the accuracy of the Born approximation, $T \simeq V$. We note that if T is known, the full Green's function G can be determined.

The Lippmann–Schwinger equation (10.8) can also be written in a more compact form. Using the above notation, we have

$$U = U_i + G_0 V U \,. \tag{10.38}$$

By iterating (10.38), we obtain

$$\begin{aligned} U &= U_i + G_0 V U_i + G_0 V G_0 V U_i + \cdots & (10.39) \\ &= U_i + G_0 (V + V G_0 V + V G_0 V G_0 V + \cdots) U_i \,. & (10.40) \end{aligned}$$

Noting that the series in parenthesis is the T-matrix, we find that

$$U = U_i + G_0 T U_i \,. \tag{10.41}$$

This result immediately yields the physical meaning of the T-matrix, namely that the quantity $T U_i$ is the source of the scattered field.

References and Additional Reading

This reference is a comprehensive textbook on scattering theory:

R. Newton, *Scattering Theory of Waves and Particles* (McGraw-Hill, New York, 1966).

For an introduction to the theory of light scattering, see:

M. Born and E. Wolf, *Principles of Optics*, 7th edition (Cambridge University Press, Cambridge, 2005), chap. 13.

The following textbooks focus on light scattering from particles, and include chapters dedicated to the general theory of scattering:

H.C. van de Hulst, *Light Scattering by Small Particles* (Dover, New York, 1981).

C.F. Bohren and D.R. Huffman, *Absorption and Scattering of Light by Small Particles* (Wiley, New York, 1983).

11
Optical Theorem

The optical theorem is a fundamental result in scattering theory. It relates the power extinguished from a plane wave incident on a scatterer to the imaginary part of the scattering amplitude in the forward direction of scattering.

11.1 Extinguished Power

We begin by recalling that the power absorbed by a scatterer is given by the integral of the inward going normal component of the current of the total field taken over a surface enclosing the scatterer. As shown in Section 2.4, the absorbed power is

$$P_a = k_0 \mathrm{Im} \int_V |U(\mathbf{r})|^2 \delta\varepsilon(\mathbf{r})\, d^3r \ , \tag{11.1}$$

where V denotes the volume of the scatterer.

The power carried by the scattered field P_s is given by

$$P_s = \int_S \mathbf{J}_s \cdot \hat{\mathbf{n}}\, d^2r \ , \tag{11.2}$$

where S is the surface bounding the volume V. Here the energy current of the scattered field is given by

$$\mathbf{J}_s = \frac{1}{2ik_0}\left(U_s^* \nabla U_s - U_s \nabla U_s^*\right) \ . \tag{11.3}$$

Applying Green's theorem to convert the surface integral to a volume integral, and making use of the identity

$$U_s^* \nabla^2 U_s - U_s \nabla^2 U_s^* = -2ik_0^2 \,\mathrm{Im}\left(U_s^* U \delta\varepsilon\right) , \tag{11.4}$$

which follows from (10.6), we find that P_s is given by the expression

$$P_s = -k_0 \,\mathrm{Im} \int_V U_s^*(\mathbf{r}) U(\mathbf{r}) \delta\varepsilon(\mathbf{r})\, d^3r \ . \tag{11.5}$$

The extinguished power P_e is defined by

$$P_e = P_a + P_s \,. \tag{11.6}$$

It represents the power depleted from the incident field due to the presence of the scatterer. Using (11.1) and (11.5) we see that

$$P_e = k_0 \, \mathrm{Im} \int_V U_i^*(\mathbf{r}) U(\mathbf{r}) \delta\varepsilon(\mathbf{r}) \, d^3 r \,. \tag{11.7}$$

The physical interpretation of (11.7) is that power is extinguished from the incident field due to interference between the total field and the incident field within the volume of the scatterer.

If the incident field is a plane wave of the form

$$U_i(\mathbf{r}) = \exp(i\mathbf{k} \cdot \mathbf{r}) \,, \tag{11.8}$$

then the integral in (11.7) is proportional to the scattering amplitude $A(\mathbf{k}, \mathbf{k})$ in the direction of incidence $\mathbf{k}$ [see Eq. (10.16)]:

$$A(\mathbf{k}, \mathbf{k}) = \frac{k_0^2}{4\pi} \int_V \exp(-i\mathbf{k} \cdot \mathbf{r}) \, U(\mathbf{r}) \delta\varepsilon(\mathbf{r}) \, d^3 r \,. \tag{11.9}$$

Thus the extinguished power is given by the expression

$$P_e = \frac{4\pi}{k_0} \mathrm{Im} A(\mathbf{k}, \mathbf{k}) \,. \tag{11.10}$$

Equation (11.10) is known as the optical theorem. It shows that the extinguished power can be deduced from the scattering amplitude A in the forward direction. In physical terms, extinction can be understood as resulting from the destructive interference between the incident field and the field scattered in the forward direction. Since the relative amplitude and phase between both fields is determined by $A(\mathbf{k}, \mathbf{k})$, this quantity encodes information about extinction.[1]

11.2 Generalized Optical Theorem

The optical theorem may be generalized to the case where the incident field consists of a superposition of plane waves. In this instance,

$$U_i(\mathbf{r}) = \int a(\mathbf{q}) \exp(i\mathbf{k}(\mathbf{q}) \cdot \mathbf{r}) \, d^2 q \,, \tag{11.11}$$

where $\mathbf{k}(\mathbf{q}) = (\mathbf{q}, k_z(q))$ and the dependence on the transverse wave vector $\mathbf{q}$ has been made explicit. The total field is then obtained by separately considering the

[1] Here we follow the approach presented in P.S. Carney, J.C. Schotland and E. Wolf, Phys. Rev. E **70**, 036611 (2004).

scattering of each plane-wave mode, and making use of the linearity of the wave equation. Thus it may be seen that

$$U(\mathbf{r}) = \int a(\mathbf{q}) U(\mathbf{r}; \mathbf{k}(\mathbf{q}))\, d^2q \,. \tag{11.12}$$

Here $U(\mathbf{r}; \mathbf{k}(\mathbf{q}))$ denotes the field that is produced from the scattering of a unit-amplitude plane wave with wave vector $\mathbf{k}(\mathbf{q})$. Using (11.7) we see that the extinguished power is of the form

$$P_e = \frac{4\pi}{k_0} \mathrm{Im} \int a^*(\mathbf{q}) a(\mathbf{q}') A(\mathbf{k}(\mathbf{q}), \mathbf{k}(\mathbf{q}'))\, d^2q d^2q' \,, \tag{11.13}$$

where the scattering amplitude is given by

$$A(\mathbf{k}(\mathbf{q}), \mathbf{k}(\mathbf{q}')) = \frac{k_0^2}{4\pi} \int_V \exp(-i\mathbf{k}(\mathbf{q}') \cdot \mathbf{r}) U(\mathbf{r}; \mathbf{k}(\mathbf{q})) \delta\varepsilon(\mathbf{r})\, d^3r \,. \tag{11.14}$$

The result in (11.13) is referred to as the generalized optical theorem.

Let us examine the case of illumination by a superposition of two plane waves. In this situation, we have

$$a(\mathbf{q}) = a_1 \delta(\mathbf{q} - \mathbf{q}_1) + a_2 \delta(\mathbf{q} - \mathbf{q}_2) \,, \tag{11.15}$$

where a_1, a_2 and $\mathbf{q}_1, \mathbf{q}_2$ are the generally complex amplitudes and transverse wave vectors of the incident plane waves with wave vectors $\mathbf{k}_1, \mathbf{k}_2$. Using (11.13) we find that the extinguished power is given by

$$P_e = \frac{4\pi}{k_0} \mathrm{Im} \left[|a_1|^2 A(\mathbf{k}_1, \mathbf{k}_1) + a_1^* a_2 A(\mathbf{k}_1, \mathbf{k}_2) + a_1 a_2^* A(\mathbf{k}_2, \mathbf{k}_1) + |a_2|^2 A(\mathbf{k}_2, \mathbf{k}_2) \right] . \tag{11.16}$$

Equation (11.16) can be applied to inverse scattering problems in which the phase of the optical field is not directly measured.

References and Additional Reading

In this chapter, we have followed the derivation of the optical theorem given in:

P.S. Carney, J.C. Schotland and E. Wolf, Phys. Rev. E **70**, 036611 (2004).
D. Lytle, P.S. Carney, J.C. Schotland and E. Wolf, Phys. Rev. E **71**, 056610 (2005).

An interesting discussion of the optical theorem is presented in:

J. Schwinger, L.L. DeRaad Jr., K.A. Milton and W. Tsai, *Classical Electrodynamics* (Perseus Book, Reading, 1998), chap. 50.

A derivation making use of far-field asymptotics is given in:

C.F. Bohren and D.R. Huffman, *Absorption and Scattering of Light by Small Particles* (Wiley, New York, 1983), chap. 3, section 3.4.

The following papers describe some applications of the optical theorem to inverse scattering:

P.S. Carney, E. Wolf and G.S. Agarwal, J. Opt. Soc. Am. A **11**, 2643 (2000).
P.S. Carney, V. Markel and J. Schotland, Phys. Rev. Lett. **86**, 5874 (2001).
A. Govyadinov, G. Panasyuk and J. Schotland, Phys. Rev. Lett. **103**, 213901 (2009).

12

Scattering in Model Systems

In this chapter, we study the problem of scattering from a point scatterer and collections of such scatterers. We also discuss the related problem of scattering from spheres of arbitrary size. These are among the few exactly solvable problems in scattering theory.

12.1 Point Scatterer

We consider the scattering of a monochromatic wave from a spherical particle whose size is small compared to the wavelength λ of the incident wave. We will refer to such a particle as a point scatterer. The corresponding dielectric function is $\varepsilon(\mathbf{r}) = 1 + \delta\varepsilon(\mathbf{r})$, with $\delta\varepsilon(\mathbf{r}) = \delta\varepsilon_s \chi(\mathbf{r})$, where

$$\chi(\mathbf{r}) = \begin{cases} 1 & |\mathbf{r}| \leq a \,, \\ 0 & |\mathbf{r}| > a \,. \end{cases} \tag{12.1}$$

Here a is the radius of the sphere and $k_0 a \ll 1$, with $k_0 = 2\pi/\lambda$. The field U satisfies the Lippmann–Schwinger equation, introduced in Chapter 10 :

$$U(\mathbf{r}) = U_i(\mathbf{r}) + k_0^2 \int G_0(\mathbf{r}, \mathbf{r}')\delta\varepsilon(\mathbf{r}')U(\mathbf{r}')\, d^3r' \,. \tag{12.2}$$

Since the scatterer is small compared to the wavelength, the field is nearly uniform inside of the sphere and we can make the approximation $U(\mathbf{r}) \simeq U(0)$ for $|\mathbf{r}| \leq a$. Equation (12.2) thus becomes

$$U(\mathbf{r}) = U_i(\mathbf{r}) + k_0^2\, \delta\varepsilon_s U(0) \int_V G_0(\mathbf{r}, \mathbf{r}')\, d^3r' \,, \tag{12.3}$$

where V is the ball of radius a centered at the origin. The above integral can be performed with the result

$$\int_V G_0(\mathbf{r},\mathbf{r}')d^3r' = \begin{cases} VG_0(\mathbf{r},0)\left[1+O\left((k_0a)^2\right)\right] & \text{if} \quad \mathbf{r}\neq 0\,, \\ \dfrac{1}{k_0^2}\left[\dfrac{1}{2}(k_0a)^2+\dfrac{i}{3}(k_0a)^3+O\left((k_0a)^4\right)\right] & \text{if} \quad \mathbf{r}=0\,, \end{cases} \tag{12.4}$$

where $V = (4/3)\pi a^3$.

To find the local field $U(0)$ inside the sphere, we put $\mathbf{r} = 0$ in (12.3) and make use of (12.4). Upon solving the resulting algebraic equation, we obtain

$$U(0) = \frac{U_i(0)}{1-\delta\varepsilon_s\left[\dfrac{1}{2}(k_0a)^2+\dfrac{i}{3}(k_0a)^3\right]} . \tag{12.5}$$

For $\mathbf{r}$ outside the scatterer, we have from (12.3) and (12.4) that

$$U(\mathbf{r}) = U_i(\mathbf{r}) + k_0^2\delta\varepsilon_s VG_0(\mathbf{r},0)U(0)\,. \tag{12.6}$$

Using (12.5) the above can be rewritten as

$$U(\mathbf{r}) = U_i(\mathbf{r}) + k_0^2\alpha G_0(\mathbf{r},0)U_i(0)\,, \tag{12.7}$$

where

$$\alpha = \frac{\alpha_0}{1-\alpha_0\left(\dfrac{3k_0^2}{8\pi a}+i\dfrac{k_0^3}{4\pi}\right)} . \tag{12.8}$$

Here α is the polarizability of the scatterer and $\alpha_0 = \delta\varepsilon_s V$. We see that $\alpha = \alpha_0$ in the limit $k_0 \to 0$, and α_0 is known as the quasi-static polarizability. We note that α includes radiative corrections to the quasi-static polarizability.

It can be seen from (10.41) and (12.7) that the T-matrix of a point scatterer located at the point $\mathbf{r}_0$ is of the form

$$T(\mathbf{r},\mathbf{r}') = t\,\delta(\mathbf{r}-\mathbf{r}_0)\delta(\mathbf{r}'-\mathbf{r}_0)\,, \tag{12.9}$$

where $t = \alpha k_0^2$. If we define $k_*^2 = 8\pi a/(3\alpha_0)$, we see that t can be rewritten as

$$t = \frac{\alpha_0 k_*^2 k_0^2}{k_*^2 - k_0^2 - i\dfrac{\alpha_0 k_*^2 k_0^3}{4\pi}} . \tag{12.10}$$

The above formula indicates that t has a resonance at $k_0 = k_*$. The existence of a resonance for a subwavelength scatterer deserves a comment. Since the condition $k_0a \ll 1$ is assumed, one can easily see that the resonance can be observed provided that the dielectric contrast $|\delta\varepsilon_s| \gg 1$.

We now examine the effect of the resonance on the scattering cross section. Suppose that the incident field is a unit-amplitude plane wave of the form

$$U_i(\mathbf{r}) = \exp(i\mathbf{k} \cdot \mathbf{r}) , \tag{12.11}$$

with $|\mathbf{k}| = k_0$. Then, using (10.16) and (12.7), we see that in the far-zone of the scatterer, the scattered field is of the form

$$U_s(\mathbf{r}) \sim A(\mathbf{k}, \mathbf{k}') \frac{\exp(ik_0 r)}{r} , \tag{12.12}$$

where $\mathbf{k}' = k_0 \hat{\mathbf{r}}$ is the outgoing wave vector and the scattering amplitude is given by

$$A(\mathbf{k}, \mathbf{k}') = t \exp[i(\mathbf{k} - \mathbf{k}') \cdot \mathbf{r}_0] . \tag{12.13}$$

Making use of the optical theorem (11.10), we find that the extinction cross section is given by the expression

$$\sigma_e = \frac{4\pi}{k_0} \mathrm{Im}\, t . \tag{12.14}$$

For a non-absorbing scatterer, the extinction and scattering cross sections coincide, and (12.14) gives the expression for σ_s. We see that scattering is isotropic for a point scatterer. The frequency-dependence of the scattering cross section is shown in Fig. 12.1. Note the peak of the cross section near the resonance frequency k_*.

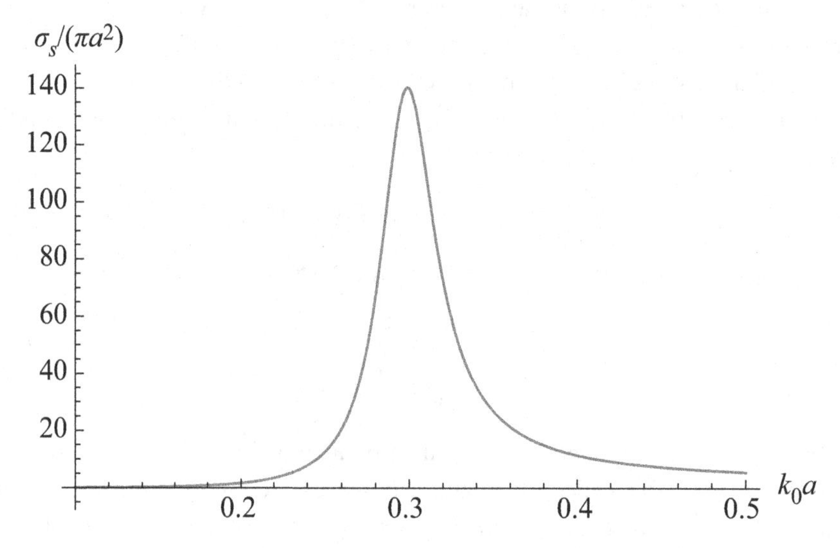

Fig. 12.1 Frequency dependence of the scattering cross section for a point scatterer with a resonance frequency $k_* a = 0.3$.

12.2 Collection of Point Scatterers

We now generalize the above results to scattering from a collection of point scatterers. The dielectric function is taken to be of the form

$$\delta\varepsilon(\mathbf{r}) = \delta\varepsilon_s \sum_{j=1}^{N} \chi(\mathbf{r} - \mathbf{r}_j) , \tag{12.15}$$

where $\mathbf{r}_1, \ldots, \mathbf{r}_N$ denote the positions of the scatterers, which for simplicity are assumed to be identical. Inserting the above expression for $\delta\varepsilon$ into the Lippmann–Schwinger equation (12.2), and making use of (12.4), we find that the field obeys the equation

$$U(\mathbf{r}) = U_i(\mathbf{r}) + \alpha_0 k_0^2 \sum_j G_0(\mathbf{r}, \mathbf{r}_j) U(\mathbf{r}_j), \tag{12.16}$$

with the local fields inside the scatterers obeying the following system of linear algebraic equations:

$$\sum_j A_{ij} U(\mathbf{r}_j) = U_i(\mathbf{r}_i) , \tag{12.17}$$

where

$$A_{ij} = \begin{cases} 1 - \alpha_0 \left(\dfrac{3k_0^2}{8\pi a} + i\dfrac{k_0^3}{4\pi} \right) & \text{if} \quad i = j , \\ -\alpha_0 k_0^2 G_0(\mathbf{r}_i, \mathbf{r}_j) & \text{if} \quad i \neq j . \end{cases} \tag{12.18}$$

Note that the diagonal elements of the matrix A account for self-interactions.

The scattering amplitude A may be immediately obtained from (10.15) and (12.16). We find that

$$A = \frac{\alpha_0 k_0^2}{4\pi} \sum_j \exp(-ik_0 \hat{\mathbf{r}} \cdot \mathbf{r}_j) U(\mathbf{r}_j) . \tag{12.19}$$

In Fig. 12.2, we show the frequency dependence of the extinction cross section, calculated from the optical theorem (11.10) and from (12.19), for a pair of point scatterers illuminated by a unit-amplitude plane wave. Two cases are considered: the wave vector of the incident plane wave is either parallel or perpendicular to the line containing the scatterers. In the parallel case, the scattering resonance is split.

12.3 Scattering from Spheres of Arbitrary Size

In this section, we consider the scattering of a plane wave from a homogeneous spherical scatterer of radius a. This is known as Mie scattering. The field

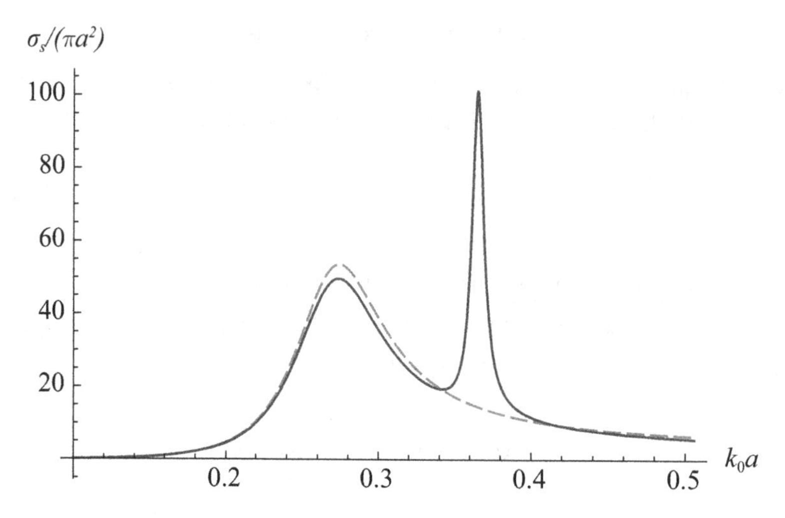

Fig. 12.2 Frequency dependence of the scattering cross section for two point scatterers with resonance frequencies $k_* a = 0.3$ and separation $2a$. Parallel illumination is shown in solid. Perpendicular illumination is shown as a dashed line.

everywhere in space satisfies

$$\nabla^2 U + k^2 U = 0 \,, \tag{12.20}$$

where k is the wavenumber. Inside the sphere, the field which we denote by U_1, obeys (12.20) with $k = nk_0$, where $n = \sqrt{\varepsilon}$ is the index of refraction. If we work in spherical coordinates with the origin of the coordinate system at the center of the sphere, we find that U_1 is given by

$$U_1(\mathbf{r}) = \sum_{l=0}^{\infty} A_l j_l(nk_0 r) P_l(\cos\theta) \,, \tag{12.21}$$

where θ is the angle between the incident wave vector and the direction of observation $\hat{\mathbf{r}}$, and A_l is the amplitude of the lth partial wave. The field outside the spheres, U_2, obeys (12.20) with $k = k_0$. If the incident field is a unit-amplitude plane wave, then we have

$$U_2(\mathbf{r}) = \exp(i\mathbf{k} \cdot \mathbf{r}) + \sum_{l=0}^{\infty} i^l (2l+1) B_l h_l^{(1)}(k_0 r) P_l(\cos\theta) \,, \tag{12.22}$$

where $\mathbf{k}$ is the incident wave vector. Note that, for later convenience, the amplitude of the lth partial wave is written as $i^l(2l+1)B_l$.

The boundary conditions on the field require the continuity of the field and its normal derivative on the surface $r = a$:

$$U_1|_{r=a} = U_2|_{r=a} \tag{12.23}$$

$$\left.\frac{\partial U_1}{\partial r}\right|_{r=a} = \left.\frac{\partial U_2}{\partial r}\right|_{r=a} . \tag{12.24}$$

Note that there are two sets of unknown coefficients and two boundary conditions that must be satisfied for each partial wave. To make further progress, we require the expansion of a plane wave into spherical waves:

$$\exp(i\mathbf{k}\cdot\mathbf{r}) = \sum_{l=0}^{\infty} i^l(2l+1)j_l(k_0 r)P_l(\cos\theta) . \tag{12.25}$$

Applying the boundary conditions, and using (12.25), we obtain the following relations among the coefficients:

$$A_l j_l(nk_0 a) = i^l(2l+1)\left[j_l(k_0 a) + B_l h_l^{(1)}(k_0 a)\right] , \tag{12.26}$$

$$nA_l j_l'(nk_0 a) = i^l(2l+1)\left[j_l'(k_0 a) + B_l h_l^{(1)\prime}(k_0 a)\right] . \tag{12.27}$$

Equations (12.26) and (12.27) are linear equations that can be solved for B_l with the result

$$B_l = \frac{j_l(nk_0 a)j_l'(k_0 a) - nj_l(k_0 a)j_l'(nk_0 a)}{nh_l^{(1)}(k_0 a)j_l'(nk_0 a) - h_l^{(1)\prime}(k_0 a)j_l(nk_0 a)} . \tag{12.28}$$

The B_l are known as the Mie coefficients. There is a related expression for A_l that we will not need.

We may now obtain the scattering amplitude for the sphere. To proceed, we require the asymptotic form of the spherical Hankel function

$$h_l^{(1)}(x) \sim (-i)^{l+1}\frac{\exp(ix)}{x} , \quad x \gg l . \tag{12.29}$$

Making use of (12.22) and the above result, we see that in the far-zone, the scattered field is given by

$$U_s(\mathbf{r}) \sim A\frac{\exp(ik_0 r)}{r} , \tag{12.30}$$

where the scattering amplitude is

$$A = \frac{1}{ik_0}\sum_{l=0}^{\infty}(2l+1)B_l P_l(\cos\theta) . \tag{12.31}$$

To calculate the scattering cross section σ_s, we note that

$$\begin{aligned}\sigma_s &= \int |A|^2 d\hat{\mathbf{r}} \\ &= \frac{2\pi}{k_0^2}\int_{-1}^{1} dx \sum_{l,l'}(2l+1)(2l'+1)B_l B_{l'}^* P_l(x)P_{l'}(x) ,\end{aligned} \tag{12.32}$$

which follows from (10.23) and (12.31). Using the orthogonality of the Legendre functions

$$\int_{-1}^{1} P_l(x)P_{l'}(x)dx = \frac{2}{2l+1}\delta_{ll'} , \tag{12.33}$$

we obtain

$$\sigma_s = \frac{4\pi}{k_0^2}\sum_{l=0}^{\infty}(2l+1)|B_l|^2 . \tag{12.34}$$

If the sphere is large, with $k_0 a \gg 1$, many terms in the above series are required. It may be shown that the Mie coefficients are exponentially small if $l > k_0 a$, due to the asymptotic behavior of spherical Bessel functions for large orders and small arguments. The frequency dependence of σ_s is shown in Fig. 12.3. We note the presence of multiple scattering resonances, known as Mie resonances, which occur at frequencies where the Mie coefficients exhibit relative maxima (small denominators). We also note that at high frequencies, the scattering cross section approaches the asymptote $\sigma_s = 2\pi a^2$, which corresponds to twice the cross-sectional area of the sphere. Note that for an absorbing sphere we would get $\sigma_e = 2\pi a^2$. The fact that the extinction cross section for a large sphere exceeds the geometrical cross section is due to diffraction.

For a small and weakly scattering sphere, with $k_0 a \ll 1$ and $n \simeq 1$, only the lowest order Mie coefficient in (12.31) is important, and it can be seen that

$$B_0 = \frac{i}{3}(k_0 a)^3(n^2 - 1) + O\left((k_0 a)^6\right) . \tag{12.35}$$

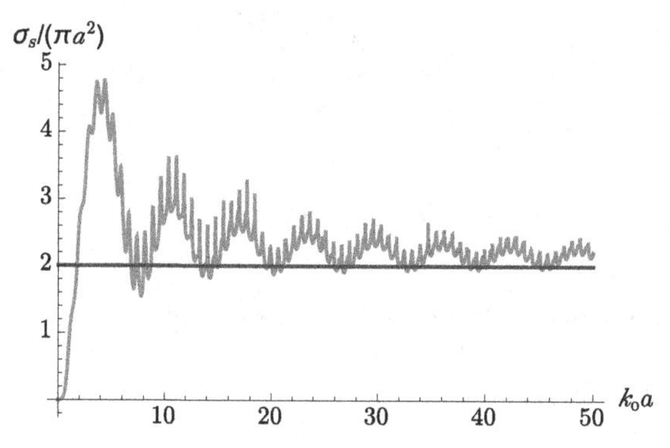

Fig. 12.3 Frequency dependence of the normalized scattering cross section $\sigma_s/(\pi a^2)$ for a spherical scatterer with index of refraction $n = 1.5$.

The differential cross section is thus given by

$$\frac{d\sigma_s}{d\Omega} = k_0^4 \left(\delta\varepsilon \frac{a^3}{3} \right)^2 , \tag{12.36}$$

where we have made use of the relation $n^2 - 1 = \delta\varepsilon$. As expected, this result agrees with (10.30), the differential cross section for Rayleigh scattering.

We can also obtain the normalized differential scattering cross section, also known as the single-scatterer phase function. It is defined by

$$p = \frac{1}{\sigma_s} \frac{d\sigma_s}{d\Omega} , \tag{12.37}$$

where, according to (10.26), the differential cross section is given by

$$\frac{d\sigma_s}{d\Omega} = |A|^2 . \tag{12.38}$$

Using (12.31) and (12.34), we find that the single-scatterer phase function is given by

$$p(\theta) = \frac{\left| \sum_{l=0}^{\infty} (2l+1) B_l P_l(\cos\theta) \right|^2}{4\pi \sum_{l=0}^{\infty} (2l+1) |B_l|^2} . \tag{12.39}$$

If the sphere is small and weakly scattering, with $k_0 a \ll 1$ and $n \simeq 1$, then only the lowest order Mie coefficient in (12.39) is important and it can be seen that $p = 1/(4\pi)$, which corresponds to isotropic scattering. In the opposite limit, where the sphere is large and strongly scattering, the phase function is peaked about the forward direction with $\theta \simeq 0$, as shown in Fig. 12.4.

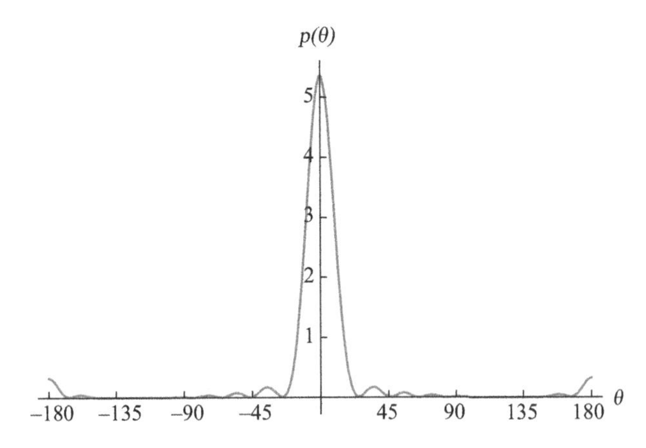

Fig. 12.4 Phase function for a sphere with $k_0 a = 20$ and index of refraction $n = 1.5$.

References and Additional Reading

This book is a basic reference on scattering from spheres:

W.T. Grandy, *Scattering of Waves from Large Spheres* (Cambridge University Press, Cambridge, 2005).

The exact solution of scattering of an electromagnetic plane wave from a homogeneous sphere was given by Mie in a seminal paper:

G. Mie, Ann. Phys. **25**, 377 (1908).

It is also recognized that L. Lorenz developed an equivalent theory in 1890. For an overview of Lorenz's works in optics, see:

H. Kragh, Appl. Opt. **30**, 4688 (1991).

Light scattering from spheres is extensively discussed in:

H.C. van de Hulst, *Light Scattering by Small Particles* (Dover, New York, 1981).

C.F. Bohren and D.R. Huffman, *Absorption and Scattering of Light by Small Particles* (Wiley, New York, 1983).

The following paper discusses light scattering by a sphere and emphasizes the so-called extinction paradox:

L. Brillouin, J. Appl. Phys. **20**, 1110 (1949).

Scattering from an ensemble of point scatterers was addressed in early papers by Foldy and Lax. A comprehensive presentation of these works can be found in:

M. Lax, Rev. Mod. Phys. **23**, 287 (1951).

13

Renormalized Perturbation Theory

In this chapter, we consider a perturbation expansion that may be used in place of the Born series introduced in Chapter 10. This so-called Rytov series allows the amplitude and phase of the scattered field to be treated on a separate footing.

13.1 Rytov Series

We consider the scattering of a monochromatic scalar wave field U. To construct the Rytov series, we consider the renormalized field ψ, which is defined by

$$\psi = \frac{U}{U_i} , \tag{13.1}$$

where, as usual, U_i is the incident field. Using the Born series (10.11), we see that ψ is given by

$$\begin{aligned}\psi(\mathbf{r}) = 1 &+ \int \Gamma^{(1)}(\mathbf{r}; \mathbf{r}_1)\delta\varepsilon(\mathbf{r}_1)\, d^3r_1 \\ &+ \frac{1}{2}\int \Gamma^{(2)}(\mathbf{r}; \mathbf{r}_1, \mathbf{r}_2)\delta\varepsilon(\mathbf{r}_1)\, \delta\varepsilon(\mathbf{r}_2)\, d^3r_1 d^3r_2 + \cdots ,\end{aligned} \tag{13.2}$$

where

$$\Gamma^{(1)}(\mathbf{r}; \mathbf{r}_1) = \frac{k_0^2}{U_i(\mathbf{r})} G_0(\mathbf{r}, \mathbf{r}_1) U_i(\mathbf{r}_1) , \tag{13.3}$$

$$\Gamma^{(2)}(\mathbf{r}; \mathbf{r}_1, \mathbf{r}_2) = \frac{2k_0^4}{U_i(\mathbf{r})} G_0(\mathbf{r}, \mathbf{r}_1) G_0(\mathbf{r}_1, \mathbf{r}_2) U_i(\mathbf{r}_2) . \tag{13.4}$$

It can be seen that ψ is a generating functional for the Γ's, which can be found by functional differentiation:

$$\Gamma^{(1)}(\mathbf{r}; \mathbf{r}_1) = \left. \frac{\delta \psi(\mathbf{r})}{\delta\, \delta\varepsilon(\mathbf{r}_1)} \right|_{\delta\varepsilon = 0} , \tag{13.5}$$

$$\Gamma^{(2)}(\mathbf{r};\mathbf{r}_1,\mathbf{r}_2) = \left.\frac{\delta^2\psi(\mathbf{r})}{\delta\,\delta\varepsilon(\mathbf{r}_1)\,\delta\,\delta\varepsilon(\mathbf{r}_2)}\right|_{\delta\varepsilon=0} . \tag{13.6}$$

We now turn to the renormalization of the Born series. This procedure is based on the ansatz that the series (13.2) can be written in the form

$$\psi(\mathbf{r}) = \exp\left[\int \Gamma_R^{(1)}(\mathbf{r};\mathbf{r}_1)\delta\varepsilon(\mathbf{r}_1)\,d^3r_1 + \frac{1}{2}\int \Gamma_R^{(2)}(\mathbf{r};\mathbf{r}_1,\mathbf{r}_2)\delta\varepsilon(\mathbf{r}_1)\delta\varepsilon(\mathbf{r}_2)\,d^3r_1 d^3r_2 + \cdots\right] , \tag{13.7}$$

where the Γ_R's are to be determined. Equation (13.7) is known as the Rytov series. Evidently, $\ln\psi$ is the generating functional for the Γ_R's:

$$\Gamma_R^{(1)}(\mathbf{r};\mathbf{r}_1) = \left.\frac{\delta\ln\psi(\mathbf{r})}{\delta\,\delta\varepsilon(\mathbf{r}_1)}\right|_{\delta\varepsilon=0} , \tag{13.8}$$

$$\Gamma_R^{(2)}(\mathbf{r};\mathbf{r}_1,\mathbf{r}_2) = \left.\frac{\delta^2\ln\psi(\mathbf{r})}{\delta\,\delta\varepsilon(\mathbf{r}_1)\,\delta\,\delta\varepsilon(\mathbf{r}_2)}\right|_{\delta\varepsilon=0} . \tag{13.9}$$

Carrying out the necessary differentiations, we find that

$$\Gamma_R^{(1)}(\mathbf{r};\mathbf{r}_1) = \Gamma^{(1)}(\mathbf{r};\mathbf{r}_1) , \tag{13.10}$$

$$\Gamma_R^{(2)}(\mathbf{r};\mathbf{r}_1,\mathbf{r}_2) = \Gamma^{(2)}(\mathbf{r};\mathbf{r}_1,\mathbf{r}_2) - \Gamma^{(1)}(\mathbf{r};\mathbf{r}_1)\Gamma^{(1)}(\mathbf{r};\mathbf{r}_2) . \tag{13.11}$$

If the Rytov series is truncated after its first term, (13.7) can be rewritten in the form

$$\ln\psi(\mathbf{r}) = \frac{k_0^2}{U_i(\mathbf{r})}\int G_0(\mathbf{r},\mathbf{r}')U_i(\mathbf{r}')\delta\varepsilon(\mathbf{r}')\,d^3r' , \tag{13.12}$$

which is referred to as the Rytov approximation. It is instructive to compare the Born and Rytov approximations. If we reexponentiate (13.12) and expand the resulting exponential, we obtain

$$U(\mathbf{r}) = U_i(\mathbf{r}) + k_0^2\int G_0(\mathbf{r},\mathbf{r}')U_i(\mathbf{r}')\delta\varepsilon(\mathbf{r}')\,d^3r' + O\left((\delta\varepsilon)^2\right) . \tag{13.13}$$

This result agrees to first order in $\delta\varepsilon$ with the Born approximation (10.12). It is important to note that terms of all orders in $\delta\varepsilon$ are present in (13.13). Thus the Rytov series can be viewed as a resummation of the Born series.

13.2 Geometrical Optics and the Radon Transform

We now show that the Rytov approximation allows for a useful separation of the amplitude and phase of the scattered field. If we put

$$U = |U|\exp(i\phi) , \quad U_i = |U_i|\exp(i\phi_i) , \tag{13.14}$$

where ϕ and ϕ_i are the phases of the total and incident fields U and U_i, respectively, we find that

$$\ln\psi = \frac{1}{2}\ln T + i\,(\phi - \phi_i)\ . \tag{13.15}$$

Here T denotes the intensity transmission coefficient, which is defined by

$$T = \frac{|U|^2}{|U_i|^2}\ . \tag{13.16}$$

Using (13.12) and (13.15), we find, within the accuracy of the Rytov approximation, that T obeys the integral equation

$$\ln T(\mathbf{r}) = \frac{k_0^2}{U_i(\mathbf{r})}\int G_0(\mathbf{r},\mathbf{r}')U_i(\mathbf{r}')\delta\varepsilon(\mathbf{r}')\,d^3r' + \text{c.c.}\ . \tag{13.17}$$

Suppose that the incident field is a plane wave in the z direction:

$$U_i(\mathbf{r}) = \exp(ik_0 z)\ . \tag{13.18}$$

Then, using the plane-wave decomposition for G_0 given by (6.13), (13.17) becomes

$$\begin{aligned} -\ln T(\mathbf{r}) = \frac{k_0^2}{4\pi^2}\mathrm{Im}\int d^3r'\int \frac{d^2q}{k_z(q)}\exp\big[i\mathbf{q}\cdot(\boldsymbol{\rho}-\boldsymbol{\rho}') \\ +i(k_z(q)-k_0)(z-z')\big]\delta\varepsilon(\mathbf{r}')\ , \end{aligned} \tag{13.19}$$

where we have assumed that T is measured in the transmission geometry with the point of observation to the right of the scatterer. In the geometrical optics limit, with $k_0 \to \infty$, it can be seen that

$$\frac{\exp\big[i(k_z(q)-k_0)(z-z')\big]}{k_z(q)} = \frac{1}{k_0} + O(q^2)\ . \tag{13.20}$$

Using this result and carrying out the integrations in (13.19), we obtain

$$-\ln T = \int \alpha(\boldsymbol{\rho}, z)\,dz\ , \tag{13.21}$$

where $\mathbf{r} = (\boldsymbol{\rho}, z)$ and the absorption α is defined by

$$\alpha = k_0\,\mathrm{Im}\,\delta\varepsilon\ . \tag{13.22}$$

We note that the identification of the absorption with the imaginary part of the dielectric function has been mentioned in Chapter 2 in connection with energy conservation (Section 2.4).

Evidently, (13.21) predicts that the intensity of the transmitted field is exponentially attenuated in a manner that is independent of z but depends on the transverse coordinate $\boldsymbol{\rho}$. This is consistent with the geometrical optics limit. That is, the wavefronts are not distorted even close to the scatterer. If we regard the incident plane wave as composed of many well-collimated beams then the transmission along a particular beam is given by

$$-\ln T = \int_L \alpha \, ds \,, \tag{13.23}$$

where the line L lies along the axis of the beam in the direction of propagation. Equation (13.23) defines the Radon transform of the function α along the line L. The Radon transform plays a fundamental role in tomography.

References and Additional Reading

The Born series in scattering theory is presented in:

R. Newton, *Scattering Theory of Waves and Particles* (McGraw-Hill, New York, 1966), chap. 9.

The Rytov series is described in:

M. Born and E. Wolf, *Principles of Optics*, 7th edition (Cambridge University Press, Cambridge, 2005), chap. 13.

M. Nieto-Vesperinas, *Scattering and Diffraction in Physical Optics*, 2nd edition (World Scientific, Singapore, 2006), chap. 3.

14

Wave Reciprocity

Reciprocity is a fundamental property of wave propagation that is very useful in the treatment of radiation and scattering problems. In this chapter, we derive reciprocity relations for the scalar field amplitude, Green's function and scattering matrix. Reciprocity relations for vector electromagnetic waves are presented in Chapter 40.

14.1 Fundamental Relation

We consider a monochromatic scalar wave field U_1 that propagates in an inhomogeneous medium and satisfies the wave equation

$$\nabla^2 U_1(\mathbf{r}) + k_0^2 \varepsilon(\mathbf{r}) U_1(\mathbf{r}) = S_1(\mathbf{r}) , \tag{14.1}$$

where S_1 is the source (localized to a volume V_1) and ε is the dielectric permittivity, as shown in Fig. 14.1(a). Likewise, let U_2 obey

$$\nabla^2 U_2(\mathbf{r}) + k_0^2 \varepsilon(\mathbf{r}) U_2(\mathbf{r}) = S_2(\mathbf{r}) , \tag{14.2}$$

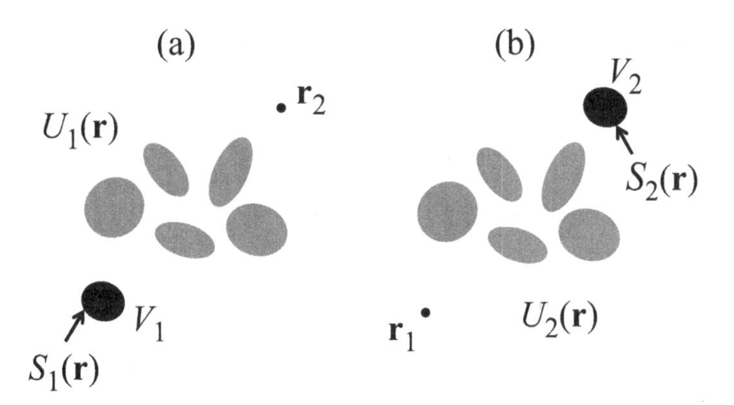

Fig. 14.1 Illustrating the two situations considered in the reciprocity theorem.

which describes the propagation of a wave in the same medium, originating from a different source S_2 (localized to a volume V_2), as shown in Fig. 14.1(b). The application of Green's theorem to a volume V bounded by a surface ∂V with unit outward normal $\hat{\mathbf{n}}$ leads to

$$\int_V \left[U_2(\mathbf{r})\nabla^2 U_1(\mathbf{r}) - U_1(\mathbf{r})\nabla^2 U_2(\mathbf{r})\right] d^3r = \int_{\partial V} \left[U_2(\mathbf{r})\frac{\partial U_1(\mathbf{r})}{\partial n} - U_1(\mathbf{r})\frac{\partial U_2(\mathbf{r})}{\partial n}\right] d^2r \,. \tag{14.3}$$

By making use of (14.1) and (14.2), the left-hand side of (14.3) becomes

$$\int_{V_1} S_1(\mathbf{r})\, U_2(\mathbf{r})\, d^3r - \int_{V_2} S_2(\mathbf{r})\, U_1(\mathbf{r})\, d^3r = \int_{\partial V} \left[U_2(\mathbf{r})\frac{\partial U_1(\mathbf{r})}{\partial n} - U_1(\mathbf{r})\frac{\partial U_2(\mathbf{r})}{\partial n}\right] d^2r. \tag{14.4}$$

The above relation is the most general form of the reciprocity theorem. Note that it holds in the case of an absorbing medium, since the dielectric function ε is not assumed to be real. We will see that the robustness of reciprocity with respect to absorption is a remarkable fact.

14.2 Local Form of the Reciprocity Theorem

In an open geometry, the surface ∂V can be taken to be a sphere centered at the origin of radius R. On the right-hand side in (14.4), the fields can be replaced by their far-field asymptotic expansions $U_{1,2}(\mathbf{r}) \sim \exp(ik_0 r)/r$, so that the integrand vanishes as $R \to \infty$. We therefore obtain the relation

$$\int_{V_1} S_1(\mathbf{r})\, U_2(\mathbf{r})\, d^3r = \int_{V_2} S_2(\mathbf{r})\, U_1(\mathbf{r})\, d^3r \,. \tag{14.5}$$

Suppose that $S_1 = \delta(\mathbf{r}-\mathbf{r}_1)$ and $S_2 = \delta(\mathbf{r}-\mathbf{r}_2)$, corresponding to two point sources with unit amplitude. Eq. (14.5) thus becomes

$$U_1(\mathbf{r}_2) = U_2(\mathbf{r}_1) \,, \tag{14.6}$$

which we refer to as the local reciprocity theorem.

14.3 Reciprocity of the Green's Function

A reciprocity relation for the Green's function follows immediately from (14.6). The Green's function G obeys

$$\nabla^2 G(\mathbf{r}, \mathbf{r}') + k_0^2 \varepsilon(\mathbf{r}) G(\mathbf{r}, \mathbf{r}') = -\delta(\mathbf{r} - \mathbf{r}') \,, \tag{14.7}$$

together with the outgoing radiation condition. For the point sources defined in the previous section, we simply have $U_1(\mathbf{r}_2) = G(\mathbf{r}_2, \mathbf{r}_1)$ and $U_2(\mathbf{r}_1) = G(\mathbf{r}_1, \mathbf{r}_2)$. Using (14.6), we deduce that

$$G(\mathbf{r}_1, \mathbf{r}_2) = G(\mathbf{r}_2, \mathbf{r}_1) , \tag{14.8}$$

which is the reciprocity theorem for the scalar Green's function.

14.4 Reciprocity of the Scattering Matrix

A scattering process can be thought of as a transformation of an incoming wave U^{in} into an outgoing wave U^{out}, described mathematically by an operator relation of the form $U^{\text{out}} = \mathcal{S}U^{\text{in}}$, where the operator $\mathcal{S}$ is called the scattering matrix or S-matrix. We will see that reciprocity leads to an interesting symmetry property of the S-matrix.

Let us consider the scattering problem depicted in Fig. 14.2. The regions $0 \leq z \leq z_1$ and $z_2 \leq z \leq L$, denoted by $\mathcal{R}^-$ and $\mathcal{R}^+$, respectively, are assumed to be vacuum, so that the field amplitude in these regions obeys the equation

$$\nabla^2 U(\mathbf{r}) + k_0^2\, U(\mathbf{r}) = 0 . \tag{14.9}$$

The regions $z < 0$ and $z > L$ may contain sources, whereas the region $z_1 < z < z_2$ contains the scattering medium.

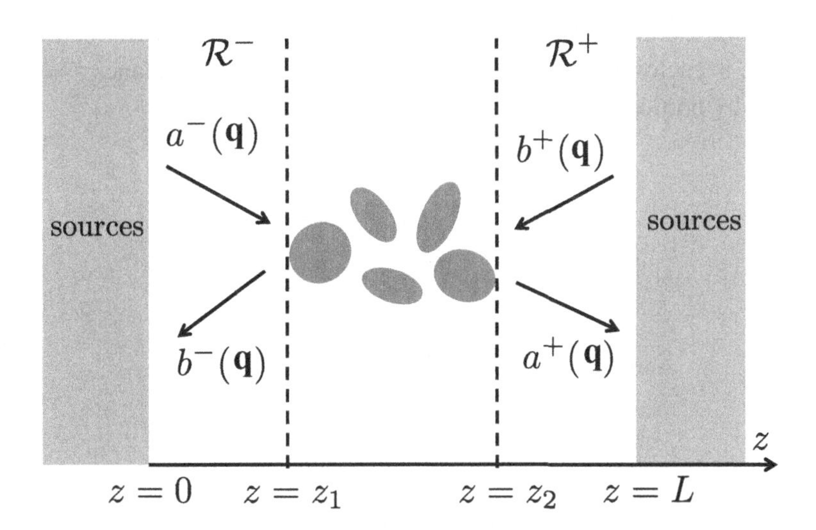

Fig. 14.2 Geometry considered for the definition of the S-matrix.

Let us denote by U^+ [U^-] the field amplitude in region $\mathcal{R}^+$ [$\mathcal{R}^-$]. Then, $U^\pm$ can be written in the form of a plane-wave expansion (see Chapter 6) of the form

$$U^{\pm}(\mathbf{r}) = \int a^{\pm}(\mathbf{q}) \exp[i\mathbf{q}\cdot\boldsymbol{\rho} + ik_z(\mathbf{q})z]\, d^2q + \int b^{\pm}(\mathbf{q}) \exp[i\mathbf{q}\cdot\boldsymbol{\rho} - ik_z(\mathbf{q})z]\, d^2q\,, \tag{14.10}$$

where $k_z(\mathbf{q}) = \sqrt{k_0^2 - q^2}$ and the integrals are carried out over both homogeneous plane waves, corresponding to $|\mathbf{q}| \leq k_0$, and evanescent waves corresponding to $|\mathbf{q}| > k_0$. The amplitudes $a^+(\mathbf{q})$ and $b^-(\mathbf{q})$ denote either outgoing waves or waves that decay exponentially from the scatterer. Likewise, the amplitudes $a^-(\mathbf{q})$ and $b^+(\mathbf{q})$ denote either incoming waves or waves that increase exponentially from the source regions towards the scatterer. Note that because in $\mathcal{R}^+$ and $\mathcal{R}^-$ the value of $|z|$ remains finite, the integrals in Eq. (14.10) are well defined.

The S-matrix $\mathbf{S}(\mathbf{q}, \mathbf{q}')$ connects the outgoing field vector $\mathbf{U}^{\text{out}} = [a^+(\mathbf{q}), b^-(\mathbf{q})]$ to the input field vector $\mathbf{U}^{\text{in}} = [a^-(\mathbf{q}), b^+(\mathbf{q})]$ by the relation

$$\mathbf{U}^{\text{out}}(\mathbf{q}) = \int \mathbf{S}(\mathbf{q}, \mathbf{q}')\, \mathbf{U}^{\text{in}}(\mathbf{q}')\, d^2q'\,. \tag{14.11}$$

Here the operator $\mathbf{S}$ is a 2×2 matrix which can be written in the form

$$\mathbf{S}(\mathbf{q}, \mathbf{q}') = \begin{bmatrix} r(\mathbf{q}, \mathbf{q}') & \tau(\mathbf{q}, \mathbf{q}') \\ t(\mathbf{q}, \mathbf{q}') & \rho(\mathbf{q}, \mathbf{q}') \end{bmatrix} \tag{14.12}$$

where the four elements r, t, τ, ρ can be understood as generalized reflection and transmission factors.

Using the general relation (14.4), we will now derive reciprocity relations for the S-matrix, and for the generalized reflection and transmission coefficients. To proceed, we define the integration volume V represented by the dotted line Fig. 14.3. This volume is enclosed by the surface ∂V defined by the two planes $z = z^-$ and $z = z^+$, and by portions of two spheres of radius R.

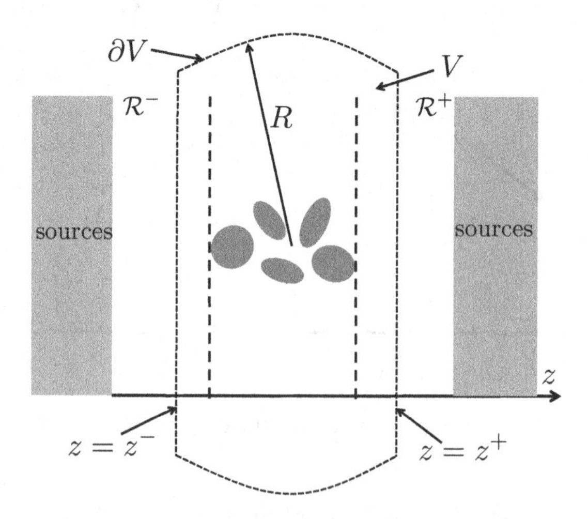

Fig. 14.3 Integration volume used for the derivation of reciprocity relations for the S-matrix.

Let U_1 and U_2 be solutions of the scattering problem defined in Fig. 14.2. An application of then relation (14.4) to the volume V, which encloses the scattering medium but does not contain the sources, gives

$$\int_{\partial V}\left[U_2(\mathbf{r})\frac{\partial U_1(\mathbf{r})}{\partial n}-U_1(\mathbf{r})\frac{\partial U_2(\mathbf{r})}{\partial n}\right]d^2r=0\,. \tag{14.13}$$

In the far-field, the above integrand vanishes on the spheres in the limit $R\to\infty$. We thereby obtain

$$\int_{z=z^-}\left[U_2(\mathbf{r})\frac{\partial U_1(\mathbf{r})}{\partial z}-U_1(\mathbf{r})\frac{\partial U_2(\mathbf{r})}{\partial z}\right]d^2r=\int_{z=z^+}\left[U_2(\mathbf{r})\frac{\partial U_1(\mathbf{r})}{\partial z}-U_1(\mathbf{r})\frac{\partial U_2(\mathbf{r})}{\partial z}\right]d^2r\,. \tag{14.14}$$

Introducing the expansion (14.10) of U_1 and U_2 into (14.14) leads to the following relation involving the S-matrix:

$$\int\left[a_1^-(\mathbf{q})\;b_1^+(\mathbf{q})\right]\left\{k_z(\mathbf{q})\,\mathbf{S}(\mathbf{q},\mathbf{q}')-k_z(\mathbf{q}')\,\mathbf{S}^T(-\mathbf{q}',-\mathbf{q})\right\}\begin{bmatrix}a_2^-(\mathbf{q}')\\ b_2^+(\mathbf{q}')\end{bmatrix}d^2q\,d^2q'=0\,. \tag{14.15}$$

Since this equality must be valid for any incoming field vector $[a_1^-(\mathbf{q})\;b_1^+(\mathbf{q})]$ and any incoming field vector $[a_2^-(\mathbf{q})\;b_2^+(\mathbf{q})]$, we conclude that

$$k_z(\mathbf{q})\,\mathbf{S}(\mathbf{q},\mathbf{q}')=k_z(\mathbf{q}')\,\mathbf{S}^T(-\mathbf{q}',-\mathbf{q})\,, \tag{14.16}$$

which is the reciprocity theorem for the scattering matrix. The presence of the factors $k_z(\mathbf{q})$ and $k_z(\mathbf{q}')$ in the relation is a consequence of the definition of the plane-wave expansion of the field. An integration over solid angles results in reciprocity relations in which these factors disappear. Nevertheless, the $\mathbf{q}$ representation is more appropriate to describe both the homogeneous and evanescent components of the field.

The reciprocity theorem for the S-matrix directly leads to reciprocity relations for the generalized reflection and transmission coefficients:

$$k_z(\mathbf{q})\,r(\mathbf{q},\mathbf{q}')=k_z(\mathbf{q}')\,r(-\mathbf{q}',-\mathbf{q})\,, \tag{14.17}$$

$$k_z(\mathbf{q})\,\rho(\mathbf{q},\mathbf{q}')=k_z(\mathbf{q}')\,\rho(-\mathbf{q}',-\mathbf{q})\,, \tag{14.18}$$

$$k_z(\mathbf{q})\,t(\mathbf{q},\mathbf{q}')=k_z(\mathbf{q}')\,\tau(-\mathbf{q}',-\mathbf{q})\,. \tag{14.19}$$

These relations hold for both the homogeneous and evanescent components of the field.

References and Additional Reading

The following papers are historical references on reciprocity:

H. Helmholtz, J. Reine Angew. Math. 67, 1 (1859).

Lord Rayleigh, *The Theory of Sound* (Dover, New York, 1945), section 294.

The following textbooks contain a chapter or section on reciprocity:

P.M. Morse and H. Feshbach, *Methods of Theoretical Physics* (McGraw-Hill, New York, 1953), part I, section 7.5.

P. Roman, *Advanced Quantum Theory* (Addison-Wesley, Reading, 1965), chap. 4 (S-matrix formalism) and 5 (symmetry and invariance).

R.G. Newton, *Scattering Theory of Waves and Particles* (McGraw-Hill, New York, 1966), p. 46.

M. Nieto-Vesperinas, *Scattering and Diffraction in Physical Optics*, 2nd edition (World Scientific, Singapore, 2006), chap. 5.

The following papers derive the reciprocity theorem in the context of light scattering:

E. Gerjuoy and D.S. Saxon, Phys. Rev. **94**, 1445 (1954).

D.S. Saxon, Phys. Rev. **100**, 1771 (1955).

M. Nieto-Vesperinas and E. Wolf, J. Opt. Soc. Am. A **3**, 2038 (1986).

R. Carminati, J.J. Sáenz, J.-J. Greffet and M. Nieto-Vesperinas, Phys. Rev. A **62**, 012712 (2000).

Exercises

II.1 Derive the far-field asymptotic expression of the free-space Green function given by (10.15), and show that the conditions of validity of this expression are $r \gg r'$ and $r \gg r'^2/\lambda$.

II.2 In Chapter 12, the integral of the scalar free-space Green function G_0 over a small sphere with radius a centered at the origin is used. By expanding the Green function in terms of the small parameter $k_0 a$, derive (12.4).

II.3 The optical theorem for a scatterer illuminated by a plane wave with wave vector $\mathbf{k}$ is given by (11.10). Can you give a physical interpretation of the optical theorem in terms of interferences between the incident and the scattered field? This interference phenomenon explains the structure of (11.10), which involves a power on the left-hand side and a field amplitude on the right-hand side.

II.4 Consider a spherical scatterer with radius a and refractive index n, illuminated by a plane wave with wavelength λ. Using the integral expression of the scattered field in the Born approximation, can you derive a condition of validity of this approximation in terms of a, n and λ?

II.5 A sphere with radius $a = 2.5\ \mu$m is illuminated by a plane wave with wavelength $\lambda = 633$ nm. Draw schematically the angular dependence of the phase function. Using simple diffraction arguments, estimate the angular width of the forward-scattering peak.

II.6 A point scatterer, placed at position $\mathbf{r}_s$ in free space, is described by its polarizability α. Construct the Green function $G(\mathbf{r}, \mathbf{r}')$ that accounts for the presence of the scatterer, in terms of α and the free-space Green function $G_0(\mathbf{r}, \mathbf{r}')$.

II.7 For the problem of the scattering of a scalar wave from a collection of point scatterers (Section 12.2), solve the set of linear equations (12.16) to obtain the local field inside the scatterers in terms of the incident field. Use this result in (12.17) to derive the expression for the T-matrix of the scatterers.

II.8 In the derivation of reciprocity relations for the scattering matrix, fill in the steps to deduce (14.15) from (14.14).

Part III

Wave Transport

15

Multiple Scattering: Average Field

In this chapter, we consider the problem of calculating the average field in a disordered medium. By making use of diagrammatic perturbation theory, we obtain an integral equation for the average Green's function that can be solved in the weak-scattering limit.

15.1 Gaussian Model

We begin by considering the propagation of a monochromatic scalar wave field in a disordered medium. The field obeys the wave equation

$$\nabla^2 U(\mathbf{r}) + k_0^2\, \varepsilon(\mathbf{r}) U(\mathbf{r}) = -S(\mathbf{r}) \,, \tag{15.1}$$

where $k_0 = \omega/c$ is the wavenumber in vacuum and S is the source. The medium is described statistically, by taking the dielectric function ε to be a random field. This statistical approach will be taken throughout the following chapters, where a disordered medium is often referred to as a random medium. For simplicity, we assume that $\varepsilon(\mathbf{r}) = 1 + \delta\varepsilon(\mathbf{r})$, where $\delta\varepsilon$ is real-valued, so that the medium is nonabsorbing. We further assume that $\delta\varepsilon$ has Gaussian statistics satisfying

$$\langle \delta\varepsilon(\mathbf{r}) \rangle = 0 \,, \tag{15.2}$$

$$\left\langle \delta\varepsilon(\mathbf{r}) \delta\varepsilon(\mathbf{r}') \right\rangle = C(\mathbf{r} - \mathbf{r}') \,, \tag{15.3}$$

where C is the two-point correlation function and $\langle \cdots \rangle$ denotes statistical averaging over realizations of the disorder. We suppose that the medium is statistically homogeneous and isotropic. That is, the correlation function $C(\mathbf{r} - \mathbf{r}')$ depends only upon the quantity $|\mathbf{r} - \mathbf{r}'|$. A case of particular interest is that of Gaussian correlations of the form

$$C(\mathbf{r}) = \frac{A}{\pi^{3/2}\ell_c^3} \exp(-r^2/\ell_c^2) \,, \tag{15.4}$$

where A is constant and ℓ_c is the correlation length. The prefactor is chosen so that in the limit $\ell_c \to 0$ we obtain $\langle \delta\varepsilon(\mathbf{r})\, \delta\varepsilon(\mathbf{r}')\rangle = A\, \delta(\mathbf{r} - \mathbf{r}')$, which corresponds to the white-noise model. We also impose the condition $k_0\ell_c \ll 1$, which further corresponds to neglecting spatial dispersion. The effect of absorption is discussed at the end of the chapter, where more general models of disorder are considered.

We note the following important properties of Gaussian random fields. Odd moments of $\delta\varepsilon$ vanish. Even moments can be expressed in terms of products of second moments according to

$$\langle \delta\varepsilon(\mathbf{r}_1) \cdots \delta\varepsilon(\mathbf{r}_n)\rangle = \sum_{\pi} \left\langle \delta\varepsilon(\mathbf{r}_{\pi(1)})\delta\varepsilon(\mathbf{r}_{\pi(2)})\right\rangle \cdots \left\langle \delta\varepsilon(\mathbf{r}_{\pi(n-1)})\delta\varepsilon(\mathbf{r}_{\pi(n)})\right\rangle \;, \tag{15.5}$$

where the sum is over all permutations π. As an example, we see that

$$\begin{aligned} \langle \delta\varepsilon(\mathbf{r}_1)\delta\varepsilon(\mathbf{r}_2)\delta\varepsilon(\mathbf{r}_3)\delta\varepsilon(\mathbf{r}_4)\rangle &= \langle \delta\varepsilon(\mathbf{r}_1)\delta\varepsilon(\mathbf{r}_2)\rangle \, \langle \delta\varepsilon(\mathbf{r}_3)\delta\varepsilon(\mathbf{r}_4)\rangle \\ + \langle \delta\varepsilon(\mathbf{r}_1)\delta\varepsilon(\mathbf{r}_3)\rangle \, \langle \delta\varepsilon(\mathbf{r}_2)\delta\varepsilon(\mathbf{r}_4)\rangle &+ \langle \delta\varepsilon(\mathbf{r}_1)\delta\varepsilon(\mathbf{r}_4)\rangle \, \langle \delta\varepsilon(\mathbf{r}_2)\delta\varepsilon(\mathbf{r}_3)\rangle \;. \end{aligned} \tag{15.6}$$

15.2 Average Field

The solution to the wave equation (15.1) is given by

$$U(\mathbf{r}) = \int G(\mathbf{r}, \mathbf{r}')S(\mathbf{r}')d^3r', \tag{15.7}$$

where the Green's function G obeys

$$\nabla^2_{\mathbf{r}} G(\mathbf{r}, \mathbf{r}') + k_0^2\, \varepsilon(\mathbf{r})G(\mathbf{r}, \mathbf{r}') = -\delta(\mathbf{r} - \mathbf{r}') \,. \tag{15.8}$$

It follows immediately that the average field is given by

$$\langle U(\mathbf{r})\rangle = \int \left\langle G(\mathbf{r}, \mathbf{r}')\right\rangle S(\mathbf{r}')d^3r' \;, \tag{15.9}$$

provided that the source term S is deterministic. Thus the average field is determined by the average Green's function.

We now turn to the problem of calculating $\langle G\rangle$. We recall from Chapter 10 that G obeys the integral equation [see (10.10)]

$$G(\mathbf{r}, \mathbf{r}') = G_0(\mathbf{r}, \mathbf{r}') + k_0^2 \int d^3r'' G_0(\mathbf{r}, \mathbf{r}'')\delta\varepsilon(\mathbf{r}'')G(\mathbf{r}'', \mathbf{r}') \;, \tag{15.10}$$

where the free-space Green's function G_0 is given by

$$G_0(\mathbf{r}, \mathbf{r}') = \frac{\exp(ik_0|\mathbf{r} - \mathbf{r}'|)}{4\pi\,|\mathbf{r} - \mathbf{r}'|} \;. \tag{15.11}$$

Equation (15.10) can be written in operator notation, introduced in Chapter 10, as

$$G = G_0 + G_0 V G \;, \tag{15.12}$$

where $V(\mathbf{r},\mathbf{r}') = \langle \mathbf{r}|V|\mathbf{r}'\rangle = k_0^2\delta\varepsilon(\mathbf{r})\delta(\mathbf{r}-\mathbf{r}')$. Upon iteration of Eq. (15.12), we obtain the multiple scattering series

$$G = G_0 + G_0VG_0 + G_0VG_0VG_0 + \cdots \tag{15.13}$$

We note that each term of the above series corresponds to successively higher orders of scattering. It will prove useful to represent the series (15.13) in diagrammatic form as

$G =$ 

(15.14)

The diagrammatic rules are as follows:

1. A solid line corresponds to the Green's function G_0.
2. A vertex corresponds to a factor of $k_0^2\delta\varepsilon$.
3. Integration is carried out over all coordinates corresponding to internal vertices.

Thus the second and third diagrams in (15.2) are given by

$$k_0^2 \int d^3r_1 G_0(\mathbf{r},\mathbf{r}_1)\delta\varepsilon(\mathbf{r}_1)G_0(\mathbf{r}_1,\mathbf{r}') , \tag{15.15}$$

$$k_0^4 \int d^3r_1 d^3r_2 G_0(\mathbf{r},\mathbf{r}_1)\delta\varepsilon(\mathbf{r}_1)G_0(\mathbf{r}_1,\mathbf{r}_2)\delta\varepsilon(\mathbf{r}_2)G_0(\mathbf{r}_2,\mathbf{r}') . \tag{15.16}$$

The average over disorder is carried out by making use of (15.5). We find that diagrams with an odd number of vertices vanish. Those with an even number of vertices are averaged by pairing vertices in all possible ways and summing the result according to (15.5). We can thereby express the average Green's function $\langle G\rangle$ diagrammatically as

$\langle G\rangle =$ 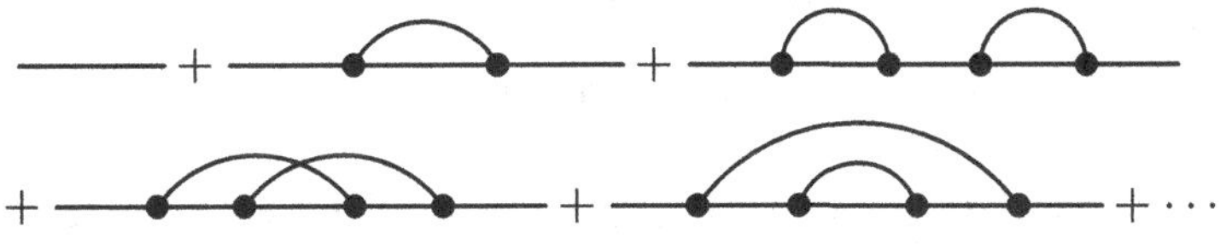

(15.17)

Here the diagrammatic rules must be supplemented by associating a closed loop with a factor of k_0^4C. We easily see that the second and third diagrams above are given by

$$k_0^4 \int d^3r_1 d^3r_2 G_0(\mathbf{r},\mathbf{r}_1)G_0(\mathbf{r}_1,\mathbf{r}_2)C(\mathbf{r}_1-\mathbf{r}_2)G_0(\mathbf{r}_2,\mathbf{r}') \tag{15.18}$$

and

$$k_0^8 \int d^3r_1 d^3r_2 d^3r_3 d^3r_4 G_0(\mathbf{r},\mathbf{r}_1)G_0(\mathbf{r}_1,\mathbf{r}_2)C(\mathbf{r}_1-\mathbf{r}_2)G_0(\mathbf{r}_2,\mathbf{r}_3) \times G_0(\mathbf{r}_3,\mathbf{r}_4)C(\mathbf{r}_3-\mathbf{r}_4)G_0(\mathbf{r}_4,\mathbf{r}') . \tag{15.19}$$

By examining (15.2), we see that there are diagrams that are disconnected in the sense that by cutting an internal line, a diagram separates into two subdiagrams. A diagram that is not disconnected is said to be connected. We define the self-energy Σ to be the sum of all connected diagrams with external lines amputated. That is,

$$\Sigma = \text{[diagram]} + \text{[diagram]} + \text{[diagram]} + \cdots \quad (15.20)$$

We note that the first three self-energy diagrams are given by

$$\Sigma_1(\mathbf{r}, \mathbf{r}') = k_0^4 G_0(\mathbf{r}, \mathbf{r}') C(\mathbf{r} - \mathbf{r}') , \quad (15.21)$$

$$\Sigma_2(\mathbf{r}, \mathbf{r}') = k_0^8 \int d^3r_1 d^3r_2 C(\mathbf{r} - \mathbf{r}_2) C(\mathbf{r}_1 - \mathbf{r}') G_0(\mathbf{r}, \mathbf{r}_1) G_0(\mathbf{r}_1, \mathbf{r}_2) G_0(\mathbf{r}_2, \mathbf{r}') , \quad (15.22)$$

$$\Sigma_3(\mathbf{r}, \mathbf{r}') = k_0^8 C(\mathbf{r} - \mathbf{r}') \int d^3r_1 d^3r_2 C(\mathbf{r}_1 - \mathbf{r}_2) G_0(\mathbf{r}, \mathbf{r}_1) G_0(\mathbf{r}_1, \mathbf{r}_2) G_0(\mathbf{r}_2, \mathbf{r}') . \quad (15.23)$$

By straightforward calculations with diagrams, it can be seen that $\langle G \rangle$ can be rewritten in the form

$$\langle G \rangle = G_0 + G_0 \Sigma G_0 + G_0 \Sigma G_0 \Sigma G_0 + \cdots . \quad (15.24)$$

Upon summation of the above series, we obtain the Dyson equation obeyed by the average Green's function

$$\langle G \rangle = G_0 + G_0 \Sigma \langle G \rangle , \quad (15.25)$$

which can be rewritten as the integral equation

$$\langle G(\mathbf{r}, \mathbf{r}') \rangle = G_0(\mathbf{r}, \mathbf{r}') + \int d^3r_1 d^3r_2 G_0(\mathbf{r}, \mathbf{r}_1) \Sigma(\mathbf{r}_1, \mathbf{r}_2) \langle G(\mathbf{r}_2, \mathbf{r}') \rangle . \quad (15.26)$$

Equation (15.26) can be solved by Fourier transformation, which follows from the fact that $\langle G(\mathbf{r}_1, \mathbf{r}_2) \rangle$ and $\Sigma(\mathbf{r}, \mathbf{r}')$ are translationally invariant in a statistically homogeneous medium. Accordingly, we define the Fourier transform of $\langle G \rangle$ as

$$\langle \widetilde{G}(\mathbf{k}, \mathbf{k}') \rangle = \int d^3r d^3r' \exp(-i\mathbf{k} \cdot \mathbf{r} + i\mathbf{k}' \cdot \mathbf{r}') \langle G(\mathbf{r}, \mathbf{r}') \rangle \quad (15.27)$$

$$= (2\pi)^3 \langle G(\mathbf{k}) \rangle \, \delta(\mathbf{k} - \mathbf{k}') , \quad (15.28)$$

which defines $\langle G(\mathbf{k}) \rangle$. Likewise, we define $\Sigma(\mathbf{k})$ and $G_0(\mathbf{k})$ by

$$\widetilde{\Sigma}(\mathbf{k}, \mathbf{k}') = (2\pi)^3 \, \Sigma(\mathbf{k}) \delta(\mathbf{k} - \mathbf{k}') , \quad \widetilde{G}_0(\mathbf{k}, \mathbf{k}') = (2\pi)^3 \, G_0(\mathbf{k}) \delta(\mathbf{k} - \mathbf{k}') . \quad (15.29)$$

Then, Fourier transforming (15.26) according to the definition (15.27), we obtain

$$\langle G(\mathbf{k})\rangle = G_0(\mathbf{k}) + G_0(\mathbf{k})\Sigma(\mathbf{k})\,\langle G(\mathbf{k})\rangle\ . \tag{15.30}$$

Here

$$G_0(\mathbf{k}) = \frac{1}{k^2 - k_0^2 - i\epsilon}\ , \tag{15.31}$$

which follows from (5.18), where the limit $\epsilon \to 0^+$ is to be taken. Solving for $\langle G(\mathbf{k})\rangle$, we find that

$$\langle G(\mathbf{k})\rangle = \frac{1}{k^2 - k_0^2 - \Sigma(\mathbf{k}) - i\epsilon}\ . \tag{15.32}$$

Equation (15.32) is an exact expression of the average Green's function. Evidently, the problem of calculating the self-energy is as difficult as calculating the average Green's function itself. To make further progress, various approximations are needed.

15.3 Weak Scattering and Effective Medium

The simplest approximation to the self-energy is known as the one-loop approximation, or weak-scattering approximation (for reasons that will become clear below), in which only the lowest order self-energy diagram is retained and $\Sigma(\mathbf{r},\mathbf{r}') \simeq \Sigma_1(\mathbf{r},\mathbf{r}')$. We also assume that $|\Sigma(\mathbf{k})| \ll k_0^2$. This condition means that the Green's function in (15.32) is peaked around k_0, and so we can require that $\mathbf{k}$ remains on shell ($|\mathbf{k}| = k_0$) in the calculation of $\Sigma(\mathbf{k})$. It will prove useful to introduce the quantities

$$k_R^2 = k_0^2 + \mathrm{Re}\,\Sigma(k_0)\ , \tag{15.33}$$

$$k_I^2 = \mathrm{Im}\,\Sigma(k_0)\ , \tag{15.34}$$

through which we define the effective wavenumber k_{eff} by

$$k_{\mathrm{eff}}^2 = k_R^2 + ik_I^2\ . \tag{15.35}$$

Now, the weak-scattering and on-shell approximations amount to stating that $k_R \simeq k_0$ and $k_I \ll k_R$, so that

$$k_{\mathrm{eff}} \simeq k_R + i\frac{k_I^2}{2k_R}\ . \tag{15.36}$$

The average Green's function can be rewritten as

$$\langle G(\mathbf{k})\rangle = \frac{1}{k^2 - k_{\mathrm{eff}}^2 - i\epsilon}\ , \tag{15.37}$$

which takes the same form as the free-space Green's function, with k_0 replaced by k_{eff}. By performing an inverse Fourier transform, we see that the average Green's function is given by

$$\langle G(\mathbf{r}, \mathbf{r}')\rangle = \frac{\exp(ik_R|\mathbf{r}-\mathbf{r}'|)}{4\pi|\mathbf{r}-\mathbf{r}'|} \exp(-|\mathbf{r}-\mathbf{r}'|/2\ell_s), \tag{15.38}$$

where $k_R = \text{Re}(k_{\text{eff}})$ and ℓ_s is the scattering mean free path defined as

$$\ell_s = \frac{k_R}{k_I^2} = \frac{1}{2\text{Im}(k_{\text{eff}})}. \tag{15.39}$$

We see that the average field propagates with an effective wavelength $\lambda_R = 2\pi/k_R$, and decays exponentially on the scale of the scattering mean free path ℓ_s.

To understand the idea of an effective medium, it is convenient to define the effective dielectric function ε_{eff} by

$$\varepsilon_{\text{eff}} = \frac{k_{\text{eff}}^2}{k_0^2}, \tag{15.40}$$

as well as the effective refractive index $n_{\text{eff}} = \sqrt{\varepsilon_{\text{eff}}}$. In the weak-scattering approximation $k_R\ell_s \gg 1$, we find that

$$\varepsilon_{\text{eff}} = \varepsilon_R\left(1 + \frac{i}{k_R\ell_s}\right), \tag{15.41}$$

$$n_{\text{eff}} = n_R\left(1 + \frac{i}{2k_R\ell_s}\right), \tag{15.42}$$

where $n_R = \sqrt{\varepsilon_R} = k_R/k_0$.

The effective wavevector can be calculated explicitly for a given model of disorder. For the Gaussian correlations given by (15.4), we find upon taking the Fourier transform of Eq. (15.21), and making use of Eq. (15.29) that

$$\Sigma(\mathbf{k}) = \frac{k_0^4 A}{\pi^{3/2}\ell_c^3} \int d^3r\, \frac{\exp(ik_0 r)}{4\pi r} \exp(-r^2/\ell_c^2) \exp(-i\mathbf{k}\cdot\mathbf{r}). \tag{15.43}$$

Performing the above integration, we obtain

$$\Sigma(\mathbf{k}) = \frac{k_0^3 A}{\pi^{3/2}\ell_c^3} \int_0^\infty dr \exp(ik_0 r) \exp(-r^2/\ell_c^2) \sin(k_0 r), \tag{15.44}$$

where we have put $k = k_0$ since the integral is calculated on-shell. Assuming $\ell_c \ll \lambda$ and expanding the integrand to first order in $k_0 r$, we get

$$\Sigma(\mathbf{k}) = \frac{k_0^4 A}{\pi^{3/2}\ell_c^3} \int_0^\infty d\rho\, (\rho + ik_0\rho^2) \exp(-\rho^2/\ell_c^2). \tag{15.45}$$

The indicated Gaussian integrals are easily evaluated, which leads to

$$k_R^2 = k_0^2 + \frac{k_0^4 A}{2\pi^{3/2}\ell_c} , \tag{15.46}$$

$$k_I^2 = \frac{k_0^5 A}{4\pi} . \tag{15.47}$$

In the weak-scattering limit $k_R \ell_s \gg 1$, we find that

$$k_{\text{eff}} = k_0 + \frac{i}{2\ell_s} \tag{15.48}$$

with

$$\ell_s = \frac{4\pi}{k_0^4 A} . \tag{15.49}$$

We note that the self-energy diagrams Σ_2 and Σ_3 can be seen to be of the order $O\left(1/(k_R\ell_s)^2\right)$ and thus may be neglected.

15.4 General Models of Disorder

So far, we have considered the Gaussian model for wave propagation in non-absorbing random media. More general types of disorder, including media consisting of discrete scatterers, can also be described with small modifications of the above theory. The key idea is that the self-energy Σ can be related to the average T-matrix of a random medium. We begin by using the Dyson equation (15.25) to define Σ as

$$\Sigma = G_0^{-1} \left(\langle G \rangle - G_0 \right) \langle G \rangle^{-1} . \tag{15.50}$$

We recall from Chapter 10 that the Green's function is given in terms of the T-matrix by

$$G = G_0 + G_0 T G_0 , \tag{15.51}$$

where the T-matrix is defined by

$$T = V + V G_0 V + \cdots . \tag{15.52}$$

Performing the average over disorder, we obtain

$$\langle G \rangle = G_0 + G_0 \langle T \rangle G_0 . \tag{15.53}$$

Making use of the above result and (15.50), we find that Σ is given by

$$\Sigma = \langle T \rangle \left(1 + G_0 \langle T \rangle\right)^{-1} , \tag{15.54}$$

which is the required relation. We note that (15.54) does not lead to a diagrammatic expansion for Σ, in contrast to the case of Gaussian disorder. We can recover the results for the self-energy of the Gaussian model by expanding (15.54) in powers of $\langle T \rangle$. We find that to lowest order in $\langle T \rangle$,

$$\Sigma = \langle T \rangle \; . \tag{15.55}$$

Recalling that $\langle \mathbf{r}|V|\mathbf{r}'\rangle = k_0^2 \delta\varepsilon(\mathbf{r})\delta(\mathbf{r}-\mathbf{r}')$ and making use of (15.52), (15.2) and (15.3), we obtain

$$\Sigma(\mathbf{r},\mathbf{r}') = k_0^4 G_0(\mathbf{r},\mathbf{r}') C(\mathbf{r}-\mathbf{r}') \, , \tag{15.56}$$

which agrees with (15.21).

Equation (15.55) for the self-energy has the advantage of being very general, and contains the absorbing medium as a special case. Making use of the on-shell approximation, we can follow the same path as in Section 15.2 and find that the average Green's function is of the form

$$\big\langle G(\mathbf{r},\mathbf{r}')\big\rangle = \frac{\exp(ik_R|\mathbf{r}-\mathbf{r}'|)}{4\pi|\mathbf{r}-\mathbf{r}'|} \exp(-|\mathbf{r}-\mathbf{r}'|/2\ell_e) \, . \tag{15.57}$$

Here the extinction mean free path,

$$\ell_e = \frac{1}{2\mathrm{Im}k_{\mathrm{eff}}}, \tag{15.58}$$

describes the decay of the average field due to scattering and absorption. In a non-absorbing medium, we see that $\ell_e = \ell_s$. In an absorbing medium, we define the absorption mean free path ℓ_a by

$$\frac{1}{\ell_e} = \frac{1}{\ell_s} + \frac{1}{\ell_a} \, . \tag{15.59}$$

References and Additional Reading

The following papers introduce the use of diagrammatic expansions in multiple scattering:

U. Frisch, Ann. Astrophys. (France) **29**, 645 (1966); Ann. Astrophys. (France) **30**, 565 (1967).

U. Frisch, in *Probabilistic Methods in Applied Mathematics*, edited by A.A. Bharucha-Reid (Academic, New York, 1968), vols. 1 and 2.

This review article offers an overview of multiple-scattering theory:

M.C.W. van Rossum and Th.M. Nieuwenhuizen, Rev. Mod. Phys. **71**, 313 (1999).

A comprehensive account of multiple-scattering theory can be found in the following textbooks:

S.M. Rytov, Y.A. Kravtsov and V.I. Tatarskii, *Principles of Statistical Radiophysics* (Springer-Verlag, Berlin, 1989), vol. 4.

P. Sheng, *Introduction to Wave Scattering, Localization, and Mesoscopic Phenomena* (Academic Press, San Diego, 1995).
E. Akkermans and G. Montambaux, *Mesoscopic Physics of Electrons and Photons* (Cambridge University Press, Cambridge, 2007). Perturbation theory for the averaged field is treated in chapter 3.

16

Multiple Scattering: Field Correlations and Radiative Transport

In this chapter, we study field correlations in random media. By making use of diagrammatic perturbation theory, we obtain an integral equation for the correlations of he Green's function that can be solved in the weak-scattering limit. This result is then used to obtain the radiative transport equation.

16.1 Field Correlations

We return to the framework of Chapter 15 for wave propagation in random media. We begin by observing that the second-order correlation function of the field is given by

$$\langle U(\mathbf{r}_1)U^*(\mathbf{r}_2)\rangle = \int d^3r_1'd^3r_2'\langle G(\mathbf{r}_1,\mathbf{r}_1')G^*(\mathbf{r}_2,\mathbf{r}_2')\rangle\, S(\mathbf{r}_1')S^*(\mathbf{r}_2')\ , \tag{16.1}$$

which follows from (15.7) and the assumption that the source S is deterministic. To calculate $\langle GG^*\rangle$, we introduce the appropriate diagrammatic perturbation theory. The diagrammatic rules are summarized in Table 16.1.

Diagrammatic symbol	Interpretation
——— (single line)	G_0 or G_0^*
●	$k_0^2\delta\varepsilon$
⌒ (arc)	$k_0^4 C$
═══ (double line)	$\langle G\rangle$ or $\langle G^*\rangle$

Table 16.1 Diagrammatic rules.

To proceed, we note that the diagrammatic expansion for the quantity GG^* can be written as

$$GG^* = \text{[diagram]} \qquad (16.2)$$

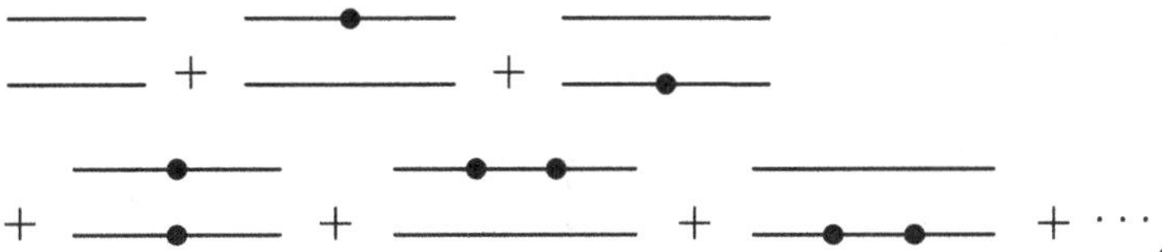

Here the upper line stands for the retarded Green's function, and the lower line for the advanced Green's function. Using (15.5) to average over the disorder, we obtain the following diagrammatic expansion for $\langle GG^* \rangle$:

$$\langle GG^* \rangle = \text{[diagram]} \qquad (16.3)$$

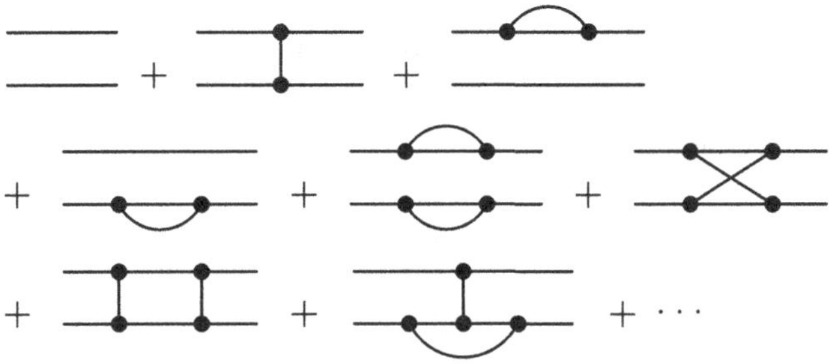

The above diagrams fall into connected and disconnected types. In a manner similar to the construction of the self-energy in Chapter 15, we define the irreducible vertex Γ as the sum of all connected diagrams with external legs amputated:

$$\Gamma = \text{[diagram]} \qquad (16.4)$$

By diagrammatic manipulation, we find that $\langle GG^* \rangle$ obeys an equation of the form

$$\langle GG^* \rangle = \langle G \rangle \langle G^* \rangle + \langle G \rangle \langle G^* \rangle \Gamma \langle G \rangle \langle G^* \rangle + \langle G \rangle \langle G^* \rangle \Gamma \langle G \rangle \langle G^* \rangle \Gamma \langle G \rangle \langle G^* \rangle + \cdots , \qquad (16.5)$$

which can be summed to obtain

$$\langle GG^* \rangle = \langle G \rangle \langle G^* \rangle + \langle G \rangle \langle G^* \rangle \Gamma \langle GG^* \rangle . \qquad (16.6)$$

Equation (16.6) is known as the Bethe–Salpeter equation. It can be expressed in real space as

$$\langle G(\mathbf{r}_1, \mathbf{r}_2) G^*(\mathbf{r}_1', \mathbf{r}_2') \rangle = \langle G(\mathbf{r}_1, \mathbf{r}_2) \rangle \langle G^*(\mathbf{r}_1', \mathbf{r}_2') \rangle + \int d^3 R_1 d^3 R_1' d^3 R_2 d^3 R_2'$$
$$\times \langle G(\mathbf{r}_1, \mathbf{R}_1) \rangle \langle G^*(\mathbf{r}_1', \mathbf{R}_1') \rangle \Gamma(\mathbf{R}_1, \mathbf{R}_2; \mathbf{R}_1', \mathbf{R}_2') \langle G(\mathbf{R}_2, \mathbf{r}_2) G^*(\mathbf{R}_2', \mathbf{r}_2') \rangle . \qquad (16.7)$$

If only the first diagram in (16.4) is retained, which is known as the ladder approximation, then

$$\Gamma(\mathbf{R}_1, \mathbf{R}_2; \mathbf{R}_1', \mathbf{R}_2') = k_0^4 C(\mathbf{R}_1 - \mathbf{R}_1') \delta(\mathbf{R}_1 - \mathbf{R}_2) \delta(\mathbf{R}_1' - \mathbf{R}_2') . \qquad (16.8)$$

It follows that (16.6) can be expressed as a sum of so-called ladder diagrams of the form

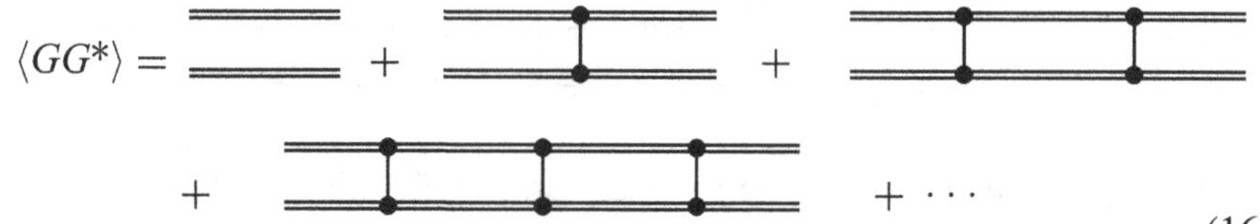

$$\langle GG^* \rangle = \quad + \quad + \quad + \quad + \cdots \tag{16.9}$$

It is easy to see that the higher-order diagrams in the expansion (16.4) of the irreducible vertex are of order $O\left(1/(k_R \ell_s)^2\right)$ and may be neglected. Thus the ladder approximation is accurate to first order in the small parameter $1/(k_R \ell_s)$. By introducing (16.8) into (16.7), we see that the Bethe–Salpeter equation in the ladder approximation becomes

$$\begin{aligned}\left\langle G(\mathbf{r}_1, \mathbf{r}_2) G^*(\mathbf{r}_1', \mathbf{r}_2') \right\rangle &= \langle G(\mathbf{r}_1, \mathbf{r}_2) \rangle \left\langle G^*(\mathbf{r}_1', \mathbf{r}_2') \right\rangle + k_0^4 \int d^3r d^3r' \\ &\times \langle G(\mathbf{r}_1, \mathbf{r}) \rangle \left\langle G^*(\mathbf{r}_1', \mathbf{r}') \right\rangle C(\mathbf{r} - \mathbf{r}') \left\langle G(\mathbf{r}, \mathbf{r}_2) G^*(\mathbf{r}', \mathbf{r}_2') \right\rangle .\end{aligned} \tag{16.10}$$

We now introduce the vertex, which plays the role of the average T-matrix for computing field correlations. We proceed by rewriting (16.5) as

$$\left\langle GG^* \right\rangle = \langle G \rangle \left\langle G^* \right\rangle + \langle G \rangle \left\langle G^* \right\rangle \Lambda \langle G \rangle \left\langle G^* \right\rangle . \tag{16.11}$$

Here the vertex Λ is defined by the series

$$\Lambda = 1 + \Gamma \langle G \rangle \left\langle G^* \right\rangle + \Gamma \langle G \rangle \left\langle G^* \right\rangle \Gamma \langle G \rangle \left\langle G^* \right\rangle + \cdots . \tag{16.12}$$

By summing the above geometric series, we obtain

$$\Lambda = 1 + \Gamma \langle G \rangle \left\langle G^* \right\rangle \Lambda . \tag{16.13}$$

We note that in real space Eq. (16.11) becomes

$$\begin{aligned}\left\langle G(\mathbf{r}_1, \mathbf{r}_2) G^*(\mathbf{r}_1', \mathbf{r}_2') \right\rangle &= \langle G(\mathbf{r}_1, \mathbf{r}_2) \rangle \left\langle G^*(\mathbf{r}_1', \mathbf{r}_2') \right\rangle + \int d^3R_1 d^3R_1' d^3R_2 d^3R_2' \\ &\times \langle G(\mathbf{r}_1, \mathbf{R}_1) \rangle \left\langle G^*(\mathbf{r}_1', \mathbf{R}_1') \right\rangle \Lambda(\mathbf{R}_1, \mathbf{R}_2; \mathbf{R}_1', \mathbf{R}_2') \langle G(\mathbf{R}_2, \mathbf{r}_2) \rangle \left\langle G^*(\mathbf{R}_2', \mathbf{r}_2') \right\rangle .\end{aligned} \tag{16.14}$$

16.2 Wigner Transform

We recall from Chapter 2 that the conservation of energy is governed by the relation

$$\nabla \cdot \mathbf{J} + k_0 \, \mathrm{Im}\varepsilon \, I = 0 , \tag{16.15}$$

where $I = |U|^2$ is the intensity and the energy current $\mathbf{J}$ is defined by

$$\mathbf{J} = \frac{1}{2ik_0} \left(U^* \nabla U - U \nabla U^* \right) . \tag{16.16}$$

Although the conservation law (16.15) determines how the intensity of the field is distributed in space, it does not describe how the intensity propagates. To obtain a local conservation law for the intensity that is resolved over both position and direction, we introduce the Wigner transform of the field $W(\mathbf{r}, \mathbf{k})$, defined by

$$W(\mathbf{r}, \mathbf{k}) = \int \frac{d^3r'}{(2\pi)^3} \exp(-i\mathbf{k} \cdot \mathbf{r}') \left\langle U(\mathbf{r} + \mathbf{r}'/2) U^*(\mathbf{r} - \mathbf{r}'/2) \right\rangle . \tag{16.17}$$

By Fourier inversion, we see that W is related to the field correlation function by

$$\left\langle U(\mathbf{r}) U^*(\mathbf{r}') \right\rangle = \int d^3k \exp(i\mathbf{k} \cdot (\mathbf{r} - \mathbf{r}')) W\left(\frac{\mathbf{r} + \mathbf{r}'}{2}, \mathbf{k}\right) . \tag{16.18}$$

Thus the Wigner transform is a measure of the spatial coherence of the field. However, here the origin of the randomness of the field is due to the medium rather than the source.

The Wigner transform has several important properties. It is real-valued and is related to the average intensity by

$$\langle I(\mathbf{r}) \rangle = \int d^3k \, W(\mathbf{r}, \mathbf{k}) . \tag{16.19}$$

In addition, the Wigner transform is related to the average current $\langle \mathbf{J}(\mathbf{r}) \rangle$ by

$$\langle \mathbf{J}(\mathbf{r}) \rangle = \int d^3k \, \mathbf{k} \, W(\mathbf{r}, \mathbf{k}) . \tag{16.20}$$

We emphasize that W is not directly measurable. Nevertheless, as indicated above, its moments are connected to observable quantities.

To further understand the physical meaning of the Wigner transform, it is instructive to consider the case of wave propagation in vacuum. The solution of the wave equation in this case is a plane wave of the form

$$U(\mathbf{r}) = A \exp(i\mathbf{p} \cdot \mathbf{r}) , \tag{16.21}$$

where A is the amplitude of the wave and $\mathbf{p}$ its wave vector. The Wigner transform is then given by

$$W(\mathbf{r}, \mathbf{k}) = |A|^2 \delta(\mathbf{k} - \mathbf{p}) , \tag{16.22}$$

and the intensity and current are

$$I = |A|^2 , \quad \mathbf{J} = |A|^2 \mathbf{p}/k_0 . \tag{16.23}$$

Note that the above W is localized in spatial frequency.

The Wigner transform of the field is related to the Wigner transform of the source according to

$$W(\mathbf{r}, \mathbf{k}) = \int d^3r' d^3k' \, \Phi(\mathbf{r} - \mathbf{r}'; \mathbf{k}, \mathbf{k}') W_0(\mathbf{r}', \mathbf{k}') , \tag{16.24}$$

which follows from (16.1) and (16.17). Here the propagator Φ is defined by

$$\Phi(\mathbf{r}-\mathbf{r}';\mathbf{k},\mathbf{k}') = \frac{1}{(2\pi)^3}\int d^3R d^3R' \exp(-i\mathbf{k}\cdot\mathbf{R}+i\mathbf{k}'\cdot\mathbf{R}') \times\langle G(\mathbf{r}+\mathbf{R}/2,\mathbf{r}'+\mathbf{R}'/2)\times G^*(\mathbf{r}-\mathbf{R}/2,\mathbf{r}'-\mathbf{R}'/2)\rangle \tag{16.25}$$

and

$$W_0(\mathbf{r},\mathbf{k}) = \int \frac{d^3r'}{(2\pi)^3}\exp(-i\mathbf{k}\cdot\mathbf{r}')S(\mathbf{r}+\mathbf{r}'/2)S^*(\mathbf{r}-\mathbf{r}'/2)\,. \tag{16.26}$$

We now turn to the derivation of the Liouville equation, which describes the spatial evolution of the Wigner transform. We begin by considering the Fourier transform of $W(\mathbf{r},\mathbf{k})$ with respect to the space variable $\mathbf{r}$, which is defined by

$$\widetilde{W}(\mathbf{q},\mathbf{k}) = \int d^3r\,\exp(-i\mathbf{q}\cdot\mathbf{r})W(\mathbf{r},\mathbf{k})\,. \tag{16.27}$$

Making use of (16.24), we find that

$$\widetilde{W}(\mathbf{q},\mathbf{k}) = \int \frac{d^3k'}{(2\pi)^3}\Phi(\mathbf{q};\mathbf{k},\mathbf{k}')\widetilde{W}_0(\mathbf{q},\mathbf{k}')\,, \tag{16.28}$$

where $\Phi(\mathbf{q};\mathbf{k},\mathbf{k}')$ is defined by

$$\Phi(\mathbf{r}-\mathbf{r}';\mathbf{k},\mathbf{k}') = \int \frac{d^3q}{(2\pi)^3}\exp(i\mathbf{q}\cdot(\mathbf{r}-\mathbf{r}'))\Phi(\mathbf{q};\mathbf{k},\mathbf{k}')\,. \tag{16.29}$$

After some calculation, it follows from (16.10) and (16.25) that within the ladder approximation, $\Phi(\mathbf{q};\mathbf{k},\mathbf{k}')$ satisfies

$$\Phi(\mathbf{q};\mathbf{k},\mathbf{k}') = \langle G(\mathbf{k}+\mathbf{q}/2)\rangle\left\langle G^*(\mathbf{k}-\mathbf{q}/2)\right\rangle\Big[\delta(\mathbf{k}-\mathbf{k}') + k_0^4\int\frac{d^3k''}{(2\pi)^3}\widetilde{C}(\mathbf{k}-\mathbf{k}'')\Phi(\mathbf{q};\mathbf{k}'',\mathbf{k}')\Big]\,. \tag{16.30}$$

The above equation for the propagator is formally equivalent to the Bethe–Salpeter equation in the ladder approximation. Next, we use the expression for the average Green's function (15.32),

$$\langle G(\mathbf{k})\rangle = \frac{1}{k^2-k_0^2-\Sigma(\mathbf{k})-i\epsilon}, \tag{16.31}$$

along with the identities

$$\langle G(\mathbf{k}+\mathbf{q}/2)\rangle\left\langle G^*(\mathbf{k}-\mathbf{q}/2)\right\rangle = \frac{\langle G^*(\mathbf{k}-\mathbf{q}/2)\rangle-\langle G(\mathbf{k}+\mathbf{q}/2)\rangle}{\langle G(\mathbf{k}+\mathbf{q}/2)\rangle^{-1}-\langle G^*(\mathbf{k}-\mathbf{q}/2)\rangle^{-1}}, \tag{16.32}$$

$$|\mathbf{k}+\mathbf{q}/2|^2-|\mathbf{k}-\mathbf{q}/2|^2 = 2\mathbf{k}\cdot\mathbf{q}\,, \tag{16.33}$$

to rewrite (16.30). We thus obtain the following equation for $\Phi(\mathbf{q};\mathbf{k},\mathbf{k}')$:

$$-2\mathbf{k}\cdot\mathbf{q}\,\Phi(\mathbf{q};\mathbf{k},\mathbf{k}') + \Delta\Sigma(\mathbf{q},\mathbf{k})\Phi(\mathbf{q};\mathbf{k},\mathbf{k}') =$$
$$\Delta G(\mathbf{q},\mathbf{k})\left[\delta(\mathbf{k}-\mathbf{k}') + k_0^4\int\frac{d^3k''}{(2\pi)^3}\widetilde{C}(\mathbf{k}-\mathbf{k}'')\Phi(\mathbf{q};\mathbf{k}'',\mathbf{k}')\right]. \quad (16.34)$$

Here

$$\Delta\Sigma(\mathbf{q},\mathbf{k}) = \Sigma(\mathbf{k}+\mathbf{q}/2) - \Sigma^*(\mathbf{k}-\mathbf{q}/2)\,, \quad (16.35)$$
$$\Delta G(\mathbf{q},\mathbf{k}) = \langle G(\mathbf{k}+\mathbf{q}/2)\rangle - \left\langle G^*(\mathbf{k}-\mathbf{q}/2)\right\rangle\,. \quad (16.36)$$

Finally, upon multiplying (16.34) by $\widetilde{W}_0(\mathbf{q},\mathbf{k}')$, integrating over $\mathbf{k}'$ and inverting the Fourier transform (16.27), we find that W obeys the equation

$$\mathbf{k}\cdot\nabla_{\mathbf{r}}W(\mathbf{r},\mathbf{k}) + \frac{1}{2i}\int\frac{d^3q}{(2\pi)^3}\exp(i\mathbf{q}\cdot\mathbf{r})\Delta\Sigma(\mathbf{q},\mathbf{k})\widetilde{W}(\mathbf{q},\mathbf{k}) =$$
$$\frac{k_0^4}{2i}\int\frac{d^3q}{(2\pi)^3}\exp(i\mathbf{q}\cdot\mathbf{r})\int\frac{d^3k'}{(2\pi)^3}\Delta G(\mathbf{q},\mathbf{k})\widetilde{C}(\mathbf{k}-\mathbf{k}')\widetilde{W}(\mathbf{q},\mathbf{k}') + S(\mathbf{r},\mathbf{k})\,, \quad (16.37)$$

where

$$S(\mathbf{r},\mathbf{k}) = \frac{1}{2i}\int\frac{d^3q}{(2\pi)^3}\exp(i\mathbf{q}\cdot\mathbf{r})\Delta G(\mathbf{q},\mathbf{k})\widetilde{W}_0(\mathbf{q},\mathbf{k})\,. \quad (16.38)$$

Equation (16.37) is known as the Liouville equation. It can be understood as a transport equation for the Wigner transform of the field. Note that the dependence on the wavevector $\mathbf{q}$ makes this equation spatially nonlocal.

16.3 Radiative Transport

In order to obtain the radiative transport equation (RTE), we consider the large scale limit $\mathbf{q}\to 0$, also known as the Kubo limit. To this end, we replace $\Delta\Sigma(\mathbf{q},\mathbf{k})$ and $\Delta G(\mathbf{q},\mathbf{k})$ by $\Delta\Sigma(0,\mathbf{k})$ and $\Delta G(0,\mathbf{k})$, respectively, in (16.37). We thus obtain

$$\mathbf{k}\cdot\nabla_{\mathbf{r}}W(\mathbf{r},\mathbf{k}) + \frac{1}{2i}\Delta\Sigma(0,\mathbf{k})W(\mathbf{r},\mathbf{k}) = \frac{k_0^4}{2i}\Delta G(0,\mathbf{k})\int\frac{d^3k'}{(2\pi)^3}\widetilde{C}(\mathbf{k}-\mathbf{k}')W(\mathbf{r},\mathbf{k}')$$
$$+\frac{1}{2i}\Delta G(0,\mathbf{k})W_0(\mathbf{r},\mathbf{k})\,. \quad (16.39)$$

Next, we rewrite the average Green's function as

$$\langle G(\mathbf{k})\rangle = \frac{1}{k^2 - k_R^2 - ik_I^2 - i\epsilon}\,, \quad (16.40)$$

where k_R and k_I are defined by (15.33) and (15.34). Making use of the identities

$$\frac{1}{k^2-\kappa^2-i\epsilon}=P\frac{1}{k^2-\kappa^2}+i\pi\delta(k^2-\kappa^2)\,, \tag{16.41}$$

$$\delta(k^2-\kappa^2)=\frac{1}{2\kappa}\left[\delta(k-\kappa)+\delta(k+\kappa)\right]\,, \tag{16.42}$$

where P denotes the Cauchy principal value, we find that in the weak-scattering limit (equivalent to $k_R\ell_s \gg 1$, with $\ell_s = k_R/k_I^2$ the scattering mean free path), $\Delta G(0,\mathbf{k})$ is given by

$$\Delta G(0,\mathbf{k})=\frac{i\pi}{k_R}\delta(k-k_R)\,. \tag{16.43}$$

Likewise, using (15.39) we find that

$$\Delta\Sigma(0,\mathbf{k})=2ik_R/\ell_s\,. \tag{16.44}$$

Using the above results, and defining the specific intensity of the field I, phase function p, scattering coefficient μ_s, and specific intensity of the source I_0 by

$$\delta(k-k_R)I(\mathbf{r},\hat{\mathbf{k}})=k_R W(\mathbf{r},\mathbf{k})\,, \tag{16.45}$$

$$\delta(k-k_R)I_0(\mathbf{r},\hat{\mathbf{k}})=\frac{\pi}{2k_R}W_0(\mathbf{r},\mathbf{k})\,, \tag{16.46}$$

$$p(\hat{\mathbf{k}},\hat{\mathbf{k}}')=\frac{\widetilde{C}(k_R(\hat{\mathbf{k}}-\hat{\mathbf{k}}'))}{\int\widetilde{C}(k_R(\hat{\mathbf{k}}-\hat{\mathbf{k}}'))d\hat{\mathbf{k}}'}\,, \tag{16.47}$$

$$\mu_s=\frac{k_0^4}{16\pi^2}\int\widetilde{C}(k_R(\hat{\mathbf{k}}-\hat{\mathbf{k}}'))d\hat{\mathbf{k}}'\,, \tag{16.48}$$

we find that (16.39) becomes

$$\hat{\mathbf{k}}\cdot\nabla_{\mathbf{r}}I+(1/\ell_s)I=\mu_s\int d\hat{\mathbf{k}}'p(\hat{\mathbf{k}},\hat{\mathbf{k}}')I(\mathbf{r},\hat{\mathbf{k}}')+I_0\,. \tag{16.49}$$

In the absence of sources, integrating the above equation over $\hat{\mathbf{k}}$, and using the energy conservation law $\nabla\cdot\mathbf{J}=0$, which holds in a nonabsorbing medium (see (2.38)), we immediately see that $\mu_s = 1/\ell_s$. We thus find that

$$\hat{\mathbf{k}}\cdot\nabla_{\mathbf{r}}I+\mu_s I=\mu_s\int d\hat{\mathbf{k}}'p(\hat{\mathbf{k}},\hat{\mathbf{k}}')I(\mathbf{r},\hat{\mathbf{k}}')+I_0\,, \tag{16.50}$$

which is the RTE in a nonabsorbing medium. The RTE describes the gains and losses of the specific intensity due to scattering. Here we note that since the medium is assumed to be statistically homogeneous and isotropic, $p(\hat{\mathbf{k}},\hat{\mathbf{k}}')$ depends only on $\hat{\mathbf{k}}\cdot\hat{\mathbf{k}}'$. Likewise, μ_s does not depend on the direction $\hat{\mathbf{k}}$.

Finally, it is interesting to note that using (16.43), (16.44) and (16.48), and observing that $\Delta\Sigma(0,\mathbf{k}) = 2i\mathrm{Im}\Sigma(\mathbf{k})$ and $\Delta G(0,\mathbf{k}) = 2i\mathrm{Im}\,\langle G(\mathbf{k})\rangle$, the equality $\mu_s = 1/\ell_s$ can be rewritten as

$$\mathrm{Im}\Sigma(\mathbf{k}) = k_0^4 \int \frac{d^3k'}{(2\pi)^3}\widetilde{C}(\mathbf{k}-\mathbf{k}')\mathrm{Im}\left\langle G(\mathbf{k}')\right\rangle . \tag{16.51}$$

This relation, which results from energy conservation, is a special case of the Ward identity, as discussed at the end of this chapter.

16.4 General Models of Disorder

We now consider the relation of the irreducible vertex Γ to the T-matrix. Here we seek a result that is analogous to (15.54), which relates the self-energy and the average T-matrix for a general model of disorder. To proceed, we use the Bethe–Salpeter equation (16.6) to define Γ according to

$$\Gamma = \left(\langle G\rangle\left\langle G^*\right\rangle\right)^{-1} - \left\langle GG^*\right\rangle^{-1} . \tag{16.52}$$

We recall that the Green's function is given in terms of the T-matrix by

$$G = G_0 + G_0 T G_0 . \tag{16.53}$$

Here the T-matrix is defined as

$$T = V + VG_0V + \cdots , \tag{16.54}$$

where $V(\mathbf{r},\mathbf{r}') = k_0^2\delta\varepsilon(\mathbf{r})\delta(\mathbf{r}-\mathbf{r}')$. Using (16.52) and (16.53), we find that Γ is given by

$$\begin{aligned}\Gamma = \left(G_0G_0^*\right)^{-1}\big[&\left(1 + G_0\langle T\rangle + G_0^*\left\langle T^*\right\rangle + \langle T\rangle\, G_0G_0^*\left\langle T^*\right\rangle\right)^{-1} \\ &- \left(1 + G_0\langle T\rangle + G_0^*\left\langle T^*\right\rangle + \left\langle TG_0G_0^*T^*\right\rangle\right)^{-1}\big] .\end{aligned} \tag{16.55}$$

We note that (16.55) does not directly lead to a diagrammatic expansion for Γ, as is the case for Gaussian disorder.

We can recover the irreducible vertex of the Gaussian model in the ladder approximation by expanding (16.55) in powers of T. We find that to the lowest order, (16.55) becomes

$$\Gamma = \left\langle TT^*\right\rangle - \langle T\rangle\left\langle T^*\right\rangle . \tag{16.56}$$

Making use of (16.54), (15.2) and (15.3), we obtain

$$\Gamma(\mathbf{R}_1,\mathbf{R}_2;\mathbf{R}_1',\mathbf{R}_2') = k_0^4 C(\mathbf{R}_1-\mathbf{R}_1')\delta(\mathbf{R}_1-\mathbf{R}_2)\delta(\mathbf{R}_1'-\mathbf{R}_2') , \tag{16.57}$$

which agrees with (16.8).

We now obtain the RTE that derives from the general irreducible vertex function, without specifying a model of disorder. We begin by observing that the Fourier transform of Γ, which is defined by

$$\widetilde{\Gamma}(\mathbf{k}_1, \mathbf{k}_2; \mathbf{k}'_1, \mathbf{k}'_2) = \int d^3R_1 d^3R_2 d^3R'_1 d^3R'_2 \exp(-i\mathbf{k}_1 \cdot \mathbf{R}_1 + i\mathbf{k}_2 \cdot \mathbf{R}_2 + i\mathbf{k}'_1 \cdot \mathbf{R}'_1 - i\mathbf{k}'_2 \cdot \mathbf{R}'_2)\Gamma(\mathbf{R}_1, \mathbf{R}_2; \mathbf{R}'_1, \mathbf{R}'_2) , \tag{16.58}$$

obeys the condition

$$\widetilde{\Gamma}(\mathbf{k}_1, \mathbf{k}_2; \mathbf{k}'_1, \mathbf{k}'_2) = (2\pi)^3\delta\left(\mathbf{k}_1 - \mathbf{k}_2 - \mathbf{k}'_1 + \mathbf{k}'_2\right)\gamma(\mathbf{k}_1, \mathbf{k}_2; \mathbf{k}'_1, \mathbf{k}'_2) , \tag{16.59}$$

which defines the vertex function γ. Here we have assumed that the medium is statistically homogeneous and made use of the translational invariance of Γ. Starting from (16.25) and making use of the Bethe–Salpeter equation (16.7) together with (16.59), we find that the Liouville equation (16.37) becomes

$$\begin{aligned} \mathbf{k} \cdot \nabla_{\mathbf{r}} W(\mathbf{r}, \mathbf{k}) &+ \frac{1}{2i}\int \frac{d^3q}{(2\pi)^3} \exp(i\mathbf{q} \cdot \mathbf{r})\Delta\Sigma(\mathbf{q}, \mathbf{k})\widetilde{W}(\mathbf{q}, \mathbf{k}) \\ &= \frac{1}{2i}\int \frac{d^3q}{(2\pi)^3} \exp(i\mathbf{q} \cdot \mathbf{r}) \int \frac{d^3k'}{(2\pi)^3}\Delta G(\mathbf{q}, \mathbf{k}) \\ &\times \gamma(\mathbf{k} + \mathbf{q}/2, \mathbf{k}' + \mathbf{q}/2; \mathbf{k} - \mathbf{q}/2, \mathbf{k}' - \mathbf{q}/2)\widetilde{W}(\mathbf{q}, \mathbf{k}') . \end{aligned} \tag{16.60}$$

Following the same steps as in Section 16.3, and using the average Green function for a general disorder model defined in Section 15.4, we obtain

$$\hat{\mathbf{k}} \cdot \nabla_{\mathbf{r}} I + \mu_e I = \mu_s \int d\hat{\mathbf{k}}' p(\hat{\mathbf{k}}, \hat{\mathbf{k}}') I(\mathbf{r}, \hat{\mathbf{k}}') + I_0 , \tag{16.61}$$

which is the general form of the RTE, valid in the presence of absorption. Here the scattering coefficient μ_s, phase function p and extinction coefficient μ_e are given by

$$\mu_s = \frac{1}{\ell_s} = \frac{1}{16\pi^2}\int d\hat{\mathbf{k}}' \gamma(k_R\hat{\mathbf{k}}, k_R\hat{\mathbf{k}}'; k_R\hat{\mathbf{k}}, k_R\hat{\mathbf{k}}') , \tag{16.62}$$

$$p(\hat{\mathbf{k}}, \hat{\mathbf{k}}') = \frac{1}{16\pi^2 \mu_s}\gamma(k_R\hat{\mathbf{k}}, k_R\hat{\mathbf{k}}'; k_R\hat{\mathbf{k}}, k_R\hat{\mathbf{k}}') , \tag{16.63}$$

$$\mu_e = \frac{1}{\ell_e} = 2\mathrm{Im}(k_{\mathrm{eff}}) . \tag{16.64}$$

16.5 Ward Identity

A Ward identity is a relation between one-particle and two-particle Green's functions. The optical theorem, introduced in Chapter 11, is a simple example of a

Ward identity. It relates the scattering amplitude, which is a one-particle Green's function, to the scattering cross section, which is a two-particle Green's function. In scattering theory, Ward identities usually result from energy conservation. Here we derive a Ward identity that connects the self-energy (one-particle quantity) to the irreducible vertex (two-particle quantity).

We begin by noting that the Liouville equation (16.60) is equivalent to the Bethe–Salpeter equation. Integrating (16.60) over $\mathbf{k}$, we see that the first term is $\nabla \cdot \mathbf{J}$, which vanishes in the absence of absorption. We thus obtain

$$\int d^3k \, \Delta\Sigma(\mathbf{q}, \mathbf{k}) \widetilde{W}(\mathbf{q}, \mathbf{k}) = \int d^3k \int \frac{d^3k'}{(2\pi)^3} \Delta G(\mathbf{q}, \mathbf{k}) \times \gamma(\mathbf{k} + \mathbf{q}/2, \mathbf{k}' + \mathbf{q}/2; \mathbf{k} - \mathbf{q}/2, \mathbf{k}' - \mathbf{q}/2) \widetilde{W}(\mathbf{q}, \mathbf{k}') . \tag{16.65}$$

The above equation must hold for the case $\mathbf{q} = 0$. Interchanging $\mathbf{k}$ and $\mathbf{k}'$ on the left-hand side, we obtain

$$\int d^3k \, \Delta\Sigma(0, \mathbf{k}) \widetilde{W}(0, \mathbf{k}) = \int d^3k \int \frac{d^3k'}{(2\pi)^3} \Delta G(0, \mathbf{k}') \gamma(\mathbf{k}', \mathbf{k}; \mathbf{k}', \mathbf{k}) \widetilde{W}(0, \mathbf{k}) . \tag{16.66}$$

The equation above must hold for any field distribution, and in particular for a statistically homogenous and isotropic field. Since in this case $W(0, \mathbf{k})$ is constant, we conclude that

$$\mathrm{Im}\Sigma(\mathbf{k}) = \int \frac{d^3k'}{(2\pi)^3} \gamma(\mathbf{k}', \mathbf{k}; \mathbf{k}', \mathbf{k}) \, \mathrm{Im} \langle G(\mathbf{k}') \rangle , \tag{16.67}$$

where we have used $\Delta\Sigma(0, \mathbf{k}) = 2i\mathrm{Im}\Sigma(\mathbf{k})$ and $\Delta G(0, \mathbf{k}) = 2i\mathrm{Im} \langle G(\mathbf{k}) \rangle$. Since the irreducible vertex Γ obeys the reciprocity relation

$$\Gamma(\mathbf{R}_1, \mathbf{R}_2; \mathbf{R}_1', \mathbf{R}_2') = \Gamma(\mathbf{R}_2, \mathbf{R}_1; \mathbf{R}_2', \mathbf{R}_1') , \tag{16.68}$$

which can be easily derived from (16.52) and the reciprocity relation for the Green function $G(\mathbf{R}_1, \mathbf{R}_2) = G(\mathbf{R}_2, \mathbf{R}_1)$, we also have $\gamma(\mathbf{k}_1, \mathbf{k}_2; \mathbf{k}_1', \mathbf{k}_2') = \gamma(\mathbf{k}_2, \mathbf{k}_1; \mathbf{k}_2', \mathbf{k}_1')$. Equation (16.67) can be rewritten as

$$\mathrm{Im}\Sigma(\mathbf{k}) = \int \frac{d^3k'}{(2\pi)^3} \gamma(\mathbf{k}, \mathbf{k}'; \mathbf{k}, \mathbf{k}') \, \mathrm{Im} \langle G(\mathbf{k}') \rangle , \tag{16.69}$$

which is the general form of the Ward identity. This equation connects the self-energy and the irreducible vertex of a non-absorbing medium. For a Gaussian disorder described by fluctuations of the dielectric function $\delta\varepsilon$, it can be easily verified that $\gamma(\mathbf{k}, \mathbf{k}'; \mathbf{k}, \mathbf{k}') = k_0^4 \widetilde{C}(\mathbf{k} - \mathbf{k}')$. Inserting this result into the Ward identity above leads to (16.51).

References and Additional Reading

A presentation of the Bethe–Salpeter equation and radiative transport can be found in:

P. Sheng, *Introduction to Wave Scattering, Localization, and Mesoscopic Phenomena* (Academic Press, San Diego, 1995), chap. 4.

E. Akkermans and G. Montambaux, *Mesoscopic Physics of Electrons and Photons* (Cambridge University Press, Cambridge, 2007), chap. 4.

The following textbooks include a derivation of the RTE from the Bethe–Salpeter equation and the Wigner transform:

S.M. Rytov, Y.A. Kravtsov, and V.I. Tatarskii, *Principles of Statistical Radiophysics* (Springer-Verlag, Berlin, 1989), vol. 4.

L.A. Apresyan and Y.A. Kravtsov, *Radiation Transfer. Statistical and Wave Aspects* (Gordon and Breach, 1996).

This review article offers an overview of multiple-scattering theory, including radiative transport:

M.C.W. Rossum and Th.M. Nieuwenhuizen, Rev. Mod. Phys. **71**, 313 (1999).

This tutorial article includes a derivation of the RTE from the Bethe–Salpeter equation and the ladder approximation:

A. Cazé and J.C. Schotland, J. Opt. Soc. Am. A **32**, 1475 (2015).

A derivation of the RTE in inhomogeneous media is given in:

J. Hoskins, J. Kraisler and J. Schotland, J. Opt. Soc. Am. A **35**, 1855 (2018).

17

Radiative Transport: Multiscale Theory

In this chapter, we derive the radiative transport equation (RTE) from the high-frequency asymptotics of the wave equation in random media. We begin by rescaling the wave equation to allow for the separation of microscopic and macroscopic scales. We then introduce the corresponding scaled Wigner transform. Finally, we make use of a multiscale asymptotic expansion to average the Wigner transform, which ultimately leads to the RTE.

17.1 High-Frequency Asymptotics

We consider the propagation of a scalar wave-field in a disordered medium. The field U obeys the wave equation

$$\nabla^2 U + k_0^2\,\varepsilon(\mathbf{r})U = 0\,, \tag{17.1}$$

where k_0 is the vacuum wavenumber and $\varepsilon(\mathbf{r}) = 1+\delta\varepsilon(\mathbf{r})$ is the dielectric function, which we take to be real-valued so that the medium is nonabsorbing. We assume that $\delta\varepsilon$ is a random variable satisfying

$$\langle\delta\varepsilon(\mathbf{r})\rangle = 0\,, \tag{17.2}$$

$$\left\langle\delta\varepsilon(\mathbf{r})\delta\varepsilon(\mathbf{r}')\right\rangle = C(\mathbf{r}-\mathbf{r}'), \tag{17.3}$$

where $\langle\cdots\rangle$ denotes a statistical average over an ensemble of realizations of the disordered medium. We also assume that the medium is statistically homogeneous and isotropic, so that the correlation function $C(\mathbf{r}-\mathbf{r}')$ depends only upon $|\mathbf{r}-\mathbf{r}'|$. To make further progress, we consider the relative sizes of the important physical scales. The solution to the wave equation (17.1) oscillates on the scale of the wavelength $\lambda = 2\pi/k_0$. However, we are interested in the behavior of the solutions on the macroscopic scale $L \gg \lambda$. We thus introduce a small parameter $\epsilon = 1/(k_0 L)$

and rescale the position $\mathbf{r}$ by $\mathbf{r} \to \mathbf{r}/\epsilon$. In addition, we assume that the randomness is sufficiently weak so that the correlation function C is $O(\epsilon)$. Thus (17.1) becomes

$$\epsilon^2 \nabla^2 U_\epsilon + [1 + \sqrt{\epsilon}\, \delta\varepsilon(\mathbf{r}/\epsilon)] U_\epsilon = 0 , \tag{17.4}$$

where $U_\epsilon(\mathbf{r}) = U(\mathbf{r}/\epsilon)$. Note that we have rescaled $\delta\varepsilon$ by $\delta\varepsilon \to \sqrt{\epsilon}\,\delta\varepsilon$ to be consistent with the $O(\epsilon)$ scaling of C. We will see that the RTE is obtained in the high-frequency limit $\epsilon \to 0$.

Next, we introduce the scaled Wigner transform $W_\epsilon(\mathbf{r}, \mathbf{k})$, which is defined by

$$W_\epsilon(\mathbf{r}, \mathbf{k}) = \int \frac{d^3 r'}{(2\pi)^3} \exp(i\mathbf{k} \cdot \mathbf{r}') U_\epsilon^*(\mathbf{r} + \epsilon\mathbf{r}'/2) U_\epsilon(\mathbf{r} - \epsilon\mathbf{r}'/2) . \tag{17.5}$$

Note that W_ϵ is not averaged over disorder, in contrast to the Wigner transform introduced in Chapter 16. The scaled Wigner transform has several important properties. It is real-valued and is related to the intensity by

$$\int W_\epsilon(\mathbf{r}, \mathbf{k}) d^3 k = |U_\epsilon(\mathbf{r})|^2 . \tag{17.6}$$

In addition, W_ϵ is related to the energy current density

$$\mathbf{J}_\epsilon(\mathbf{r}) = \frac{\epsilon}{2i} \left[U_\epsilon^*(\mathbf{r}) \nabla U_\epsilon(\mathbf{r}) - U_\epsilon(\mathbf{r}) \nabla U_\epsilon^*(\mathbf{r}) \right] \tag{17.7}$$

by

$$\int \mathbf{k} W_\epsilon(\mathbf{r}, \mathbf{k}) d^3 k = \mathbf{J}_\epsilon(\mathbf{r}) . \tag{17.8}$$

It is instructive to consider the example of a high-frequency plane wave of the form $U_\epsilon(\mathbf{r}) = \exp(i\mathbf{p} \cdot \mathbf{r}/\epsilon)$. Then, $W_\epsilon(\mathbf{r}, \mathbf{k}) = \delta(\mathbf{k} - \mathbf{p})$, which is localized in frequency.

We now turn to the derivation of the Liouville equation, which is a conservation law for W_ϵ. It will prove useful to introduce the function $\phi_\epsilon(\mathbf{r}_1, \mathbf{r}_2) = U_\epsilon(\mathbf{r}_1) U_\epsilon^*(\mathbf{r}_2)$. Evidently, ϕ obeys the pair of wave equations

$$\epsilon^2 \nabla_{\mathbf{r}_1}^2 \phi_\epsilon + \left[1 + \sqrt{\epsilon}\, \delta\varepsilon(\mathbf{r}_1/\epsilon)\right] \phi_\epsilon = 0 , \tag{17.9}$$

$$\epsilon^2 \nabla_{\mathbf{r}_2}^2 \phi_\epsilon + \left[1 + \sqrt{\epsilon}\, \delta\varepsilon(\mathbf{r}_2/\epsilon)\right] \phi_\epsilon = 0 , \tag{17.10}$$

which, upon subtraction, yields

$$\epsilon^2 (\nabla_{\mathbf{r}_1}^2 - \nabla_{\mathbf{r}_2}^2) \phi_\epsilon + \sqrt{\epsilon} \left[\delta\varepsilon(\mathbf{r}_1/\epsilon) - \delta\varepsilon(\mathbf{r}_2/\epsilon)\right] \phi_\epsilon = 0 . \tag{17.11}$$

We now introduce the change of variables

$$\mathbf{r}_1 = \mathbf{r} - \epsilon\mathbf{r}'/2 , \quad \mathbf{r}_2 = \mathbf{r} + \epsilon\mathbf{r}'/2 . \tag{17.12}$$

Equation (17.11) thus becomes

$$\epsilon \nabla_{\mathbf{r}} \cdot \nabla_{\mathbf{r}'} \phi_\epsilon + \frac{k_0^2}{2}\sqrt{\epsilon}\left[\delta\varepsilon\left(\frac{1}{\epsilon}\left(\mathbf{r}+\epsilon\mathbf{r}'/2\right)\right) - \delta\varepsilon\left(\frac{1}{\epsilon}\left(\mathbf{r}-\epsilon\mathbf{r}'/2\right)\right)\right]\phi_\epsilon = 0 . \tag{17.13}$$

Next, we multiply the above equation by $\exp(i\mathbf{k}\cdot\mathbf{r}')$, integrate with respect to $\mathbf{r}'$ and make use of the definition (17.5) of the Wigner transform. We find, after some calculation, that W_ϵ obeys the Liouville equation

$$\mathbf{k}\cdot\nabla_{\mathbf{r}} W_\epsilon + \frac{1}{\sqrt{\epsilon}} L W_\epsilon = 0 , \tag{17.14}$$

where

$$L W_\epsilon(\mathbf{r},\mathbf{k}) = \frac{i k_0^2}{2}\int \frac{d^3 p}{(2\pi)^3} \exp(-i\mathbf{p}\cdot\mathbf{r}/\epsilon)\,\widetilde{\delta\varepsilon}(\mathbf{p}) \times\left[W_\epsilon(\mathbf{r},\mathbf{k}+\mathbf{p}/2) - W_\epsilon(\mathbf{r},\mathbf{k}-\mathbf{p}/2)\right] . \tag{17.15}$$

Here $\widetilde{\delta\varepsilon}$ denotes the Fourier transform of $\delta\varepsilon$.

17.2 Multiscale Expansion

We now consider the asymptotics of the Wigner transform in the high-frequency limit $\epsilon \to 0$. To this end, we introduce a multiscale expansion for the Wigner transform of the form

$$W_\epsilon(\mathbf{r},\mathbf{k}) = W_0(\mathbf{r},\mathbf{R},\mathbf{k}) + \sqrt{\epsilon}\, W_1(\mathbf{r},\mathbf{R},\mathbf{k}) + \epsilon\, W_2(\mathbf{r},\mathbf{R},\mathbf{k}) + \cdots , \tag{17.16}$$

where $\mathbf{R} = \mathbf{r}/\epsilon$ is a fast variable. We then make the replacement

$$\nabla_{\mathbf{r}} \to \nabla_{\mathbf{r}} + \frac{1}{\epsilon}\nabla_{\mathbf{R}} \tag{17.17}$$

and substitute (17.16) into (17.14), thus obtaining

$$\epsilon\mathbf{k}\cdot\nabla_{\mathbf{r}} W_\epsilon + \mathbf{k}\cdot\nabla_{\mathbf{R}} W_\epsilon + \sqrt{\epsilon} L W_\epsilon = 0 . \tag{17.18}$$

Collecting terms of the same order in ϵ, we find that at $O(1)$,

$$\mathbf{k}\cdot\nabla_{\mathbf{R}} W_0 = 0 . \tag{17.19}$$

Since the above result must hold for all wavevectors $\mathbf{k}$, we see that W_0 can depend only upon the variables $\mathbf{r}$ and $\mathbf{k}$.

At $O(\sqrt{\epsilon})$, we have

$$\mathbf{k}\cdot\nabla_{\mathbf{R}} W_1 + L W_0 = 0 . \tag{17.20}$$

Equation (17.20) can be solved by Fourier transformation with the result

$$\widetilde{W}_1(\mathbf{r}, \mathbf{q}, \mathbf{k}) = \frac{k_0^2}{2} \frac{\widetilde{\delta\varepsilon}(\mathbf{q}) \left[W_0(\mathbf{r}, \mathbf{k} + \mathbf{q}/2) - W_0(\mathbf{r}, \mathbf{k} - \mathbf{q}/2) \right]}{\mathbf{q} \cdot \mathbf{k} + i\theta} , \tag{17.21}$$

where

$$\widetilde{W}_1(\mathbf{r}, \mathbf{q}, \mathbf{k}) = \int d^3R \, \exp(i\mathbf{q} \cdot \mathbf{R}) W_1(\mathbf{r}, \mathbf{R}, \mathbf{k}) \tag{17.22}$$

and θ is a regularization parameter that will eventually be set to zero.

At $O(\epsilon)$, we find that

$$\mathbf{k} \cdot \nabla_{\mathbf{r}} W_0 + \mathbf{k} \cdot \nabla_{\mathbf{R}} W_2 + L W_1 = 0 \, . \tag{17.23}$$

More generally, the recursion relation

$$\mathbf{k} \cdot \nabla_{\mathbf{r}} W_{n-2} + \mathbf{k} \cdot \nabla_{\mathbf{R}} W_n + L W_{n-1} = 0 \tag{17.24}$$

is obeyed for $n > 1$.

The RTE may be derived by averaging (17.23) over realizations of the disordered medium. To do so, we impose the condition $\langle \nabla_{\mathbf{R}} W_2 \rangle = 0$, which closes the hierarchy (17.24) at $n = 2$, and corresponds to the assumption that W_2 is statistically stationary in the fast variable $\mathbf{R}$. Equation (17.23) thus becomes

$$\mathbf{k} \cdot \nabla_{\mathbf{r}} W_0 + \frac{ik_0^2}{2} \int \frac{d^3p}{(2\pi)^3} \exp(-i\mathbf{p} \cdot \mathbf{r}/\epsilon) \tag{17.25}$$

$$\times \left\langle \widetilde{\delta\varepsilon}(\mathbf{p}) \left[W_1(\mathbf{r}, \mathbf{R}, \mathbf{k} + \mathbf{p}/2) - W_1(\mathbf{r}, \mathbf{R}, \mathbf{k} - \mathbf{p}/2) \right] \right\rangle = 0 \, . \tag{17.26}$$

The next step is to substitute the formula (17.21) for $\widetilde{W}_1$ (after performing a Fourier transform) into (17.25) and use the relation

$$\left\langle \widetilde{\delta\varepsilon}(\mathbf{p}) \widetilde{\delta\varepsilon}(\mathbf{q}) \right\rangle = (2\pi)^3 \tilde{C}(\mathbf{p}) \delta(\mathbf{p} + \mathbf{q}) \, , \tag{17.27}$$

which follows from (17.2). In performing the indicated average, we assume that W_0 is deterministic, that is, $\left\langle \widetilde{\delta\varepsilon}(\mathbf{p}) \widetilde{\delta\varepsilon}(\mathbf{q}) W_0 \right\rangle = \left\langle \widetilde{\delta\varepsilon}(\mathbf{p}) \widetilde{\delta\varepsilon}(\mathbf{q}) \right\rangle W_0$. Accordingly, we find that the second term on the left-hand side of (17.25) becomes

$$\begin{aligned}
&\frac{ik_0^4}{4} \int \frac{d^3p}{(2\pi)^3} \tilde{C}(\mathbf{p}) \left[\frac{W_0(\mathbf{r}, \mathbf{k}) - W_0(\mathbf{r}, \mathbf{k} + \mathbf{p})}{-\mathbf{p} \cdot (\mathbf{k} + \mathbf{p}/2) + i\theta} - \frac{W_0(\mathbf{r}, \mathbf{k}) - W_0(\mathbf{r}, \mathbf{k} - \mathbf{p})}{\mathbf{p} \cdot (\mathbf{k} - \mathbf{p}/2) - i\theta} \right] \\
&= \frac{ik_0^4}{4} \int \frac{d^3p}{(2\pi)^3} \tilde{C}(\mathbf{p}) \left(W_0(\mathbf{r}, \mathbf{k} + \mathbf{p}) - W_0(\mathbf{r}, \mathbf{k}) \right) \\
&\qquad \times \left[\frac{1}{\mathbf{p} \cdot (\mathbf{k} + \mathbf{p}/2) - i\theta} - \frac{1}{\mathbf{p} \cdot (\mathbf{k} + \mathbf{p}/2) + i\theta} \right] \\
&= -\frac{k_0^4}{4} \int \frac{d^3p}{(2\pi)^3} \tilde{C}(\mathbf{p} - \mathbf{k}) \left(W_0(\mathbf{r}, \mathbf{p}) - W_0(\mathbf{r}, \mathbf{k}) \right) \frac{2\theta}{\frac{1}{4} \left(p^2 - k^2 \right)^2 + \theta^2} \, .
\end{aligned} \tag{17.28}$$

Using the fact that

$$\lim_{\theta\to 0^+} \frac{\theta}{x^2+\theta^2} = \pi\delta(x) \tag{17.29}$$

and (17.28), we find that (17.25) becomes

$$\begin{aligned}\mathbf{k}\cdot\nabla_{\mathbf{r}} W_0 &+ \frac{1}{16\pi^2}\int d^3p\,\tilde{C}(\mathbf{p}-\mathbf{k})\delta\left(\frac{p^2}{2}-\frac{k^2}{2}\right)W_0(\mathbf{r},\mathbf{k}) \\ &= \frac{1}{16\pi^2}\int d^3p\,\tilde{C}(\mathbf{p}-\mathbf{k})\delta\left(\frac{p^2}{2}-\frac{k^2}{2}\right)W_0(\mathbf{r},\mathbf{p})\,. \end{aligned}\tag{17.30}$$

Evidently, the presence of the delta function in (17.30) indicates that $W(\mathbf{r},\mathbf{k})$ depends only on the direction $\hat{\mathbf{k}}$, making it possible to set $|\mathbf{k}| = k_0$. It is then convenient to define the quantities

$$\begin{aligned}\delta(k-k_0)I(\mathbf{r},\hat{\mathbf{k}}) &= W_0(\mathbf{r},\mathbf{k})\,, \\ p(\hat{\mathbf{k}},\hat{\mathbf{k}}') &= \frac{\tilde{C}(k_0(\hat{\mathbf{k}}-\hat{\mathbf{k}}'))}{\int \tilde{C}(k_0(\hat{\mathbf{k}}-\hat{\mathbf{k}}'))d\hat{\mathbf{k}}'}\,, \end{aligned}\tag{17.31}$$

$$\mu_s = \frac{k_0^4}{16\pi^2}\int \tilde{C}(k_0(\hat{\mathbf{k}}-\hat{\mathbf{k}}'))d\hat{\mathbf{k}}'\,, \tag{17.32}$$

whose physical dimensions have been restored. The quantities I, p and μ_s are, respectively, the specific intensity, the phase function and the scattering coefficient introduced in Chapter 16. Finally, making use of the identity

$$\delta\left(\frac{p^2}{2}-\frac{k^2}{2}\right) = \frac{1}{k}\delta(p-k) \tag{17.33}$$

and the above definitions, we obtain the RTE

$$\hat{\mathbf{k}}\cdot\nabla_{\mathbf{r}} I + \mu_s I = \mu_s\int d\hat{\mathbf{k}}'\,p(\hat{\mathbf{k}},\hat{\mathbf{k}}')I(\mathbf{r},\hat{\mathbf{k}}')\,. \tag{17.34}$$

This equation agrees with (16.50) derived in Chapter 16 for a non-absorbing medium.

In conclusion, two different approaches to the derivation of the RTE have been presented. The first derivation is based on diagrammatic perturbation theory, and the second makes use of a multiscale asymptotic expansion. The Wigner transform plays a central role in both approaches, as do the assumptions of Gaussian disorder and statistical homogeneity. An additional common assumption is the separation of microscopic and macroscopic scales. Several points of departure should also be noted. In diagrammatic perturbation theory, the counterpart of the closure relation $\langle\nabla_{\mathbf{R}} W_2\rangle = 0$ is not evident. The same is true of the deterministic nature of W_0. Likewise, the analog of the ladder approximation in the multiscale expansion is not clear. Nevertheless, the fact that both approaches lead to the RTE is striking.

References and Additional Reading

Derivations of the RTE based on multiscale asymptotics can be found in:

L. Ryzhik, G. Papanicolaou and J.B. Keller, Wave Motion **24**, 327 (1996).
G. Bal, T. Komorowski and L. Ryzhik, Kinet. Relat. Models **3**, 529 (2010).

This tutorial article includes a derivation of the RTE based on multiscale asymptotics, together with a derivation based on the Bethe–Salpeter equation and diagrammatic expansions:

A. Cazé and J.C. Schotland, J. Opt. Soc. Am. A **32**, 1475 (2015).

This paper discusses self-averaging in radiative transport theory:

G. Bal, SIAM J. Multiscale Model. Simul. **2** 398 (2004).

This paper proposes a model for the acousto-optic effect in random media using the RTE:

J. Hoskins and J. Schotland, Phys. Rev. E **95**, 033002 (2017).

A general presentation of asymptotic analysis is given in the following textbook:

P. Miller, *Applied Asymptotic Analysis* (American Mathematical Society, Providence, 2006).

18

Discrete Scatterers and Spatial Correlations

In this chapter, we present a formalism to compute the scattering mean free path and phase function for a medium composed of a discrete set of scatterers, accounting for the effect of spatial correlations in the positions of the scatterers. We also introduce the transport mean free path, which plays an important role in radiative transport theory within the diffusion approximation.

18.1 T-matrix of a Discrete Set of Scatterers

The T-matrix has been introduced in Chapter 10. Here we derive the T-matrix for a medium made of a discrete set of scatterers.

We consider a monochromatic field U which obeys the Lippman-Schwinger equation

$$U = U_0 + G_0 \, V \, U \, , \tag{18.1}$$

where V is the scattering potential, G_0 the free-space Green function, and U_0 the incident field. The Born series, which accounts for multiple scattering, is obtained by iteration:

$$U = U_0 + G_0 \, V \, U_0 + G_0 \, V \, G_0 \, V \, U_0 + G_0 \, V \, G_0 \, V \, G_0 \, V \, U_0 + \cdots \, . \tag{18.2}$$

The T-matrix is defined as

$$T = V + V G_0 V + V G_0 V G_0 V + \cdots \, , \tag{18.3}$$

and characterizes the scattering medium as a whole. It allows us to rewrite (18.1) in the form

$$U = U_0 + G_0 \, T \, U_0 \, . \tag{18.4}$$

Note that in real space, the T matrix can be written as

$$T(\mathbf{r}_1, \mathbf{r}_2) = V(\mathbf{r}_1)\,\delta(\mathbf{r}_1 - \mathbf{r}_2) + V(\mathbf{r}_1)\,G_0(\mathbf{r}_1, \mathbf{r}_2)\,V(\mathbf{r}_2) + \int d^3r' V(\mathbf{r}_1)\,G_0(\mathbf{r}_1, \mathbf{r}')\,V(\mathbf{r}')\,G_0(\mathbf{r}', \mathbf{r}_2)\,V(\mathbf{r}_2) + \cdots . \tag{18.5}$$

Suppose the scattering medium is composed of a discrete set of identical scatterers. The potential V can then be written as

$$V(\mathbf{r}) = \sum_j V_j(\mathbf{r}), \tag{18.6}$$

where $V_j(\mathbf{r})$ is the potential due to a scatterer located at point $\mathbf{r}_j$. For homogeneous particles with dielectric constant ε, we have

$$V_j(\mathbf{r}) = k_0^2(\varepsilon - 1)\,\chi(\mathbf{r} - \mathbf{r}_j), \tag{18.7}$$

where $\chi(\mathbf{r} - \mathbf{r}_j) = 1$ if $\mathbf{r}$ lies inside the particle and $\chi(\mathbf{r} - \mathbf{r}_j) = 0$ otherwise. Using (18.5) and (18.6), we find that the T-matrix of the medium becomes

$$T(\mathbf{r}_1, \mathbf{r}_2) = \sum_j V_j(\mathbf{r}_1)\,\delta(\mathbf{r}_1 - \mathbf{r}_2) + \sum_{j,k} V_k(\mathbf{r}_1)\,G_0(\mathbf{r}_1, \mathbf{r}_2)\,V_j(\mathbf{r}_2) + \sum_{j,k,l} \int V_l(\mathbf{r}_1)\,G_0(\mathbf{r}_1, \mathbf{r}')\,V_k(\mathbf{r}')\,G_0(\mathbf{r}', \mathbf{r}_2)\,V_j(\mathbf{r}_2)\,d^3r' + \cdots . \tag{18.8}$$

We now introduce the T-matrix t_j of the jth scatterer, which is given by

$$t_j(\mathbf{r}_1, \mathbf{r}_2) = V_j(\mathbf{r}_1)\,\delta(\mathbf{r}_1 - \mathbf{r}_2) + V_j(\mathbf{r}_1)\,G_0(\mathbf{r}_1, \mathbf{r}_2)\,V_j(\mathbf{r}_2) + \int d^3r' V_j(\mathbf{r}_1)\,G_0(\mathbf{r}_1, \mathbf{r}')\,V_j(\mathbf{r}')\,G_0(\mathbf{r}', \mathbf{r}_2)\,V_j(\mathbf{r}_2) + \cdots . \tag{18.9}$$

We see that the T-matrix of the medium is of the form

$$T = \sum_j t_j + \sum_{j,k\neq j} t_k\,G_0\,t_j + \sum_{j,k\neq j,l\neq k} t_l\,G_0\,t_k\,G_0\,t_j + \cdots , \tag{18.10}$$

which can be seen by inserting (18.9) into (18.10). We also note that substituting (18.10) into (18.4) leads to the following expression for the field:

$$U = U_0 + \sum_j G_0\,t_j\,U_0 + \sum_{j,k\neq j} G_0\,t_k\,G_0\,t_j\,U_0 + \sum_{j,k\neq j,l\neq k} G_0\,t_l\,G_0\,t_k\,G_0\,t_j\,U_0 + \cdots , \tag{18.11}$$

in which the multiple-scattering process appears explicitly as a series of scattering events involving individual scatterers.

18.2 Irreducible Vertex

In Chapter 16, the radiative transport equation (RTE) has been derived, starting from the Bethe–Salpeter equation. A central quantity in the theory is the irreducible vertex $\Gamma(\mathbf{r}_1, \mathbf{r}_2; \mathbf{r}_1', \mathbf{r}_2')$. In a statistically homogeneous medium, its Fourier transform

$$\widetilde{\Gamma}(\mathbf{k}_1, \mathbf{k}_2; \mathbf{k}_1', \mathbf{k}_2') = \int \exp(-i\mathbf{k}_1 \cdot \mathbf{r}_1 + i\mathbf{k}_2 \cdot \mathbf{r}_2 + i\mathbf{k}_1' \cdot \mathbf{r}_1' - i\mathbf{k}_2' \cdot \mathbf{r}_2') \times \Gamma(\mathbf{r}_1, \mathbf{r}_2; \mathbf{r}_1', \mathbf{r}_2')\, d^3r_1 d^3r_2 d^3r_1' d^3r_2' \quad (18.12)$$

is given by

$$\widetilde{\Gamma}(\mathbf{k}_1, \mathbf{k}_2; \mathbf{k}_1', \mathbf{k}_2') = (2\pi)^3 \delta\left(\mathbf{k}_1 - \mathbf{k}_2 - \mathbf{k}_1' + \mathbf{k}_2'\right) \gamma(\mathbf{k}_1, \mathbf{k}_2; \mathbf{k}_1', \mathbf{k}_2'), \quad (18.13)$$

which is a consequence of translational invariance. The above equation defines the vertex γ.

In Chapter 16, a connection is established between the irreducible vertex Γ and the T-matrix of the medium [see 16.55)], which to lowest order in T is of the form

$$\Gamma = \left\langle TT^* \right\rangle - \langle T \rangle \left\langle T^* \right\rangle . \quad (18.14)$$

We consider a collection of N identical scatterers in a volume V, and derive an expression for the irreducible vertex Γ in terms of the T-matrix of an individual scatterer. We denote by t the T-matrix of a scatterer located at the origin, so that the T-matrix of a scatterer at position $\boldsymbol{r}_j$ is

$$t_j(\boldsymbol{r}_1, \boldsymbol{r}_2) = t(\boldsymbol{r}_1 - \boldsymbol{r}_j, \boldsymbol{r}_2 - \boldsymbol{r}_j) . \quad (18.15)$$

Using the first term in the expansion (18.10), which corresponds, as we shall see below, to a low-density approximation, we obtain

$$\left\langle TT^* \right\rangle = \sum_{j,k} \left\langle t_j t_k^* \right\rangle = \sum_j \left\langle t_j t_j^* \right\rangle + \sum_{j,k \neq j} \left\langle t_j t_k^* \right\rangle . \quad (18.16)$$

The first term on the right-hand side above is

$$\sum_j \left\langle t_j(\boldsymbol{r}_1, \boldsymbol{r}_2) t_j^*(\boldsymbol{r}_1', \boldsymbol{r}_2') \right\rangle = \sum_j \int P(\boldsymbol{r}_j) t(\mathbf{r}_1 - \mathbf{r}_j, \mathbf{r}_2 - \mathbf{r}_j) t^*(\mathbf{r}_1' - \mathbf{r}_j, \mathbf{r}_2' - \mathbf{r}_j) d^3 r_j , \quad (18.17)$$

where $P(\boldsymbol{r}_j)$ is the probability density of having a scatterer at the point $\boldsymbol{r}_j$, and the integral is over the volume V. For a uniform distribution, $P(\boldsymbol{r}_j) = 1/V$ and

$$\sum_j \left\langle t_j(\boldsymbol{r}_1, \boldsymbol{r}_2) t_j^*(\boldsymbol{r}_1', \boldsymbol{r}_2') \right\rangle = \frac{N}{V} \int t(\mathbf{r}_1 - \mathbf{r}, \mathbf{r}_2 - \mathbf{r}) t^*(\mathbf{r}_1' - \mathbf{r}, \mathbf{r}_2' - \mathbf{r}) d^3 r . \quad (18.18)$$

The second term on the right-hand side in (18.16) is

$$\sum_{j,k\neq j} \left\langle t_j(\boldsymbol{r}_1,\boldsymbol{r}_2)t_k^*(\boldsymbol{r}_1',\boldsymbol{r}_2')\right\rangle = \sum_{j,k\neq j} \int P(\boldsymbol{r}_j,\boldsymbol{r}_k)t(\mathbf{r}_1-\mathbf{r}_j,\mathbf{r}_2-\mathbf{r}_j) \times t^*(\mathbf{r}_1'-\mathbf{r}_k,\mathbf{r}_2'-\mathbf{r}_k)d^3r_jd^3r_k\,, \tag{18.19}$$

where $P(\boldsymbol{r}_j,\boldsymbol{r}_k)$ is the probability density of having one scatterer at position $\boldsymbol{r}_j$ and another scatterer at position $\boldsymbol{r}_k$. It is common to write $P(\boldsymbol{r}_j,\boldsymbol{r}_k) = P(\boldsymbol{r}_j)P(\boldsymbol{r}_k)g(\boldsymbol{r}_j,\boldsymbol{r}_k)$, where g is the pair correlation function. Alternatively, $P(\boldsymbol{r}_j,\boldsymbol{r}_k) = P(\boldsymbol{r}_j)P(\boldsymbol{r}_k)[1+h(\boldsymbol{r}_j,\boldsymbol{r}_k)]$, where h is known as the total correlation function. We find that

$$\sum_{j,k\neq j} \left\langle t_j(\boldsymbol{r}_1,\boldsymbol{r}_2)t_k^*(\boldsymbol{r}_1',\boldsymbol{r}_2')\right\rangle = \frac{N(N-1)}{V^2}\int [1+h(\boldsymbol{r},\boldsymbol{r}')]t(\mathbf{r}_1-\mathbf{r},\mathbf{r}_2-\mathbf{r}) \times t^*(\mathbf{r}_1'-\mathbf{r}',\mathbf{r}_2'-\mathbf{r}')d^3rd^3r'\,. \tag{18.20}$$

Again using the first term in the expansion of the T-matrix (18.10), we also have

$$\langle T\rangle\left\langle T^*\right\rangle = \sum_{j,k}\langle t_j\rangle\left\langle t_k^*\right\rangle\,. \tag{18.21}$$

The right-hand side above is

$$\sum_{j,k}\left\langle t_j(\boldsymbol{r}_1,\boldsymbol{r}_2)\right\rangle\left\langle t_j^*(\boldsymbol{r}_1',\boldsymbol{r}_2')\right\rangle = \frac{N^2}{V^2}\int t(\mathbf{r}_1-\mathbf{r},\mathbf{r}_2-\mathbf{r})d^3r\int t^*(\mathbf{r}_1'-\mathbf{r},\mathbf{r}_2'-\mathbf{r})d^3r\,. \tag{18.22}$$

Finally, from (18.14), (18.16), (18.18), (18.20), (18.21) and (18.22), and assuming $N(N-1)/V^2 \simeq N^2/V^2$ in the large N limit, we obtain

$$\Gamma(\mathbf{r}_1,\mathbf{r}_2;\mathbf{r}_1',\mathbf{r}_2') = \rho\int t(\mathbf{r}_1-\mathbf{r},\mathbf{r}_2-\mathbf{r})t^*(\mathbf{r}_1'-\mathbf{r},\mathbf{r}_2'-\mathbf{r})\,d^3r + \rho^2\int t(\mathbf{r}_1-\mathbf{r},\mathbf{r}_2-\mathbf{r})t^*(\mathbf{r}_1'-\mathbf{r}',\mathbf{r}_2'-\mathbf{r}')\,h(\mathbf{r},\mathbf{r}')\,d^3rd^3r'\,, \tag{18.23}$$

where $\rho = N/V$ is the number density of scatterers. We note that having used only the first term of the expansion (18.10) in (18.16) and (18.21) leads to an expression for Γ which is second order in ρ, and which thus holds at low densities.

It will prove useful to record the expression for the vertex in Fourier space. Defining the Fourier transform of the T-matrix of a scatterer by

$$t(\mathbf{k}_1,\mathbf{k}_2) = \int t(\mathbf{r}_1,\mathbf{r}_2)\,\exp(-i\mathbf{k}_1\cdot\mathbf{r}_1+i\mathbf{k}_2\cdot\mathbf{r}_2)\,d^3r_1d^3r_2\,, \tag{18.24}$$

and making use of (18.12) and (18.13), (18.23) can be transformed into

$$\gamma(\mathbf{k}_1, \mathbf{k}_2; \mathbf{k}_1', \mathbf{k}_2') = \rho\, t(\mathbf{k}_1, \mathbf{k}_2) t^*(\mathbf{k}_1', \mathbf{k}_2') + \rho^2\, t(\mathbf{k}_1, \mathbf{k}_2) t^*(\mathbf{k}_1', \mathbf{k}_2')\, h(\mathbf{k}_1 - \mathbf{k}_1') \,. \tag{18.25}$$

Here we have assumed that $h(\mathbf{r}, \mathbf{r}')$ depends only on $\mathbf{r} - \mathbf{r}'$ due to translational invariance.

18.3 Independent Scattering

The expressions for the scattering mean path ℓ_s and phase function p have been derived in Chapter 16. The scattering mean free path is given by

$$\frac{1}{\ell_s} = \frac{1}{16\pi^2} \int \gamma(k_R \hat{\mathbf{k}}, k_R \hat{\mathbf{k}}'; k_R \hat{\mathbf{k}}, k_R \hat{\mathbf{k}}')\, d\hat{\mathbf{k}}' \,, \tag{18.26}$$

where k_R is the real part of the effective wavevector defined in Chapter 15. The phase function is given by

$$p(\hat{\mathbf{k}}, \hat{\mathbf{k}}') = \frac{\ell_s}{16\pi^2}\, \gamma(k_R \hat{\mathbf{k}}, k_R \hat{\mathbf{k}}'; k_R \hat{\mathbf{k}}, k_R \hat{\mathbf{k}}') \,. \tag{18.27}$$

In the absence of correlations in the positions of the scatterers, and in the weak-scattering regime, we have

$$\gamma(k_R \hat{\mathbf{k}}, k_R \hat{\mathbf{k}}'; k_R \hat{\mathbf{k}}, k_R \hat{\mathbf{k}}') = \rho |t(k_R \hat{\mathbf{k}}, k_R \hat{\mathbf{k}}')|^2 \,, \tag{18.28}$$

which corresponds to the first term in (18.25). This defines the independent scattering approximation (ISA), in which the transport parameters depend only on the properties of individual scatterers. In the ISA, the scattering mean free path becomes

$$\frac{1}{\ell_s} = \frac{\rho}{16\pi^2} \int |t(k_R \hat{\mathbf{k}}, k_R \hat{\mathbf{k}}')|^2\, d\hat{\mathbf{k}}' \tag{18.29}$$

and the phase function is simply

$$p(\hat{\mathbf{k}}, \hat{\mathbf{k}}') = \frac{\rho\, \ell_s}{16\pi^2} |t(k_R \hat{\mathbf{k}}, k_R \hat{\mathbf{k}}')|^2 \,. \tag{18.30}$$

Noting that the scattering cross section σ_s of an individual scatterer is (see Chapter 10)

$$\sigma_s = \frac{1}{16\pi^2} \int |t(k_R \hat{\mathbf{k}}, k_R \hat{\mathbf{k}}')|^2\, d\hat{\mathbf{k}}' \,, \tag{18.31}$$

the expression for the scattering mean free path can be rewritten as

$$\ell_s = \frac{1}{\rho\, \sigma_s} \,. \tag{18.32}$$

This expression of ℓ_s is a feature of the ISA, which is valid in the weak-scattering limit $k_0\ell_s \gg 1$ and in the absence of correlations in the positions of the scatterers.

18.4 Structure Factor

The structure factor $S(\mathbf{q})$ of a collection of N scatterers distributed in a volume V is defined as

$$S(\mathbf{q}) = \frac{1}{N}\left\langle \left| \sum_{j=1}^{N} \exp(-i\mathbf{q}\cdot\mathbf{r}_j) \right|^2 \right\rangle, \tag{18.33}$$

where $\mathbf{r}_j$ is the position of the jth scatterer. We now derive a useful relationship connecting the structure factor $S(\mathbf{q})$ and the pair correlation function.

The number density of the particles is defined as

$$\rho(\mathbf{r}) = \sum_{j=1}^{N} \delta(\mathbf{r}-\mathbf{r}_j) \tag{18.34}$$

and satisfies $\int \rho(\mathbf{r})d^3r = N$. The one-particle probability density is $P(\mathbf{r}) = \langle\rho(\mathbf{r})\rangle/N$. For a uniform distribution of particles, we have $\langle\rho(\mathbf{r})\rangle = \rho = N/V$ and $P(\mathbf{r}) = 1/V$. A two-particle distribution function can also be introduced as

$$\sum_{j=1}^{N}\sum_{k=1,k\neq j}^{N} \delta(\mathbf{r}-\mathbf{r}_j)\delta(\mathbf{r}'-\mathbf{r}_k), \tag{18.35}$$

which, after averaging and normalization, yields the pair correlation function

$$g(\mathbf{r},\mathbf{r}') = \frac{1}{\langle\rho(\mathbf{r})\rangle\langle\rho(\mathbf{r}')\rangle}\left\langle \sum_{j=1}^{N}\sum_{k=1,k\neq j}^{N} \delta(\mathbf{r}-\mathbf{r}_j)\delta(\mathbf{r}'-\mathbf{r}_k)\right\rangle, \tag{18.36}$$

such that $P(\mathbf{r},\mathbf{r}') = P(\mathbf{r})P(\mathbf{r}')g(\mathbf{r},\mathbf{r}')$. We can then rewrite the structure factor as

$$\begin{aligned} S(\mathbf{q}) &= 1 + \frac{1}{N}\left\langle \sum_{j=1}^{N}\sum_{k=1,k\neq j}^{N} \exp[-i\mathbf{q}\cdot(\mathbf{r}_j-\mathbf{r}_k)]\right\rangle \\ &= 1 + \frac{1}{N}\int\int \sum_{j=1}^{N}\sum_{k=1,k\neq j}^{N} \exp[-i\mathbf{q}\cdot(\mathbf{r}_j-\mathbf{r}_k)] \\ &\quad \times P(\mathbf{r}_j)\,P(\mathbf{r}_k)\,g(\mathbf{r}_j,\mathbf{r}_k)d^3r_j d^3r_k\,. \end{aligned} \tag{18.37}$$

For a uniform distribution of particles, this becomes

$$S(\mathbf{q}) = 1 + \frac{N-1}{V^2}\int\int \exp[-i\mathbf{q}\cdot(\mathbf{r}-\mathbf{r}')]\,g(\mathbf{r},\mathbf{r}')d^3r d^3r'\,. \tag{18.38}$$

In the limit of an infinite system, we can assume that g only depends on $\mathbf{R} = \mathbf{r} - \mathbf{r}'$. The equation above can be written as

$$S(\mathbf{q}) = 1 + \frac{N-1}{V} \int \exp(-i\mathbf{q} \cdot \mathbf{R})\, g(\mathbf{R}) d^3 R\,. \qquad (18.39)$$

In terms of the total correlation function $h = g - 1$, the structure factor is given by

$$S(\mathbf{q}) = 1 + \frac{N-1}{V} \int \exp(-i\mathbf{q} \cdot \mathbf{R})\, [1 + h(\mathbf{R})] d^3 R\,, \qquad (18.40)$$

or equivalently

$$S(\mathbf{q}) = 1 + \frac{N-1}{V} \int \exp(-i\mathbf{q} \cdot \mathbf{R})\, h(\mathbf{R}) d^3 R + \frac{N-1}{V} (2\pi)^3\, \Delta(\mathbf{q})\,. \qquad (18.41)$$

Here $\Delta(\mathbf{q})$ is (up to a factor of $(2\pi)^3$) the Fourier transform of a function with value unity inside V and vanishing otherwise. In the limit of an infinite system, we put $(N-1)/V \simeq \rho$ and $\Delta(\mathbf{q}) \simeq \delta(\mathbf{q})$. We find that

$$S(\mathbf{q}) = 1 + \rho\, h(\mathbf{q}) + (2\pi)^3\, \rho\, \delta(\mathbf{q})\,. \qquad (18.42)$$

An equivalent expression for the structure factor is sometimes useful in practice. In a statistically homogeneous and isotropic medium, the pair correlation function depends only on $R = |\mathbf{R}|$, and the integral in (18.41) can be simplified. Introducing the angle χ between $\mathbf{q}$ and $\mathbf{R}$, we obtain for any radially symmetric function $f(R)$:

$$\begin{aligned} \int f(R) \exp(-i\mathbf{q} \cdot \mathbf{R})\, d^3 r &= 2\pi \int_0^\infty dR f(R) \int_0^\pi d\chi\, \exp(-iqR\cos\chi)\, R^2 \sin\chi \\ &= 4\pi \int_0^\infty R^2 f(R) \frac{\sin(qR)}{qR}\, dR\,. \end{aligned}$$

Using this result and setting $(N-1)/V \simeq \rho$, (18.41) becomes

$$S(q) = 1 + 4\pi\rho \int_0^\infty h(R) \frac{\sin(qR)}{qR} R^2\, dR + (2\pi)^3\, \rho\, \delta(q)\,. \qquad (18.43)$$

18.5 Correlations

In a medium with spatial correlations, the influence of the pair correlation function cannot be ignored, as it is in the ISA. The corresponding expressions for the scattering mean free path and phase function are usually given in terms of the total correlation function. From (18.25) and (18.26), we obtain

$$\frac{1}{\ell_s} = \frac{1}{16\pi^2} \int \rho\, |t(k_R\hat{\mathbf{k}}, k_R\hat{\mathbf{k}}')|^2 \left[1 + \rho\, h(k_R\hat{\mathbf{k}} - k_R\hat{\mathbf{k}}')\right] d\hat{\mathbf{k}}'\,. \qquad (18.44)$$

Similarly, from (18.25) and (18.27), we see that

$$p(\hat{\mathbf{k}}, \hat{\mathbf{k}}') = \frac{\ell_s}{16\pi^2} \rho \, |t(k_R\hat{\mathbf{k}}, k_R\hat{\mathbf{k}}')|^2 \left[1 + \rho \, h(k_R\hat{\mathbf{k}} - k_R\hat{\mathbf{k}}')\right] . \tag{18.45}$$

Making use of (18.42), we find that for an infinite medium, the scattering mean free path and the phase function can be written in terms of the structure factor as

$$\frac{1}{\ell_s} = \frac{1}{16\pi^2} \int \rho \, |t(k_R\hat{\mathbf{k}}, k_R\hat{\mathbf{k}}')|^2 \, \tilde{S}(\mathbf{q}) \, d\hat{\mathbf{k}}' \tag{18.46}$$

and

$$p(\hat{\mathbf{k}}, \hat{\mathbf{k}}') = \frac{\ell_s}{16\pi^2} \rho \, |t(k_R\hat{\mathbf{k}}, k_R\hat{\mathbf{k}}')|^2 \, \tilde{S}(\mathbf{q}) \, , \tag{18.47}$$

where $\mathbf{q} = k_R\hat{\mathbf{k}} - k_R\hat{\mathbf{k}}'$. In these expressions, we have defined

$$\tilde{S}(\mathbf{q}) = S(\mathbf{q}) - (2\pi)^3 \, \rho \, \delta(\mathbf{q}) \tag{18.48}$$

as the structure factor with the forward scattering contribution subtracted.

For isotropic scatterers, the T-matrix depends only on $q = k_R|\hat{\mathbf{k}} - \hat{\mathbf{k}}'|$, and for a statistically homogeneous and isotropic distribution of scatterers, the structure factor also depends on q. The integral in (18.46) can be performed by noting that $q^2 = 4k_R^2 \sin^2(\theta/2)$, $d\hat{\mathbf{k}}' = 2\pi \sin\theta d\theta$ and $q \, dq = k_R^2 \sin\theta d\theta$, where θ is the angle between $\hat{\mathbf{k}}'$ and $\hat{\mathbf{k}}$. This leads to the formula

$$\frac{1}{\ell_s} = \frac{\rho}{8\pi k_R^2} \int_0^{2k_R} q \, |t(q)|^2 \, \tilde{S}(q) \, dq \, . \tag{18.49}$$

This expression is sometimes written in terms of the form factor $F(q) = (k_R^2/16\pi^2)|t(q)|^2$:

$$\frac{1}{\ell_s} = \frac{2\pi \, \rho}{k_R^4} \int_0^{2k_R} q \, F(q) \, \tilde{S}(q) \, dq \, . \tag{18.50}$$

18.6 Transport Mean Free Path

We now show that the radiative transport equation (RTE) implicitly contains another length scale ℓ_t, known as the transport mean free path. In terms of the specific intensity $I(\mathbf{r}, \hat{\mathbf{k}})$, the average energy current is defined by

$$\langle \boldsymbol{J}(\mathbf{r}) \rangle = \int I(\mathbf{r}, \hat{\mathbf{k}}) \, \hat{\mathbf{k}} \, d\hat{\mathbf{k}} \, . \tag{18.51}$$

An equation for $\langle \boldsymbol{J}(\mathbf{r}) \rangle$ can be found by multiplying the RTE (16.61) by $\hat{\mathbf{k}}$ and then integrating over $\hat{\mathbf{k}}$:

$$\int [\hat{\mathbf{k}} \cdot \nabla I(\mathbf{r}, \hat{\mathbf{k}})] \, \hat{\mathbf{k}} \, d\hat{\mathbf{k}} = -\frac{1}{\ell_s} \langle \boldsymbol{J}(\mathbf{r}) \rangle + \frac{1}{\ell_s} \int \boldsymbol{H}(\hat{\mathbf{k}}') \, I(\mathbf{r}, \hat{\mathbf{k}}') \, d\hat{\mathbf{k}}' \, . \tag{18.52}$$

Here the function $\boldsymbol{H}$ is given by

$$\boldsymbol{H}(\hat{\mathbf{k}}') = \int p(\hat{\mathbf{k}}, \hat{\mathbf{k}}')\,\hat{\mathbf{k}}\,d\hat{\mathbf{k}} = \frac{\ell_s}{16\pi^2}\int \gamma(k_R\hat{\mathbf{k}}, k_R\hat{\mathbf{k}}', k_R\hat{\mathbf{k}}, k_R\hat{\mathbf{k}}')\,\hat{\mathbf{k}}\,d\hat{\mathbf{k}}\,. \tag{18.53}$$

For a statistically homogeneous and isotropic medium, the Fourier transform of the irreducible vertex γ is a function of $|\hat{\mathbf{k}} - \hat{\mathbf{k}}'|$ only. Likewise, the phase function is a function of $|\hat{\mathbf{k}} - \hat{\mathbf{k}}'|$, or equivalently of $\hat{\mathbf{k}} \cdot \hat{\mathbf{k}}'$. It is then easy to see that

$$\boldsymbol{H}(\hat{\mathbf{k}}') = g\,\hat{\mathbf{k}}'\,, \tag{18.54}$$

where g is the anisotropy factor, defined by

$$g = \int \hat{\mathbf{k}} \cdot \hat{\mathbf{k}}'\,p(\hat{\mathbf{k}} \cdot \hat{\mathbf{k}}')\,d\hat{\mathbf{k}}\,. \tag{18.55}$$

Introducing (18.54) into (18.52), we obtain the following expression for the energy current:

$$\boldsymbol{J}(\mathbf{r}) = -\frac{\ell_s}{1-g}\int \hat{\mathbf{k}} \cdot \nabla I(\mathbf{r}, \hat{\mathbf{k}})\,\hat{\mathbf{k}}\,d\hat{\mathbf{k}}\,. \tag{18.56}$$

The prefactor in the above integral is a length, known as the transport mean free path, which is defined by the relation

$$\ell_t = \frac{\ell_s}{1-g}\,. \tag{18.57}$$

The transport mean free path is an important length scale in the treatment of radiative transport within the diffusion approximation.

In the ISA, the phase function is given by (18.30) and the anisotropy factor g characterizes the scattering of an individual scatterer. In fact, g can be understood as the angularly averaged cosine of the scattering angle (the angle between $\hat{\mathbf{k}}'$ and $\hat{\mathbf{k}}$). For isotropic scattering, $g = 0$, while $g \simeq 1$ for strong forward scattering. Nearly isotropic scattering is observed for scatterers much smaller than the wavelength, while strong forward scattering is observed for scatterers much larger than the wavelength.

In the presence of spatial correlations, the value of g, and as a consequence of ℓ_t, can be substantially modified. To obtain an expression for ℓ_t, we first rewrite (18.55) as an integral over q. This leads to

$$g = 1 - \ell_s\frac{\pi\,\rho}{k_R^6}\int_0^{2k_R} q^3\,F(q)\,\tilde{S}(q)\,dq\,, \tag{18.58}$$

from which the expression for the transport mean free path is readily deduced:

$$\frac{1}{\ell_t} = \frac{\pi\,\rho}{k_R^6}\int_0^{2k_R} q^3\,F(q)\,\tilde{S}(q)\,dq\,. \tag{18.59}$$

References and Additional Reading

These review articles provide a clear presentation of the T-matrix formalism for wave scattering:

A. Lagendijk and B.A. van Tiggelen, Phys. Rep. **270**, 143 (1996).

P. de Vries, D.V. van Coevorden and A. Lagendijk, Rev. Mod. Phys. **70**, 447 (1998).

M.C.W. van Rossum and Th.M. Nieuwenhuizen, Rev. Mod. Phys. **71**, 313 (1999).

These papers introduced the expression of the transport mean free path derived in this chapter to analyze light scattering in dense colloidal suspensions:

S. Fraden and G. Maret, Phys. Rev. Lett. **65**, 512 (1990).

X. Qiu, X.L. Wu, J.Z. Xue, D.J. Pine, D.A. Weitz and P.M. Chaikin, Phys. Rev. Lett. **65**, 516 (1990).

These papers demonstrate experimentally the impact of spatial correlations on multiple scattering of light:

L.F. Rojas-Ochoa, J.M. Mendez-Alcaraz, J.J. Sáenz, P. Schurtenberger and F. Scheffold, Phys. Rev. Lett. **93**, 073903 (2004).

M. Reufer, L.F. Rojas-Ochoa, S. Eiden, J.J. Sáenz and F. Scheffold, Appl. Phys. Lett. **91**, 171904 (2007).

P.D. García, R. Sapienza, A. Blanco and C. López, Adv. Mater. **19**, 2597 (2007).

This textbook provides a general view of models of disorder:

J.M. Ziman, *Models of Disorder: The Theoretical Physics of Homogeneously Disordered Systems* (Cambridge University Press, Cambridge, 1979).

19

Time-Dependent Radiative Transport and Energy Velocity

In this chapter, we study the theory of time-dependent radiative transport in a static medium. For nonresonant scattering, we derive a time-dependent radiative transport equation (RTE). In the presence of resonances, we obtain a generalization of this result that allows us to discuss the dependence of the energy velocity on the dwell time associated with the resonances.[1]

19.1 Two-Frequency Bethe–Salpeter Equation

The Bethe–Salpeter equation has been introduced in Chapter 16 as a means to derive the radiative transport of a time-harmonic field. In order to establish a transport equation for general time-dependent fields, we start by considering the correlation function $\langle U(\mathbf{r}_1,\omega)U^*(\mathbf{r}_1',\omega')\rangle$ of two fields at two different frequencies. Following the steps leading to the single-frequency Bethe–Salpeter equation (Section 16.1), we obtain

$$\begin{aligned}\langle U(\mathbf{r}_1,\omega)U^*(\mathbf{r}_1',\omega')\rangle &= \langle U(\mathbf{r}_1,\omega)\rangle\langle U^*(\mathbf{r}_1',\omega')\rangle \\ &\quad + \int \langle G(\mathbf{r}_1,\mathbf{R}_1,\omega)\rangle\langle G^*(\mathbf{r}_1',\mathbf{R}_1',\omega')\rangle \\ &\quad \times \Gamma(\mathbf{R}_1,\mathbf{R}_2,\omega;\mathbf{R}_1',\mathbf{R}_2',\omega') \\ &\quad \times \langle U(\mathbf{R}_2,\omega)U^*(\mathbf{R}_2',\omega')\rangle\, d^3R_1 d^3R_2 d^3R_1' d^3R_2' , \end{aligned} \tag{19.1}$$

where $\Gamma(\mathbf{R}_1,\mathbf{R}_2,\omega;\mathbf{R}_1',\mathbf{R}_2',\omega')$ is the two-frequency irreducible vertex. Since our goal is to derive a RTE, we omit the first term on the right-hand side of (19.1), which only contributes a source term (see the derivation of the time-independent RTE in Chapter 16). We note that throughout this chapter, we will assume that the scattering medium is statistically homogeneous and isotropic.

[1] The treatment of resonant scattering follows the approach introduced in B.A. van Tiggelen, A. Lagendijk, M.P. van Albada and A. Tip, Phys. Rev. B **45**, 12233 (1992).

We introduce the spatial Fourier transform of the field correlation function according to

$$\langle U(\mathbf{k}_1,\omega)U^*(\mathbf{k}_1',\omega')\rangle = \int \langle U(\mathbf{r}_1,\omega)U^*(\mathbf{r}_1',\omega')\rangle \exp(-i\mathbf{k}_1\cdot\mathbf{r}_1 + i\mathbf{k}_1'\cdot\mathbf{r}_1')\, d^3r_1 d^3r_1' . \tag{19.2}$$

As a consequence of translational invariance, the Fourier transform of the average Green function can be written as

$$\langle G(\mathbf{r},\mathbf{R},\omega)\rangle = \int \langle G(\mathbf{k},\omega)\rangle \exp(i\mathbf{k}\cdot(\mathbf{r}-\mathbf{R}))\,\frac{d^3k}{(2\pi)^3} , \tag{19.3}$$

which defines $\langle G(\mathbf{k},\omega)\rangle$. The Fourier transform of the irreducible vertex is

$$\Gamma(\mathbf{R}_1,\mathbf{R}_2,\omega;\mathbf{R}_1',\mathbf{R}_2',\omega') = \int \tilde{\Gamma}(\mathbf{K}_1,\mathbf{K}_2,\omega;\mathbf{K}_1',\mathbf{K}_2',\omega')$$
$$\times \exp(i\mathbf{K}_1\cdot\mathbf{R}_1 - i\mathbf{K}_2\cdot\mathbf{R}_2 - i\mathbf{K}_1'\cdot\mathbf{R}_1' + i\mathbf{K}_2'\cdot\mathbf{R}_2')\,\frac{d^3K_1}{(2\pi)^3}\frac{d^3K_2}{(2\pi)^3}\frac{d^3K_1'}{(2\pi)^3}\frac{d^3K_2'}{(2\pi)^3} , \tag{19.4}$$

where the signs in the exponential have been chosen for convenience. Due to translational invariance, the irreducible vertex in Fourier space must be of the form

$$\Gamma(\mathbf{R}_1+\Delta\mathbf{R},\mathbf{R}_2+\Delta\mathbf{R},\omega;\mathbf{R}_1'+\Delta\mathbf{R},\mathbf{R}_2'+\Delta\mathbf{R},\omega') = \Gamma(\mathbf{R}_1,\mathbf{R}_2,\omega;\mathbf{R}_1',\mathbf{R}_2',\omega') \tag{19.5}$$

for any arbitrary displacement vector $\Delta\mathbf{R}$. In Fourier space, this means that the irreducible vertex is of the form

$$\tilde{\Gamma}(\mathbf{K}_1,\mathbf{K}_2,\omega;\mathbf{K}_1',\mathbf{K}_2',\omega') = (2\pi)^3\,\gamma(\mathbf{K}_1,\mathbf{K}_2,\omega;\mathbf{K}_1',\mathbf{K}_2',\omega')\,\delta(\mathbf{K}_1-\mathbf{K}_2-\mathbf{K}_1'+\mathbf{K}_2') , \tag{19.6}$$

which defines the two-frequency vertex γ. We now insert (19.3) and (19.4) into (19.1), and make use of the relation (19.6). We then make the change of variables

$$\mathbf{k} = \frac{\mathbf{K}_1+\mathbf{K}_1'}{2} , \quad \mathbf{q} = \mathbf{K}_1 - \mathbf{K}_1' , \tag{19.7}$$

$$\Omega = \frac{\omega+\omega'}{2} , \quad \varpi = \omega - \omega' , \tag{19.8}$$

which has unit Jacobian. We also introduce the quantity

$$\widetilde{W}(\mathbf{q},\mathbf{k},\varpi,\Omega) = \frac{1}{(2\pi)^6}\langle U(\mathbf{k}+\mathbf{q}/2,\Omega+\varpi/2)U^*(\mathbf{k}-\mathbf{q}/2,\Omega-\varpi/2)\rangle , \tag{19.9}$$

which defines the Fourier transform of the space-time Wigner transform. It can be seen that

$$\widetilde{W}(\mathbf{q},\mathbf{k},\varpi,\Omega) = \int \exp(i\varpi t)\,\widetilde{W}(\mathbf{q},\mathbf{k},t,\Omega)dt , \tag{19.10}$$

where

$$\widetilde{W}(\mathbf{q}, \mathbf{k}, t, \Omega) = \frac{1}{(2\pi)^6} \int \exp(i\Omega t')\langle U(\mathbf{k}{+}\mathbf{q}/2, t{+}t'/2)U^*(\mathbf{k}{-}\mathbf{q}/2, t{-}t'/2)\rangle \, dt' \, . \tag{19.11}$$

The above result extends the definition of the Wigner transform introduced in Chapter 16 (see (16.17) and (16.27)) to the time domain. We thus see that in terms of $\widetilde{W}$, the Bethe–Salpeter equation (19.1) takes the form

$$\begin{aligned}\widetilde{W}(\mathbf{q}, \mathbf{k}, \varpi, \Omega) = \int & \langle G(\mathbf{k} + \mathbf{q}/2, \Omega + \varpi/2)\rangle\langle G^*(\mathbf{k} - \mathbf{q}/2, \Omega - \varpi/2)\rangle \\ & \times\gamma(\mathbf{k} + \mathbf{q}/2, \mathbf{K} + \mathbf{q}/2, \Omega + \varpi/2; \mathbf{k} - \mathbf{q}/2, \mathbf{K} - \mathbf{q}/2, \Omega - \varpi/2) \\ & \times\widetilde{W}(\mathbf{q}, \mathbf{K}, \varpi, \Omega)\, \frac{d^3K}{(2\pi)^3} \, .\end{aligned} \tag{19.12}$$

The average Green function obeys the Dyson equation. In Fourier space, the Dyson equation is given by [see (15.32)]

$$\langle G(\mathbf{k}, \omega)\rangle = \frac{1}{k^2 - k_0^2 - \Sigma(\mathbf{k}, \omega) - i\epsilon} \, , \tag{19.13}$$

where Σ is the self-energy, $k = |\mathbf{k}|$ and $k_0 = \omega/c$. Thus we can rewrite the product of averaged Green functions in (19.12) as

$$\begin{aligned}&\langle G(\mathbf{k} + \mathbf{q}/2, \Omega + \varpi/2)\rangle\langle G^*(\mathbf{k} - \mathbf{q}/2, \Omega - \varpi/2)\rangle = \\ &\frac{\langle G(\mathbf{k} + \mathbf{q}/2, \Omega + \varpi/2)\rangle - \langle G^*(\mathbf{k} - \mathbf{q}/2, \Omega - \varpi/2)\rangle}{2\varpi\Omega/c^2 - 2\mathbf{k}\cdot\mathbf{q} + \Sigma(\mathbf{k} + \mathbf{q}/2, \Omega + \varpi/2) - \Sigma^*(\mathbf{k} - \mathbf{q}/2, \Omega - \varpi/2)} \, .\end{aligned} \tag{19.14}$$

Inserting this result into (19.12) leads to the two-frequency Bethe–Salpeter equation in Fourier space:

$$\begin{aligned}&\left[2\varpi\Omega/c^2 - 2\mathbf{k}\cdot\mathbf{q} + \Sigma(\mathbf{k}^+, \Omega^+) - \Sigma^*(\mathbf{k}^-, \Omega^-)\right] \widetilde{W}(\mathbf{q}, \mathbf{k}, \varpi, \Omega) = \\ &\quad \left[\langle G(\mathbf{k}^+, \Omega^+)\rangle - \langle G^*(\mathbf{k}^-, \Omega^-)\rangle\right] \int \gamma(\mathbf{k}^+, \mathbf{K}^+, \Omega^+; \mathbf{k}^-, \mathbf{K}^-, \Omega^-) \\ &\qquad\qquad \times\widetilde{W}(\mathbf{q}, \mathbf{k}, \varpi, \Omega)\, \frac{d^3K}{(2\pi)^3} \, .\end{aligned} \tag{19.15}$$

Here we have introduced the notation $\mathbf{k}^\pm = \mathbf{k}\pm\mathbf{q}/2$, $\mathbf{K}^\pm = \mathbf{K}\pm\mathbf{q}/2$ and $\Omega^\pm = \Omega \pm \varpi/2$. Equation (19.15) is the starting point of the derivation of the time-dependent RTE.

19.2 Time-Dependent Radiative Transport Equation

In this section, we derive the asymptotic form of (19.15) that holds at large space and time scales. That is, we consider the limit $q \to 0$ and $\varpi \to 0$. We find that the term involving the self-energy in (19.15) can be approximated by

$$\begin{aligned}\Sigma(\mathbf{k}^+, \Omega^+) - \Sigma^*(\mathbf{k}^-, \Omega^-) &\simeq \Sigma(\mathbf{k}, \Omega) - \Sigma^*(\mathbf{k}, \Omega) + \frac{\varpi}{2}\frac{\partial \Sigma}{\partial \Omega}(\mathbf{k}, \Omega) \\ &+ \frac{\varpi}{2}\frac{\partial \Sigma^*}{\partial \Omega}(\mathbf{k}, \Omega) = 2i \operatorname{Im} \Sigma(\mathbf{k}, \Omega) \\ &+ \varpi \operatorname{Re}\frac{\partial \Sigma}{\partial \Omega}(\mathbf{k}, \Omega) . \end{aligned} \tag{19.16}$$

The first-order term in the frequency ϖ is needed to account for the effect of resonances in the medium, as will become clear. We will also consider weak scattering, as in Chapter 15, and assume that the self-energy is taken on-shell, with $|\mathbf{k}| = \Omega/c$. As a result, Σ depends only on Ω. The contribution of the averaged Green function in (19.15) becomes, according to (19.13),

$$\langle G(\mathbf{k}, \Omega)\rangle = \frac{1}{k^2 - k_R^2(\Omega) - ik_I^2(\Omega) - i\epsilon} , \tag{19.17}$$

where we have defined k_R and k_I such that $k_R^2(\Omega) + ik_I^2(\Omega) = (\Omega/c)^2 + \Sigma(\Omega)$. Using this expression, the prefactor in the integral appearing in (19.15) can be approximated by

$$\begin{aligned}\langle G(\mathbf{k}^+, \Omega^+)\rangle - \langle G^*(\mathbf{k}^-, \Omega^-)\rangle &\simeq \langle G(\mathbf{k}, \Omega)\rangle - \langle G^*(\mathbf{k}, \Omega)\rangle \\ &= 2i \operatorname{Im}\frac{1}{k^2 - k_R^2(\Omega) - ik_I^2(\Omega) - i\epsilon} \\ &\simeq 2i\pi\, \delta(k^2 - k_R^2(\Omega)) . \end{aligned} \tag{19.18}$$

Here we have used the identity (16.41) along with the weak-scattering approximation $k_R \gg k_I$ (see Chapter 15). The irreducible vertex can be simplified in a similar manner, using the slowly varying envelope approximation

$$\begin{aligned}\gamma(\mathbf{k}^+, \mathbf{K}^+, \Omega^+; \mathbf{k}^-, \mathbf{K}^-, \Omega^-) &\simeq \gamma(\mathbf{k}, \mathbf{K}, \Omega; \mathbf{k}, \mathbf{K}, \Omega) \\ &+ \frac{\varpi}{2}\frac{\partial \gamma}{\partial \Omega'}(\mathbf{k}, \mathbf{K}, \Omega'; \mathbf{k}, \mathbf{K}, \Omega)\Big|_{\Omega'=\Omega} \\ &- \frac{\varpi}{2}\frac{\partial \gamma}{\partial \Omega'}(\mathbf{k}, \mathbf{K}, \Omega; \mathbf{k}, \mathbf{K}, \Omega')\Big|_{\Omega'=\Omega} . \end{aligned} \tag{19.19}$$

Finally, making use of (19.16), (19.18) and (19.19), the two-frequency Bethe–Salpeter equation (19.15) becomes

$$\left[\frac{2\varpi\Omega}{c^2} - 2\mathbf{k}\cdot\mathbf{q} + 2i\,\mathrm{Im}\,\Sigma(\Omega) + \varpi\,\mathrm{Re}\frac{\partial\Sigma}{\partial\Omega}(\Omega)\right]\widetilde{W}(\mathbf{q},\mathbf{k},\varpi,\Omega) =$$
$$2i\pi\,\delta(k^2 - k_R^2(\Omega))\int\left[\gamma(\mathbf{k},\mathbf{K},\Omega;\mathbf{k},\mathbf{K},\Omega) + \frac{\varpi}{2}\frac{\partial\gamma}{\partial\Omega'}(\mathbf{k},\mathbf{K},\Omega';\mathbf{k},\mathbf{K},\Omega)\big|_{\Omega'=\Omega}\right.$$
$$\left. - \frac{\varpi}{2}\frac{\partial\gamma}{\partial\Omega'}(\mathbf{k},\mathbf{K},\Omega;\mathbf{k},\mathbf{K},\Omega')\big|_{\Omega'=\Omega}\right]\widetilde{W}(\mathbf{q},\mathbf{k},\varpi,\Omega)\,\frac{d^3K}{(2\pi)^3}\,. \tag{19.20}$$

The above result can be understood as a precursor to the time-dependent generalization of the RTE. More explicitly, by introducing the specific intensity $I(\mathbf{q},\hat{\mathbf{k}},\varpi,\Omega)$ as in Chapter 16, and the relation

$$\widetilde{W}(\mathbf{q},\mathbf{k},\varpi,\Omega) = I(\mathbf{q},\hat{\mathbf{k}},\varpi,\Omega)\,\delta(k^2 - k_R^2)\,, \tag{19.21}$$

where $\hat{\mathbf{k}} = \mathbf{k}/k_R$, and rewriting the integral in (19.20) using

$$\int f(\mathbf{k})\,\delta(k^2 - p^2)\,d^3k = \frac{p}{2}\int f(\hat{\mathbf{k}})d\hat{\mathbf{k}}\,, \tag{19.22}$$

we obtain

$$\left\{-i\varpi\left[\frac{\Omega}{c^2} + \frac{1}{2}\mathrm{Re}\frac{\partial\Sigma}{\partial\Omega}(\Omega)\right] + i\mathbf{k}\cdot\mathbf{q}\right\}I(\mathbf{q},\hat{\mathbf{k}},\varpi,\Omega) = -\mathrm{Im}\,\Sigma(\Omega)\,I(\mathbf{q},\hat{\mathbf{k}},\varpi,\Omega)$$
$$+\frac{k_R(\Omega)}{16\pi^2}\int\left[\gamma(\hat{\mathbf{k}},\hat{\mathbf{k}}',\Omega;\hat{\mathbf{k}},\hat{\mathbf{k}}',\Omega) + \frac{\varpi}{2}\frac{\partial\gamma}{\partial\Omega'}(\hat{\mathbf{k}},\hat{\mathbf{k}}',\Omega';\hat{\mathbf{k}},\hat{\mathbf{k}}',\Omega)\big|_{\Omega'=\Omega}\right.$$
$$\left. - \frac{\varpi}{2}\frac{\partial\gamma}{\partial\Omega'}(\hat{\mathbf{k}},\hat{\mathbf{k}}',\Omega;\hat{\mathbf{k}},\hat{\mathbf{k}}',\Omega')\big|_{\Omega'=\Omega}\right]I(\mathbf{q},\hat{\mathbf{k}}',\varpi,\Omega)\,d\hat{\mathbf{k}}'\,. \tag{19.23}$$

The first term on the left-hand side corresponds to a time derivative in the frequency domain. Introducing the group velocity $v_g(\Omega)$, defined by $1/v_g(\Omega) = dk_R(\Omega)/d\Omega$, it is easy to see that

$$\frac{\Omega}{c^2} + \frac{1}{2}\mathrm{Re}\frac{\partial\Sigma}{\partial\Omega}(\Omega) = \frac{k_R(\Omega)}{v_g(\Omega)}\,. \tag{19.24}$$

We can then rewrite (19.23) as

$$\left[-i\varpi\,\frac{k_R(\Omega)}{v_g(\Omega)} + i\mathbf{k}\cdot\mathbf{q}\right]I(\mathbf{q},\hat{\mathbf{k}},\varpi,\Omega) = -\mathrm{Im}\,\Sigma(\Omega)\,I(\mathbf{q},\hat{\mathbf{k}},\varpi,\Omega)$$
$$+\frac{k_R(\Omega)}{16\pi^2}\int\left[\gamma(\hat{\mathbf{k}},\hat{\mathbf{k}}',\Omega;\hat{\mathbf{k}},\hat{\mathbf{k}}',\Omega) + \frac{\varpi}{2}\frac{\partial\gamma}{\partial\Omega'}(\hat{\mathbf{k}},\hat{\mathbf{k}}',\Omega';\hat{\mathbf{k}},\hat{\mathbf{k}}',\Omega)\big|_{\Omega'=\Omega}\right.$$
$$\left. - \frac{\varpi}{2}\frac{\partial\gamma}{\partial\Omega'}(\hat{\mathbf{k}},\hat{\mathbf{k}}',\Omega;\hat{\mathbf{k}},\hat{\mathbf{k}}',\Omega')\big|_{\Omega'=\Omega}\right]I(\mathbf{q},\hat{\mathbf{k}}',\varpi,\Omega)\,d\hat{\mathbf{k}}'\,. \tag{19.25}$$

The above result is the general form of the time-dependent RTE written in the frequency domain. Here $I(\mathbf{q}, \hat{\mathbf{k}}, \varpi, \Omega)$ is the Fourier transform of the time-dependent specific intensity $I(\mathbf{r}, \hat{\mathbf{k}}, t, \Omega)$ with respect to $\mathbf{r}$ and t, Ω being the central optical frequency.

19.3 Nonresonant Scattering

For nonresonant scattering, the dependence of the optical properties of the medium on frequency can be neglected, and the derivatives of γ with respect to Ω can be taken to be zero in (19.25). Moreover, we can write $k_R = n_{\text{eff}}(\Omega)\Omega/c$, with n_{eff} the real part of the effective refractive index. Assuming that n_{eff} weakly depends on Ω, the group velocity equals the phase velocity $v_p(\Omega) = c/n_{\text{eff}}(\Omega)$. Equation (19.25) thus becomes

$$\left[-i\varpi \frac{k_R(\Omega)}{v_p(\Omega)} + ik \cdot \mathbf{q}\right] I(\mathbf{q}, \hat{\mathbf{k}}, \varpi, \Omega) = -\text{Im}\,\Sigma(\Omega)\, I(\mathbf{q}, \hat{\mathbf{k}}, \varpi, \Omega) + \frac{k_R(\Omega)}{16\pi^2} \int \gamma(\hat{\mathbf{k}}, \hat{\mathbf{k}}', \Omega; \hat{\mathbf{k}}, \hat{\mathbf{k}}', \Omega)\, I(\mathbf{q}, \hat{\mathbf{k}}', \varpi, \Omega)\, d\hat{\mathbf{k}}' . \tag{19.26}$$

Using the definition of the phase function $p(\hat{\mathbf{k}}, \hat{\mathbf{k}}', \Omega)$, and of the scattering and extinction coefficients $\mu_s(\Omega)$ and $\mu_e(\Omega)$ introduced in Chapter 16, the above equation can be rewritten as

$$\left[\frac{-i\varpi}{v_p(\Omega)} + ik \cdot \mathbf{q}\right] I(\mathbf{q}, \hat{\mathbf{k}}, \varpi, \Omega) = -\mu_e(\Omega)\, I(\mathbf{q}, \hat{\mathbf{k}}, \varpi, \Omega) + \mu_s(\Omega) \int p(\hat{\mathbf{k}}, \hat{\mathbf{k}}', \Omega)\, I(\mathbf{q}, \hat{\mathbf{k}}', \varpi, \Omega)\, d\hat{\mathbf{k}}' . \tag{19.27}$$

Performing the Fourier transforms with respect to ϖ (time) and $\mathbf{q}$ (space), according to (19.10) and (16.27), we obtain the time-dependent RTE

$$\left(\frac{1}{v_p(\Omega)}\frac{\partial}{\partial t} + i\hat{\mathbf{k}} \cdot \nabla_{\mathbf{r}}\right) I(\mathbf{r}, \hat{\mathbf{k}}, t, \Omega) + \mu_e(\Omega)\, I(\mathbf{r}, \hat{\mathbf{k}}, t, \Omega) = \mu_s(\Omega) \int p(\hat{\mathbf{k}}, \hat{\mathbf{k}}', \Omega)\, I(\mathbf{r}, \hat{\mathbf{k}}', t, \Omega)\, d\hat{\mathbf{k}}' . \tag{19.28}$$

The above result extends the time-independent RTE derived in Chapters 16 and 17 to the setting of nonresonant media.

19.4 Resonant Scattering

In the presence of scattering resonances, the full frequency dependence in (19.25) must be taken into account. Here we consider the case of a medium composed of discrete resonant scatterers, and derive the specific form of (19.25) for this case. In the Kubo limit, the irreducible vertex is given by (19.19). Assuming that all scatterers are identical, and following the procedure described in Section 18.2, we easily see that

$$\gamma(\mathbf{k}, \mathbf{K}, \Omega; \mathbf{k}, \mathbf{K}, \Omega') = \rho\, t(\mathbf{k}, \mathbf{K}, \Omega)\, t^*(\mathbf{k}, \mathbf{K}, \Omega')\,, \tag{19.29}$$

to lowest order in the number density ρ. Here $t(\mathbf{k}, \mathbf{K}, \Omega)$ is the Fourier transform of the T-matrix of a single scatterer at the frequency Ω, as defined in (18.24). For simplicity, we will consider point scatterers, for which the T-matrix in the Fourier space is independent of the wavevector, as a consequence of (12.9). We have, in this case,

$$\gamma(\mathbf{k}, \mathbf{K}, \Omega; \mathbf{k}, \mathbf{K}, \Omega') = \rho\, t(\Omega)\, t^*(\Omega')\,. \tag{19.30}$$

In Chapter 15, the self-energy is shown to be $\Sigma = \langle T \rangle$ to lowest order in the T-matrix [see (15.55)]. Following again the procedure provided in Section 18.2, we can also show that

$$\Sigma(\mathbf{k}, \Omega) = \rho\, t(\Omega)\,, \tag{19.31}$$

to lowest order in ρ. Introducing the above expressions for Σ and γ into (19.25) leads to

$$\begin{aligned} &\left[-i\varpi \frac{k_R(\Omega)}{v_g(\Omega)} + i\mathbf{k}\cdot\mathbf{q}\right] I(\mathbf{q}, \hat{\mathbf{k}}, \varpi, \Omega) + \rho\, \mathrm{Im}\, t(\Omega)\, I(\mathbf{q}, \hat{\mathbf{k}}, \varpi, \Omega) \\ &= \frac{k_R(\Omega)}{16\pi^2} \int \left\{\rho\, |t(\Omega)|^2 + i\varpi\rho\, \mathrm{Im}\left[\frac{\partial t(\Omega)}{\partial \Omega} t^*(\Omega)\right]\right\} I(\mathbf{q}, \hat{\mathbf{k}}', \varpi, \Omega)\, d\hat{\mathbf{k}}'\,, \end{aligned} \tag{19.32}$$

which is the time-dependent RTE for resonant point scatterers, written in Fourier space. Note that due to dispersion, an RTE of the form (19.28) cannot be written for resonant scattering, except for the case of point scatterers.

19.5 Energy Velocity

The energy density u and the energy current $\boldsymbol{J}$ can be obtained from the specific intensity according to

$$u(\mathbf{r}, t, \Omega) = \frac{1}{v_E(\Omega)} \int I(\mathbf{r}, \hat{\mathbf{k}}, t, \Omega)\, d\hat{\mathbf{k}}\,, \tag{19.33}$$

$$\boldsymbol{J}(\mathbf{r}, t, \Omega) = \int I(\mathbf{r}, \hat{\mathbf{k}}, t, \Omega)\, \hat{\mathbf{k}}\, d\hat{\mathbf{k}}\,. \tag{19.34}$$

These expressions follow from (16.19), (16.20) and (19.21), and are valid up to a constant factor that we do not specify. The first relation defines the energy velocity $v_E(\Omega)$. Note that the expressions remain unchanged after Fourier transformation with respect to time and space, so that they directly apply to $I(\mathbf{q}, \hat{\mathbf{k}}, \varpi, \Omega)$.

A relationship between u and $\boldsymbol{J}$ can be obtained by integration of (19.32) with respect to $\hat{\mathbf{k}}$:

$$
\begin{aligned}
i\mathbf{q} \cdot \boldsymbol{J}(\mathbf{q}, \varpi, \Omega) - i\varpi\, u(\mathbf{q}, \varpi, \Omega)\, v_E(\Omega) \left\{ \frac{1}{v_g(\Omega)} + \frac{\rho}{4\pi} \operatorname{Im} \left[\frac{\partial t(\Omega)}{\partial \Omega} t^*(\Omega) \right] \right\} \\
= \left[-\frac{\rho}{k_R(\Omega)} \operatorname{Im} t(\Omega) + \frac{\rho}{4\pi} |t(\Omega)|^2 \right] u(\mathbf{q}, \varpi, \Omega)\, v_E(\Omega)\,.
\end{aligned}
\tag{19.35}
$$

The right-hand side describes absorption losses, and vanishes in the case of non-absorbing scatterers. Indeed, for such scatterers, we have $\operatorname{Im} t(\Omega) = (k_R(\Omega)/4\pi)|t(\Omega)|^2$, which is a consequence of the optical theorem. In the absence of absorption, the relation governing energy conservation takes the form

$$
i\mathbf{q} \cdot \boldsymbol{J}(\mathbf{q}, \varpi, \Omega) - i\varpi\, u(\mathbf{q}, \varpi, \Omega) = 0\,. \tag{19.36}
$$

From (19.35), we deduce that

$$
\frac{1}{v_E(\Omega)} = \frac{1}{v_g(\Omega)} + \frac{\rho}{4\pi} \operatorname{Im} \left[\frac{\partial t(\Omega)}{\partial \Omega} t^*(\Omega) \right]. \tag{19.37}
$$

The above expression for the energy velocity exhibits a frequency dependence that accounts for internal resonances in the scatterers, through the frequency dependence of the T-matrix. We note that resonances can lead to substantial reductions of the energy velocity compared to the speed of light in vacuum.

The expression (19.37) has an appealing physical interpretation. Let us introduce the transport time $\tau^* = \ell_t/v_E$, where ℓ_t is the transport mean free path. We can then write $\tau^* = \tau_p + \tau_{dw}$, where $\tau_p = \ell_t/v_g$ is the propagation time of the average field along a path of length ℓ_t. The additional time τ_{dw}, which is referred to as the dwell time, corresponds to the delay produced by the scattering resonances. In terms of these transport parameters, the energy velocity can be expressed as

$$
v_E(\Omega) = \left[\frac{1}{v_g(\Omega)} + \frac{\tau_{dw}}{\ell_t} \right]^{-1}. \tag{19.38}
$$

The role of the dwell time in dynamic multiple-scattering of light is an important difference with electronic transport in mesoscopic physics. Regimes in which the energy velocity is dominated by this contribution have been demonstrated with strongly resonant scatterers such as cold atoms.

References and Additional Reading

The derivation of the energy velocity, along the lines followed in this chapter, was initially presented in:

B.A. van Tiggelen, A. Lagendijk, M.P. van Albada and A. Tip, Phys. Rev. B **45**, 12233 (1992).

Resonant multiple scattering of light is reviewed in the following paper:

A. Lagendijk and B.A. van Tiggelen, Phys. Rep. **270**, 143 (1996).

These papers report on experimental evidence of substantial changes of the energy velocity in multiple scattering of light:

M.P. van Albada, B.A. van Tiggelen, A. Lagendijk and A. Tip, Phys. Rev. Lett. **66**, 3132 (1991).

G. Labeyrie, E. Vaujour, C. A. Müller, D. Delande, C. Miniatura, D.Wilkowski and R. Kaiser, Phys. Rev. Lett. **91**, 223904 (2003).

R. Sapienza, P.D. García, J. Bertolotti, M.D. Martín, Á. Blanco, L. Vina, C. Lópáez and D.S. Wiersma, Phys. Rev. Lett. **99**, 233902 (2007).

Exercises

III.1 The diagrammatic formalism has been introduced in Chapter 15 to compute the average Green's function. Starting from (15.2) and (15.2), perform the diagrammatic calculations needed to obtain (15.24).

III.2 In Chapter 16, diagrammatic methods are used to compute field correlation functions. Starting from (16.3) and (16.4), perform the diagrammatic calculations needed to obtain (16.5).

III.3 Perform the calculation leading to (16.30) from (16.25), making use of the Bethe–Salpeter equation (16.7) and the ladder approximation (16.8). It may be useful to consult the appendix in A. Cazé and J. Schotland, J. Opt. Soc. Am. A **32**, 1475 (2015).

III.4 The T-matrix of a discrete ensemble of scatterers has been introduced in Chapter 18. Show that by inserting (18.9) into (18.10), one recovers the series in (18.8). This justifies the expression for the multiply-scattered field in terms of T matrices of individual scatterers.

III.5 Use a Mie theory calculator to compute the angular distribution of scattered power for a glass sphere with refractive index $n = 1.5$ and radius R ranging from 10 nm to 10 μm, located in vacuum, and illuminated with unpolarized light at a wavelength $\lambda = 600$ nm. Observe the change in the power when R increases. Compute the value of the anisotropy factor g in the regimes $R \ll \lambda$, $R \lesssim \lambda$ and $R \gg \lambda$.

III.6 Standing in a fog, you estimate the visibility to be 100 m. The fog is made of water droplets with radius $R = 50$ μm and a density low enough to assume independent scattering.

(a) Calculate the scattering cross-section of the droplets.
(b) Explain why the visibility coincides (in order of magnitude) with the scattering mean free path.
(c) Deduce the numerical value of the mass of water suspended in 1 m^3 of air (mass density of water: 1000 kg m^{-3}).

III.7 A scattering material is composed of particles with diameter $d = 2.5\ \mu$m in a matrix with a low refractive index n. (We assume $n = 1$ for simplicity.) The independent scattering approximation is assumed to be valid. The illumination wavelength is $\lambda = 500$ nm.
(a) For a single particle, what is the order of magnitude of the angular aperture θ of the forward scattering lobe? Hint: This is a diffraction lobe.
(b) Can you estimate the anisotropy factor g?
(c) The scattering mean free path is $\ell_s = 50\ \mu$m. Can you deduce the value of the transport mean path ℓ_t?

III.8 In the derivation of the mean free path in correlated media (Chapter 18), derive (18.25) from (18.23).

III.9 Consider an ensemble of N identical spherical scatterers in a volume V, illuminated by a monochromatic plane wave with incident wavevector $\mathbf{k}_i$. In the single scattering regime, write the far-field expression of the scattered field U_s in a direction defined by the scattered wavevector $\mathbf{k}_s$, assuming that scatterers are described by their T-matrix $t(q)$, with $q = |\mathbf{k}_s - \mathbf{k}_i|$. Show that the averaged scattered intensity $\langle |U_s|^2 \rangle$ is translational to the average structure factor $S(q)$ defined in Chapter 18. Can you give a physical meaning to the structure factor in terms of interferences?

III.10 To lowest order in the T-matrix of a disordered medium, the self-energy is given by $\Sigma = \langle T \rangle$ [see (15.55)]. For a set of identical discrete scatterers, show that the self-energy at frequency Ω in Fourier space is

$$\Sigma(\mathbf{k}, \Omega) = \rho\, t(\mathbf{k}, \mathbf{k}, \Omega) \tag{19.39}$$

to lowest order in the number density ρ, where $t(\mathbf{k}, \mathbf{k}', \Omega)$ is the T-matrix of a single scatterer. This proves (19.31). Hint: Follow the procedure described in Section 18.2 for the irreducible vertex.

Part IV

Radiative Transport and Diffusion

20

Radiative Transport: Boundary Conditions and Integral Representations

In this chapter, we present some basic facts about the radiative transport equation (RTE). In particular, we describe the boundary conditions under which the RTE has a unique solution. We also derive a basic integral representation for such solutions and, as a consequence, obtain the relation governing reciprocity of Green's functions for the RTE.

20.1 Time-Independent Radiative Transport

The RTE is a conservation law that governs the gains and losses of the specific intensity due to scattering and absorption. We note that the RTE is a generalization of the transport equation (3.27) of geometrical optics, in which rays change direction due to scattering. The specific intensity $I(\mathbf{r}, \hat{\mathbf{k}})$ at the position $\mathbf{r}$ in the direction $\hat{\mathbf{k}}$. obeys the time-independent RTE derived in Chapter 16 [see (16.61)],

$$\hat{\mathbf{k}} \cdot \nabla I + \mu_e I = \mu_s \int d\hat{\mathbf{k}}' p(\hat{\mathbf{k}}, \hat{\mathbf{k}}') I(\mathbf{r}, \hat{\mathbf{k}}') + S \,, \tag{20.1}$$

where μ_e and μ_s are the extinction and scattering coefficients of the medium, and S is the source. It follows from (16.63) that the phase function p is nonnegative and obeys the symmetry and normalization conditions

$$p(\hat{\mathbf{k}}, \hat{\mathbf{k}}') = p(-\hat{\mathbf{k}}', -\hat{\mathbf{k}}) \,, \qquad \int p(\hat{\mathbf{k}}, \hat{\mathbf{k}}') d\hat{\mathbf{k}}' = 1 \,, \tag{20.2}$$

for all $\hat{\mathbf{k}}$ and $\hat{\mathbf{k}}'$.

Various models for the phase function are used in practice. We will often assume that the phase function $p(\hat{\mathbf{k}}, \hat{\mathbf{k}}')$ depends only upon the angle between $\hat{\mathbf{k}}$ and $\hat{\mathbf{k}}'$, which arises in the setting of a statistically homogeneous and isotropic medium.

The case $p = 1/(4\pi)$ is referred to as isotropic scattering. In general, we can expand p as

$$p(\hat{\mathbf{k}}, \hat{\mathbf{k}}') = \frac{1}{4\pi} \sum_{l=0}^{\infty} (2l+1) A_l P_l(\hat{\mathbf{k}} \cdot \hat{\mathbf{k}}') , \tag{20.3}$$

where A_l are suitable coefficients and P_l are Legendre polynomials. The Henyey–Greenstein phase function

$$p(\hat{\mathbf{k}}, \hat{\mathbf{k}}') = \frac{1}{4\pi} \frac{1 - g^2}{(1 - 2g\hat{\mathbf{k}} \cdot \hat{\mathbf{k}}' + g^2)^{3/2}} \tag{20.4}$$

is particularly well known. It can be obtained by setting $A_l = g^l$. The parameter $g \in [-1, 1]$ is known as the anisotropy factor. When $g \approx 1$, we say that the scattering is in the forward direction. Likewise, when $g \approx -1$ the scattering is said to be in a backward direction. The case $g = 0$ corresponds to isotropic scattering. It follows from Eq. (20.4) that [see also (18.55)]

$$g = \int \hat{\mathbf{k}} \cdot \hat{\mathbf{k}}' p(\hat{\mathbf{k}}, \hat{\mathbf{k}}') d\hat{\mathbf{k}}' . \tag{20.5}$$

The Delta-Eddington phase function

$$p(\hat{\mathbf{k}}, \hat{\mathbf{k}}') = \frac{1 - g}{4\pi} + g\delta(\hat{\mathbf{k}} - \hat{\mathbf{k}}') \tag{20.6}$$

is obtained by adding a term that is strongly peaked in the forward direction to a constant background. It corresponds to putting $A_l = (1 - g)\delta_{l0} + g$.

For scattering that is strongly peaked in the forward direction, an expansion of the right-hand side of the RTE in angular moments may be performed. To proceed, we expand the specific intensity $I(\mathbf{r}, \hat{\mathbf{k}}')$ in $\hat{\mathbf{k}}'$ about $\hat{\mathbf{k}}' = \hat{\mathbf{k}}$ according to

$$I(\mathbf{r}, \hat{\mathbf{k}}') = I(\mathbf{r}, \hat{\mathbf{k}}) + (\hat{\mathbf{k}}' - \hat{\mathbf{k}}) \cdot \nabla_{\hat{\mathbf{k}}} I(\mathbf{r}, \hat{\mathbf{k}}) + \frac{1}{2} \left[(\hat{\mathbf{k}}' - \hat{\mathbf{k}}) \cdot \nabla_{\hat{\mathbf{k}}} \right]^2 I(\mathbf{r}, \hat{\mathbf{k}}) + \cdots . \tag{20.7}$$

Substituting the above into the right-hand side of (20.1), we obtain the Fokker–Planck form of the RTE

$$\hat{\mathbf{k}} \cdot \nabla I + \mu_e I = \frac{\mu_s}{2} (1 - g) \Delta_{\hat{\mathbf{k}}} I , \tag{20.8}$$

where $\Delta_{\hat{\mathbf{k}}}$ is the Laplacian on the two-dimensional unit sphere. Note that since $g \simeq 1$, the higher-order terms on the right-hand side of (20.1) can be neglected.

20.2 Boundary Conditions and Uniqueness

The boundary conditions for the RTE are typical of those for transport equations. That is, they are specified only on a portion of the boundary. The following notation

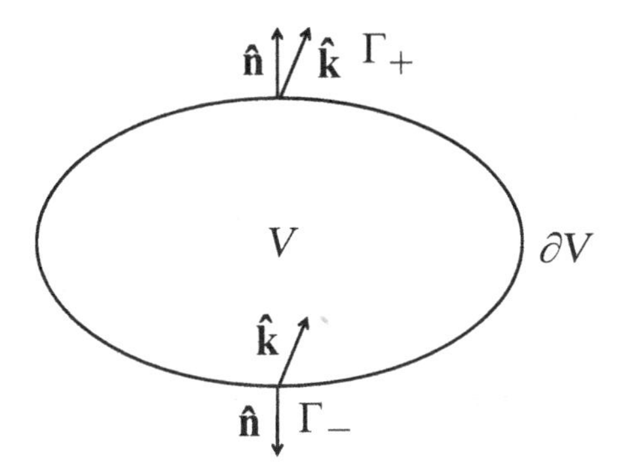

Fig. 20.1 Illustrating the incoming and outgoing boundaries.

will prove useful. Let V be a three-dimensional volume bounded by a surface ∂V with outward unit normal $\hat{\mathbf{n}}$. We define the ingoing and outgoing boundaries $\Gamma_\pm$ as

$$\Gamma_\pm = \{(\mathbf{r}, \hat{\mathbf{k}}) : \mathbf{r} \in \partial V \text{ and } \pm \hat{\mathbf{k}} \cdot \hat{\mathbf{n}} > 0\}, \tag{20.9}$$

as shown in Fig. 20.1. The RTE together with appropriate incoming boundary conditions is of the form

$$\hat{\mathbf{k}} \cdot \nabla I + \mu_e I - LI = S \quad \text{in} \quad V \times S^2, \tag{20.10}$$

$$I = f \quad \text{on} \quad \Gamma_-, \tag{20.11}$$

where S^2 is the two-dimensional unit sphere, f is a surface source and S is a volume source. The scattering operator L is defined by

$$LI(\mathbf{r}, \hat{\mathbf{k}}) = \mu_s \int p(\hat{\mathbf{k}}, \hat{\mathbf{k}}') I(\mathbf{r}, \hat{\mathbf{k}}') d\hat{\mathbf{k}}'. \tag{20.12}$$

We now show that the solution to the RTE subject to incoming boundary conditions is unique. To proceed, we suppose that I_1 and I_2 are solutions to (20.10) and obey the boundary conditions (20.11). Let $I = I_1 - I_2$. We claim that $I \equiv 0$, which implies that $I_1 = I_2$. To prove the claim, we first observe that I obeys

$$\hat{\mathbf{k}} \cdot \nabla I + \mu_e I - LI = 0 \quad \text{in} \quad V \times S^2, \tag{20.13}$$

$$I = 0 \quad \text{on} \quad \Gamma_-. \tag{20.14}$$

Next, we multiply both sides of (20.13) by I and integrate over $V \times S^2$. We thus obtain

$$\int_{V \times S^2} \left[\frac{1}{2} \nabla \cdot \left(\hat{\mathbf{k}} I^2 \right) + \mu_e I^2 - ILI \right] d^3 r d\hat{\mathbf{k}} = 0. \tag{20.15}$$

We now invoke the inequality

$$\int ILI \, d\hat{\mathbf{k}} d\hat{\mathbf{k}}' \leq \mu_s \int I^2 d\hat{\mathbf{k}}, \tag{20.16}$$

which follows from

$$\int d\hat{\mathbf{k}}d\hat{\mathbf{k}}' p(\hat{\mathbf{k}},\hat{\mathbf{k}}')\left[I(\mathbf{r},\hat{\mathbf{k}}) - I(\mathbf{r},\hat{\mathbf{k}}')\right]^2 \geq 0\,. \tag{20.17}$$

Equation (20.15) thus becomes

$$\int_{V\times S^2}\left[\frac{1}{2}\nabla\cdot\left(\hat{\mathbf{k}}I^2\right) + \mu_a I^2\right] d^3r d\hat{\mathbf{k}} \leq 0\,, \tag{20.18}$$

where $\mu_a = \mu_e - \mu_s$. Integrating by parts and applying the boundary condition (20.14), we obtain

$$\frac{1}{2}\int_{\Gamma_+} \hat{\mathrm{n}}\cdot\hat{\mathbf{k}}\, I^2 d^2r d\hat{\mathbf{k}} + \int \mu_a I^2 d^3r d\hat{\mathbf{k}} \leq 0\,. \tag{20.19}$$

Since $\hat{\mathrm{n}}\cdot\hat{\mathbf{k}} > 0$ in Γ_+ and μ_a is nonnegative, we see that both terms on the left-hand side of (20.19) are nonnegative. Thus $I \equiv 0$. This completes the proof.

20.3 Green's Functions and Integral Representations

Here we derive an integral representation for solutions to the RTE. We begin by defining the Green's function G for the RTE, which satisfies

$$\begin{aligned}\hat{\mathbf{k}}\cdot\nabla_{\mathbf{r}} G(\mathbf{r},\hat{\mathbf{k}};\mathbf{r}',\hat{\mathbf{k}}') + \mu_e G(\mathbf{r},\hat{\mathbf{k}};\mathbf{r}',\hat{\mathbf{k}}') - LG(\mathbf{r},\hat{\mathbf{k}};\mathbf{r}',\hat{\mathbf{k}}')\\ = \delta(\mathbf{r}-\mathbf{r}')\delta(\hat{\mathbf{k}}-\hat{\mathbf{k}}').\end{aligned} \tag{20.20}$$

The Green's function also obeys homogeneous incoming boundary conditions. Since the transport operator $\hat{\mathbf{k}}\cdot\nabla$ is not self-adjoint, it will prove useful to introduce the adjoint RTE

$$-\hat{\mathbf{k}}\cdot\nabla\tilde{I}(\mathbf{r},\hat{\mathbf{k}}) + \mu_e\tilde{I}(\mathbf{r},\hat{\mathbf{k}}) - L\tilde{I}(\mathbf{r},\hat{\mathbf{k}}) = \tilde{S}(\mathbf{r},\hat{\mathbf{k}}) \quad \text{in} \quad V\times S^2\,, \tag{20.21}$$

$$\tilde{I} = \tilde{f} \quad \text{on} \quad \Gamma_+\,. \tag{20.22}$$

We note that the above boundary condition is of the outgoing type.

We now obtain a useful identity. Multiplying (20.10) by $\tilde{I}$ and (20.21) by I, subtracting the results, and integrating, we find that

$$\int_{V\times S^2}\left(\tilde{I}\hat{\mathbf{k}}\cdot\nabla I + I\hat{\mathbf{k}}\cdot\nabla\tilde{I}\right) d^3r d\hat{\mathbf{k}} = \int_{V\times S^2}\left(S\tilde{I} - \tilde{S}I\right) d^3r d\hat{\mathbf{k}}\,. \tag{20.23}$$

Next, we note that the integrand on the left-hand side of (20.23) can be written as the divergence of the vector field $\hat{\mathbf{k}}I\tilde{I}$. Upon applying the divergence theorem, we obtain

$$\int_{\partial V\times S^2} \hat{\mathbf{k}}\cdot\hat{\mathbf{n}} I\tilde{I} d^2r d\hat{\mathbf{k}} = \int_{V\times S^2}\left(S\tilde{I} - \tilde{S}I\right) d^3r d\hat{\mathbf{k}}\,. \tag{20.24}$$

As an application of (20.24), we obtain an integral representation for solutions to the RTE. To proceed, we assume that $\tilde{f} = 0$ and $S(\mathbf{r}, -\hat{\mathbf{k}}) = \delta(\mathbf{r} - \mathbf{r}')\delta(\hat{\mathbf{k}} - \hat{\mathbf{k}}')$. It follows that $\tilde{I}(\mathbf{r}, -\hat{\mathbf{k}}) = G(\mathbf{r}, -\hat{\mathbf{k}}; \mathbf{r}', -\hat{\mathbf{k}}')$. Making using of (20.24), we find that

$$I(\mathbf{r}, \hat{\mathbf{k}}) = \int_{V \times S^2} G(\mathbf{r}', -\hat{\mathbf{k}}'; \mathbf{r}, -\hat{\mathbf{k}}) S(\mathbf{r}', \hat{\mathbf{k}}') d^3 r' d\hat{\mathbf{k}}' - \int_{\partial V \times S^2} \hat{\mathbf{k}} \cdot \hat{\mathrm{n}} I \tilde{I} d^2 r' d\hat{\mathbf{k}}' . \tag{20.25}$$

Applying the boundary condition (20.11) and the reciprocity relation (20.28), which is established below, we obtain

$$\begin{aligned} I(\mathbf{r}, \hat{\mathbf{k}}) = &\int_{V \times S^2} G(\mathbf{r}, \hat{\mathbf{k}}; \mathbf{r}', \hat{\mathbf{k}}') S(\mathbf{r}', \hat{\mathbf{k}}') d^3 r' d\hat{\mathbf{k}}' \\ &+ \int_{\Gamma_-} |\hat{\mathbf{k}} \cdot \hat{\mathrm{n}}| G(\mathbf{r}, \hat{\mathbf{k}}; \mathbf{r}', \hat{\mathbf{k}}') f(\mathbf{r}', \hat{\mathbf{k}}') d^2 r' d\hat{\mathbf{k}}' . \end{aligned} \tag{20.26}$$

The above result expresses the solution to the RTE in terms of the Green's function G and the surface and volume sources f and S. It is a basic result in radiative transport theory which is analogous to the Kirchoff integral formula (5.5).

20.4 Reciprocity

As a second application of the identity (20.24), we demonstrate the reciprocity of the Green's function for the RTE. Suppose that $f, \tilde{f} = 0$ and

$$S(\mathbf{r}, \hat{\mathbf{k}}) = \delta(\mathbf{r} - \mathbf{r}_1)\delta(\hat{\mathbf{k}} - \hat{\mathbf{k}}_1) , \quad \tilde{S}(\mathbf{r}, -\hat{\mathbf{k}}) = \delta(\mathbf{r} - \mathbf{r}_2)\delta(\hat{\mathbf{k}} + \hat{\mathbf{k}}_2) . \tag{20.27}$$

Then the right-hand side of (20.24) vanishes and, upon carrying out the integrations, we find that

$$G(\mathbf{r}_1, \hat{\mathbf{k}}_1; \mathbf{r}_2, \hat{\mathbf{k}}_2) = G(\mathbf{r}_2, -\hat{\mathbf{k}}_2; \mathbf{r}_1, -\hat{\mathbf{k}}_1) . \tag{20.28}$$

This result is the reciprocity theorem for the specific intensity created by directional point sources.

An additional reciprocity relation can be derived by putting $S, \tilde{S} = 0$ and

$$f(\mathbf{r}, \hat{\mathbf{k}}) = \delta(\mathbf{r} - \mathbf{r}_1)\delta(\hat{\mathbf{k}} - \hat{\mathbf{k}}_1) , \quad \tilde{f}(\mathbf{r}, \hat{\mathbf{k}}) = \delta(\mathbf{r} - \mathbf{r}_2)\delta(\hat{\mathbf{k}} - \hat{\mathbf{k}}_2) . \tag{20.29}$$

It follows that the left-hand side of (20.24) vanishes, and we find that

$$\hat{\mathrm{n}}(\mathbf{r}_1) \cdot \hat{\mathbf{k}}_1 G(\mathbf{r}_1, \hat{\mathbf{k}}_1; \mathbf{r}_2, \hat{\mathbf{k}}_2) = \hat{\mathrm{n}}(\mathbf{r}_2) \cdot \hat{\mathbf{k}}_2 G(\mathbf{r}_2, -\hat{\mathbf{k}}_2; \mathbf{r}_1, -\hat{\mathbf{k}}_1) . \tag{20.30}$$

This is the reciprocity theorem for the specific intensity created by directional and local illumination on the boundary of the medium.

References and Additional Reading

The radiative transport equation was historically introduced in astrophysics by Chandrasekhar. The historical phenomenological derivation of the RTE is presented in:

S. Chandrasekhar, *Radiative Transfer* (Dover, New York, 1960).

A similar equation, known as the one-speed transport equation, was introduced in the context of neutron transport. The theoretical framework is presented in:

K.M. Case and P.F. Zweifel, *Linear Transport Theory* (Addison-Wesley, Reading, 1967).

The parallel between the phenomenological approach to the radiative transport equation, and the wave approach based on multiple scattering theory, is highlighted in this book:

L.A. Apresyan and Yu.A. Kravtsov, *Radiative Transfer: Statistical and Wave Aspects* (Gordon and Breach, Amsterdam, 1996).

The following book chapter provides an illuminating presentation of the boundary value problem for the radiative transport equation:

K.M. Case, On Boundary Value Problems of Linear Transport Theory, in *Proceedings of the Symposium in Applied Mathematics*, vol. 1, R. Bellman, G. Birkhoff, and I. Abu-Shumays, eds. (American Mathematical Society, Providence, 1969), p. 17.

Reciprocity relations for the radiative transport equation were initially published in:

K.M. Case, Rev. Mod. Phys. **29**, 651 (1957).

21

Elementary Solutions of the Radiative Transport Equation

In this chapter, we consider solutions of the radiative transport equation (RTE) for infinite media. In particular, we study the problem of isotropic scattering in homogeneous media. We also develop a perturbative method for solving the RTE known as the collision expansion.

21.1 Ballistic Propagation

We recall from Chapter 20 that the time-independent RTE is of the form

$$\hat{\mathbf{k}} \cdot \nabla I + \mu_e I - \mu_s \int p(\hat{\mathbf{k}}, \hat{\mathbf{k}}') I(\mathbf{r}, \hat{\mathbf{k}}') d\hat{\mathbf{k}}' = S \, , \tag{21.1}$$

where $I(\mathbf{r}, \hat{\mathbf{k}})$ is the specific intensity at the position $\mathbf{r}$ in direction $\hat{\mathbf{k}}$. If the source term S is independent of the direction $\hat{\mathbf{k}}$, it is referred to as isotropic. If the phase function p is constant, then the RTE is said to describe isotropic scattering. The case $p = 0$, which corresponds to the absence of scattering, is referred to as ballistic propagation. The specific intensity in this case, denoted by I_0, is given by

$$I_0(\mathbf{r}, \hat{\mathbf{k}}) = \int G_0(\mathbf{r}, \hat{\mathbf{k}}; \mathbf{r}', \hat{\mathbf{k}}') S(\mathbf{r}', \hat{\mathbf{k}}') d^3r' d\hat{\mathbf{k}}' \, , \tag{21.2}$$

where the ballistic Green's function G_0 obeys the equation

$$\hat{\mathbf{k}} \cdot \nabla_{\mathbf{r}} G_0(\mathbf{r}, \hat{\mathbf{k}}; \mathbf{r}', \hat{\mathbf{k}}') + \mu_e G_0(\mathbf{r}, \hat{\mathbf{k}}; \mathbf{r}', \hat{\mathbf{k}}') = \delta(\mathbf{r} - \mathbf{r}') \delta(\hat{\mathbf{k}} - \hat{\mathbf{k}}') \, . \tag{21.3}$$

To find G_0 in an infinite homogeneous medium (μ_e is independent of position), we make use of translational invariance. We thus expand G_0 into plane waves of the form

$$G_0(\mathbf{r}, \hat{\mathbf{k}}; \mathbf{r}', \hat{\mathbf{k}}') = \int \frac{d^3q}{(2\pi)^3} \exp(i\mathbf{q} \cdot (\mathbf{r} - \mathbf{r}')) \tilde{G}_0(\mathbf{q}; \hat{\mathbf{k}}, \hat{\mathbf{k}}') \, . \tag{21.4}$$

Upon substitution of (21.4) into (21.3), we find that

$$\tilde{G}_0(\mathbf{q};\hat{\mathbf{k}},\hat{\mathbf{k}}') = \frac{1}{\mu_e + i\hat{\mathbf{k}}\cdot\mathbf{q}}\delta(\hat{\mathbf{k}}-\hat{\mathbf{k}}') . \tag{21.5}$$

We can make further progress by writing

$$\frac{1}{\mu_e + i\hat{\mathbf{k}}\cdot\mathbf{q}} = \int_0^\infty \exp(-(\mu_e + i\hat{\mathbf{k}}\cdot\mathbf{q})R)dR . \tag{21.6}$$

Using the above result and carrying out the Fourier transform in (21.4), we obtain

$$G_0(\mathbf{r},\hat{\mathbf{k}};\mathbf{r}',\hat{\mathbf{k}}') = \frac{\exp(-\mu_e|\mathbf{r}-\mathbf{r}'|)}{|\mathbf{r}-\mathbf{r}'|^2}\delta\left(\hat{\mathbf{k}}' - \frac{\mathbf{r}-\mathbf{r}'}{|\mathbf{r}-\mathbf{r}'|}\right)\delta(\hat{\mathbf{k}}-\hat{\mathbf{k}}') . \tag{21.7}$$

Note that, as expected, propagation occurs only along the line connecting $\mathbf{r}$ and $\mathbf{r}'$.

It is instructive to consider the case of an isotropic point source of the form $S = \delta(\mathbf{r}-\mathbf{r}_0)$, where $\mathbf{r}_0$ is the position of the source. Making use of (21.2), we see that the corresponding specific intensity is given by

$$I_0(\mathbf{r},\hat{\mathbf{k}}) = \frac{\exp(-\mu_e|\mathbf{r}-\mathbf{r}_0|)}{|\mathbf{r}-\mathbf{r}_0|^2}\delta\left(\hat{\mathbf{k}} - \frac{\mathbf{r}-\mathbf{r}_0}{|\mathbf{r}-\mathbf{r}_0|}\right) . \tag{21.8}$$

Thus the energy density u_0 is

$$\begin{aligned} u_0(\mathbf{r}) &= \frac{1}{v_E}\int I_0(\mathbf{r},\hat{\mathbf{k}})d\hat{\mathbf{k}} \\ &= \frac{1}{v_E}\frac{\exp(-\mu_e|\mathbf{r}-\mathbf{r}_0|)}{|\mathbf{r}-\mathbf{r}_0|^2} , \end{aligned} \tag{21.9}$$

which has the expected $1/r^2$ decay.

21.2 Collision Expansion

We consider an infinite inhomogeneous medium, where μ_s and μ_e are both generally position dependent. The solution to the RTE (21.1) obeys the integral equation

$$I(\mathbf{r},\hat{\mathbf{k}}) = I_0(\mathbf{r},\hat{\mathbf{k}}) + \int d^3r'd\hat{\mathbf{k}}'d\hat{\mathbf{k}}''G_0(\mathbf{r},\hat{\mathbf{k}};\mathbf{r}',\hat{\mathbf{k}}')\mu_s(\mathbf{r}')p(\hat{\mathbf{k}}',\hat{\mathbf{k}}'')I(\mathbf{r}',\hat{\mathbf{k}}'') , \tag{21.10}$$

where the unscattered (ballistic) specific intensity I_0 is given by (21.2). Here the Green function G_0 obeys (21.3). Following the previous development, it is readily seen that G_0 is given by

$$G_0(\mathbf{r},\hat{\mathbf{k}};\mathbf{r}',\hat{\mathbf{k}}') = g(\mathbf{r},\mathbf{r}')\delta\left(\hat{\mathbf{k}}' - \frac{\mathbf{r}-\mathbf{r}'}{|\mathbf{r}-\mathbf{r}'|}\right)\delta(\hat{\mathbf{k}}-\hat{\mathbf{k}}') , \tag{21.11}$$

where

$$g(\mathbf{r},\mathbf{r}') = \frac{1}{|\mathbf{r}-\mathbf{r}'|^2}\exp\left[-\int_0^{|\mathbf{r}-\mathbf{r}'|}\mu_e\left(\mathbf{r}'+\ell\frac{\mathbf{r}-\mathbf{r}'}{|\mathbf{r}-\mathbf{r}'|}\right)d\ell\right] , \tag{21.12}$$

Note that if a narrow collimated beam of unit intensity is incident on the medium at the point $\mathbf{r}_0$ in the direction $\hat{\mathbf{k}}_0$, then $I_0(\mathbf{r},\hat{\mathbf{k}})$ is given by

$$I_0(\mathbf{r},\hat{\mathbf{k}}) = G_0(\mathbf{r},\hat{\mathbf{k}};\mathbf{r}_0,\hat{\mathbf{k}}_0) . \tag{21.13}$$

To derive the collision expansion, we iterate (21.10) starting from $I = I_0$ and thus obtain

$$I(\mathbf{r},\hat{\mathbf{k}}) = I_0(\mathbf{r},\hat{\mathbf{k}}) + I_1(\mathbf{r},\hat{\mathbf{k}}) + I_2(\mathbf{r},\hat{\mathbf{k}}) + \cdots , \tag{21.14}$$

where each term of the series is given by

$$I_n(\mathbf{r},\hat{\mathbf{k}}) = \int d^3r'd\hat{\mathbf{k}}'d\hat{\mathbf{k}}''G_0(\mathbf{r},\hat{\mathbf{k}};\mathbf{r}',\hat{\mathbf{k}}')\mu_s(\mathbf{r}')p(\hat{\mathbf{k}}',\hat{\mathbf{k}}'')I_{n-1}(\mathbf{r}',\hat{\mathbf{k}}'') , \tag{21.15}$$

with $n = 1, 2, \ldots$. The above series is the analog of the Born series for the RTE. Indeed, each term accounts for successively higher orders of scattering. We note that the specific intensity is necessarily nonnegative, since each term in the series is nonnegative, provided that the series converges.

It is instructive to examine the expression for I_1, which is the contribution to the specific intensity from single scattering. We obtain from (21.15) and (21.13) that I_1 is given by

$$I_1(\mathbf{r},\hat{\mathbf{k}}) = \int d^3r'd\hat{\mathbf{k}}'d\hat{\mathbf{k}}''G_0(\mathbf{r},\hat{\mathbf{k}};\mathbf{r}',\hat{\mathbf{k}}')\mu_s(\mathbf{r}')p(\hat{\mathbf{k}}',\hat{\mathbf{k}}'')I_0(\mathbf{r}',\hat{\mathbf{k}}'') . \tag{21.16}$$

Carrying out the above integral, we obtain

$$\begin{aligned} I_1(\mathbf{r},\hat{\mathbf{k}}) = p(\hat{\mathbf{k}}_0,\hat{\mathbf{k}})\int_0^\infty dRR^2 g(\mathbf{r},\mathbf{r}_0+R\hat{\mathbf{k}}_0)g(\mathbf{r}_0+R\hat{\mathbf{k}}_0,\mathbf{r}_0) \\ \times\delta\left(\frac{\mathbf{r}-\mathbf{r}_0-R\hat{\mathbf{k}}_0}{|\mathbf{r}-\mathbf{r}_0-R\hat{\mathbf{k}}_0|}-\hat{\mathbf{k}}\right)\mu_s(\mathbf{r}_0+R\hat{\mathbf{k}}_0) . \end{aligned} \tag{21.17}$$

Using (21.17), the energy density u_1 is given by

$$\begin{aligned} u_1(\mathbf{r}) &= \frac{1}{v_E}\int I_1(\mathbf{r},\hat{\mathbf{k}})d\hat{\mathbf{k}} \\ &= \frac{1}{v_E}p\left(\hat{\mathbf{k}}_0,\frac{\mathbf{r}-\mathbf{r}_0-R\hat{\mathbf{k}}_0}{|\mathbf{r}-\mathbf{r}_0-R\hat{\mathbf{k}}_0|}\right)\int_0^\infty dRR^2 g(\mathbf{r},\mathbf{r}_0+R\hat{\mathbf{k}}_0)g(\mathbf{r}_0+R\hat{\mathbf{k}}_0,\mathbf{r}_0) \\ &\quad\times\mu_s(\mathbf{r}_0+R\hat{\mathbf{k}}_0) . \end{aligned} \tag{21.18}$$

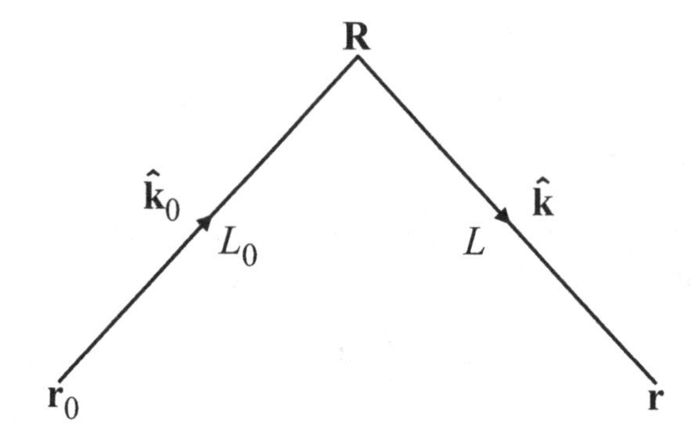

Fig. 21.1 Illustrating the geometry of the broken ray.

The integral in (21.17) cannot be evaluated in general. However, in special geometries, it can be computed explicitly. For the case of isotropic scattering in a homogeneous medium, we find that

$$I_1(\mathbf{r}, \mathbf{k}) = \frac{\mu_s}{4\pi v_E |\mathbf{r} - \mathbf{r}_0|} \exp\left[- \int_0^{L_0} \mu_e(\mathbf{r}_0 + t\hat{\mathbf{k}}_0)dt - \int_0^{L} \mu_e(\mathbf{R} + t\hat{\mathbf{k}})dt \right] . \tag{21.19}$$

Here we have assumed, without loss of generality, that $\mathbf{k}_0$, $\mathbf{k}$ and $\mathbf{r} - \mathbf{r}_0$ lie in the same plane, $\mathbf{R}$ denoting the point of intersection of the rays in the directions $\mathbf{k}_0$ and $\mathbf{k}$, $L_0 = |\mathbf{R} - \mathbf{r}_0|$, and $L = |\mathbf{R} - \mathbf{r}|$. Note that the argument of the exponential corresponds to the integral of μ_e along the broken ray that begins at $\mathbf{r}_0$, passes through $\mathbf{R}$, and terminates at $\mathbf{r}$, as shown in Fig. 21.1.

The terms in the collision expansion can be classified by their smoothness. The ballistic term I_0 is the most singular; there are two angular delta functions in (21.13). The first-order term is less singular, since there is only a single delta function in (21.17). Terms of higher order are of increasing smoothness.

21.3 Isotropic Scattering

We now turn to the problem of isotropic scattering in an infinite homogeneous medium. In this situation, the phase function $p = 1/(4\pi)$ and the Green's function G for the RTE satisfies the equation

$$\hat{\mathbf{k}} \cdot \nabla_{\mathbf{r}} G(\mathbf{r}, \hat{\mathbf{k}}; \mathbf{r}', \hat{\mathbf{k}}') + \mu_e G(\mathbf{r}, \hat{\mathbf{k}}; \mathbf{r}', \hat{\mathbf{k}}') - \frac{\mu_s}{4\pi} \int d\hat{\mathbf{k}}'' G(\mathbf{r}, \hat{\mathbf{k}}''; \mathbf{r}', \hat{\mathbf{k}}') = \delta(\mathbf{r} - \mathbf{r}')\delta(\hat{\mathbf{k}} - \hat{\mathbf{k}}'), \tag{21.20}$$

where μ_s and μ_e are independent of position. It will prove useful to expand G into plane waves of the form

$$G(\mathbf{r}, \hat{\mathbf{k}}; \mathbf{r}', \hat{\mathbf{k}}') = \int \frac{d^3q}{(2\pi)^3} \exp(i\mathbf{q} \cdot (\mathbf{r} - \mathbf{r}'))G(\mathbf{q}; \hat{\mathbf{k}}, \hat{\mathbf{k}}') . \tag{21.21}$$

Substituting (21.21) into (21.20), we find that

$$G(\mathbf{q}; \hat{\mathbf{k}}, \hat{\mathbf{k}}') = \frac{\delta(\hat{\mathbf{k}} - \hat{\mathbf{k}}')}{\mu_e + i\hat{\mathbf{k}} \cdot \mathbf{q}} + \frac{\mu_s}{4\pi} \frac{g(\mathbf{q}, \hat{\mathbf{k}})}{\mu_e + i\hat{\mathbf{k}}' \cdot \mathbf{q}} , \tag{21.22}$$

where

$$g(\mathbf{q}, \hat{\mathbf{k}}) = \int d\hat{\mathbf{k}}' G(\mathbf{q}; \hat{\mathbf{k}}, \hat{\mathbf{k}}') . \tag{21.23}$$

If we integrate (21.22) with respect to $\hat{\mathbf{k}}'$ and use the result

$$\int \frac{d\hat{\mathbf{k}}}{\mu_e + i\hat{\mathbf{k}} \cdot \mathbf{q}} = \frac{4\pi}{k} \tan^{-1}\left(\frac{q}{\mu_e}\right) , \tag{21.24}$$

we see that g is given by

$$g(\mathbf{q}, \hat{\mathbf{k}}) = \frac{1}{\left(\mu_e + i\hat{\mathbf{k}} \cdot \mathbf{q}\right)\left[1 - \frac{\mu_s}{q} \tan^{-1}\left(\frac{q}{\mu_e}\right)\right]} . \tag{21.25}$$

Making use of the above result, along with (21.22) and (21.4), we find that the Green function is of the form

$$G(\mathbf{r}, \hat{\mathbf{k}}; \mathbf{r}', \hat{\mathbf{k}}') = G_0(\mathbf{r}, \hat{\mathbf{k}}; \mathbf{r}', \hat{\mathbf{k}}') + G_1(\mathbf{r}, \hat{\mathbf{k}}; \mathbf{r}', \hat{\mathbf{k}}') , \tag{21.26}$$

where

$$\begin{aligned} G_1(\mathbf{r}, \hat{\mathbf{k}}; \mathbf{r}', \hat{\mathbf{k}}') = {} & \frac{\mu_s}{4\pi} \int \frac{d^3q}{(2\pi)^3} \exp(i\mathbf{q} \cdot (\mathbf{r} - \mathbf{r}')) \\ & \times \frac{1}{\left(\mu_e + i\hat{\mathbf{k}} \cdot \mathbf{q}\right)\left(\mu_e + i\hat{\mathbf{k}}' \cdot \mathbf{q}\right)\left[1 - \frac{\mu_s}{q} \tan^{-1}\left(\frac{q}{\mu_e}\right)\right]} \end{aligned} \tag{21.27}$$

and G_0 is the ballistic Green function given by (21.11).

The expression (21.26) for the Green's function is rather formidable. A considerably simpler formula for the energy density may be obtained. It follows immediately from (21.7) and (21.10) that the energy density

$$u(\mathbf{r}) = \frac{\mathbf{1}}{\mathbf{v_E}} \int \mathbf{I}(\mathbf{r}, \hat{\mathbf{k}}) \mathbf{d}\hat{\mathbf{k}} \tag{21.28}$$

obeys the integral equation

$$u(\mathbf{r}) = u_0(\mathbf{r}) + \int d^3r' K(\mathbf{r} - \mathbf{r}') u(\mathbf{r}') , \tag{21.29}$$

where

$$K(\mathbf{r} - \mathbf{r}') = \frac{\mu_s}{4\pi} \frac{\exp(-\mu_e|\mathbf{r} - \mathbf{r}'|)}{|\mathbf{r} - \mathbf{r}'|^2} \tag{21.30}$$

$$= \mu_s \int \frac{d^3q}{(2\pi)^3} \exp(i\mathbf{q} \cdot \mathbf{r}) \frac{\tan^{-1}\left(\frac{q}{\mu_e}\right)}{q} \tag{21.31}$$

and

$$u_0(\mathbf{r}) = \frac{1}{v_E} \int d\hat{\mathbf{k}} I_0(\mathbf{r}, \hat{\mathbf{k}}) \; . \tag{21.32}$$

If we substitute the series

$$\frac{\tan^{-1}(x)}{x} = 1 - \frac{x^2}{3} + \cdots \tag{21.33}$$

into (21.30), we obtain

$$K(\mathbf{r}) = \frac{\mu_s}{\mu_e} \left(1 + \frac{1}{3\mu_e^2} \nabla^2\right) \delta(\mathbf{r}) \; . \tag{21.34}$$

Using this result, we find that (21.29) becomes

$$-\frac{1}{3\mu_e} \nabla^2 u + \mu_a u = S \; , \tag{21.35}$$

where $S = \mu_e u_0$ and we have assumed that $\mu_a \ll \mu_s$. Thus the energy density obeys a diffusion equation with an absorption term. It is readily seen that the solution to (21.35) for a point source is given by

$$u(\mathbf{r}) = \frac{\exp(-|\mathbf{r} - \mathbf{r}_0|/\ell_d)}{4\pi |\mathbf{r} - \mathbf{r}_0|} \; , \tag{21.36}$$

where $\ell_d = 1/\sqrt{3\mu_a \mu_e}$ is the diffusion length, and the source is taken to have unit amplitude and is located at the point $\mathbf{r}_0$. In Chapter 24, we will show that the energy density obeys a diffusion equation even when the scattering is anisotropic.

References and Additional Reading

An extensive presentation of elementary solutions of the RTE can be found in:

S. Chandrasekhar, *Radiative Transfer* (Dover, New York, 1960).

A. Peraiah, *An Introduction to Radiative Transfer* (Cambridge University Press, Cambridge, 2002).

Exact solutions of the RTE (referred to as the one-speed transport equation) in special cases, including isotropic scattering, are discussed in:

K.M. Case and P.F. Zweifel, *Linear Transport Theory* (Addison-Wesley, Reading, 1967), chap. 5 and 6.

J.J. Duderstadt and W.R. Martin, *Transport Theory* (Wiley, New York, 1979), chap. 2.

The following book contains a description of exact and approximate methods for solving the RTE:

L.A. Apresyan and Yu.A. Kravtsov, *Radiative Transfer: Statistical and Wave Aspects* (Gordon and Breach, Amsterdam, 1996), chap. 4.

The following book provides a clear presentation of methods to solve the RTE, including numerical methods:

G.E. Thomas and K. Stamnes, *Radiative Transfer in the Atmosphere and Ocean* (Cambridge University Press, Cambridge, 1999).

The following papers discuss single-scattering in radiative transport theory in the context of the broken-ray transform:

L. Florescu, J. Schotland and V. Markel Phys. Rev. E **79**, 036607 (2009).

L. Florescu, V. Markel and J. Schotland Inverse Probl. **27**, 025002 (2011).

22

Problems with Planar and Azimuthal Symmetry

In this chapter, we consider the problem of radiative transport with planar and azimuthal symmetry. In this setting, the radiative transport equation (RTE) becomes spatially and angularly one-dimensional, and finding its solutions can be formulated as a generalized eigenproblem. The corresponding eigenfunctions are singular and can be used to calculate the Green function.

22.1 Singular Eigenfunctions

The time-independent RTE in an infinite homogeneous medium is of the form

$$\hat{\mathbf{k}} \cdot \nabla I + \mu_e I = \mu_s \int p(\hat{\mathbf{k}}, \hat{\mathbf{k}}')I(\mathbf{r}, \hat{\mathbf{k}}')d\hat{\mathbf{k}}' , \tag{22.1}$$

which holds in the absence of a source. In a system with planar symmetry, the specific intensity is also azimuthally symmetric (invariant under rotations about the normal direction). Thus the RTE becomes

$$\mu \frac{\partial I}{\partial x} + \mu_e I = 2\pi \mu_s \int_{-1}^{1} p(\mu, \mu')I(x, \mu')d\mu', \tag{22.2}$$

where x is the coordinate along the normal direction and $\mu = \cos\theta$, with θ the angle between $\hat{\mathbf{k}}$ and the x-axis. Here

$$p(\mu, \mu') = \frac{1}{2\pi} \int_0^{2\pi} p(\mu, \phi; \mu', \phi')d\phi' , \tag{22.3}$$

where ϕ and ϕ' are azimuthal angles. If the medium is isotropically scattering, then $p = 1/(4\pi)$ and (22.2) takes the form of the one-dimensional RTE

$$\mu \frac{\partial I}{\partial x} + \mu_e I = \frac{\mu_s}{2} \int_{-1}^{1} I(x, \mu)d\mu . \tag{22.4}$$

It will prove useful to nondimensionalize (22.4) by taking the unit of length to be $1/\mu_e$. Equation (22.4) thus becomes

$$\mu\frac{\partial I}{\partial x} + I = \frac{a}{2}\int_{-1}^{1} I(x,\mu)d\mu \ . \tag{22.5}$$

Here the albedo $a = \mu_s/\mu_e = \mu_s/(\mu_a+\mu_s)$ is a measure of the "whiteness" of the medium. We note that $0 \le a \le 1$ and $a = 1$ for a nonabsorbing medium, which corresponds to $\mu_a = 0$.

To solve (22.5), we make the ansatz

$$I(x,\mu) = \exp(-x/\nu)\phi(\mu) \ , \tag{22.6}$$

where ν is to be determined. Substituting (22.6) into (22.5), we find that ϕ obeys the integral equation

$$(\nu-\mu)\phi(\mu) = \frac{a\nu}{2}\int_{-1}^{1}\phi(\mu)d\mu \ . \tag{22.7}$$

Equation (22.7) can be viewed as a generalized eigenproblem of the form

$$A\phi = \nu B\phi \ , \tag{22.8}$$

where the operators A and B are defined by

$$A\phi = \mu\phi \ , \quad B\phi = \phi - \frac{a}{2}\int_{-1}^{1}\phi(\mu)d\mu \ . \tag{22.9}$$

Here ϕ is a generalized eigenfunction and ν is the corresponding eigenvalue. It is easily seen that A and B are noncommuting self-adjoint operators on $L^2([-1,1])$.

Since (22.7) is invariant under rescaling of ϕ, it follows that we can choose any convenient normalization of ϕ, and in particular we will set

$$\int_{-1}^{1}\phi(\mu)d\mu = 1 \ . \tag{22.10}$$

Thus (22.7) becomes

$$(\nu-\mu)\phi(\mu) = \frac{a\nu}{2} \ . \tag{22.11}$$

There are two cases to consider in solving (22.11) for ϕ. First, suppose that $\nu \notin [-1,1]$. Then

$$\phi(\mu) = \frac{a\nu}{2}\frac{1}{\nu-\mu} \ . \tag{22.12}$$

To obtain the eigenvalue ν, we integrate (22.12) and use the normalization condition (22.10). We find that $\Lambda(\nu) = 0$, where

$$\Lambda(\nu) = 1 - a\nu\tanh^{-1}\left(\frac{1}{\nu}\right) \ . \tag{22.13}$$

By using the fact that

$$\tanh^{-1}(x) = x + \frac{x^3}{3} + \frac{x^5}{5} + \cdots , \tag{22.14}$$

it is easily seen that there are two roots of the form $\nu_\pm = \pm\nu_0$, where for $a < 1$, ν_0 is given by

$$\nu_0 = \frac{1}{\sqrt{3(1-a)}}\left[1 + \frac{2}{5}(1-a) + O\left((1-a)^3\right)\right] . \tag{22.15}$$

We will refer to $\nu_\pm$ as discrete eigenvalues. The corresponding eigenfunctions are given by

$$\phi_\pm(\mu) = \frac{a\nu_\pm}{2}\frac{1}{\nu_\pm - \mu} . \tag{22.16}$$

Next, we suppose that $\nu \in [-1, 1]$. In this case, we must allow for the fact that μ and ν can be equal. To proceed, we make the ansatz

$$\phi(\mu) = \frac{a\nu}{2}P\frac{1}{\nu - \mu} + \lambda(\nu)\delta(\mu - \nu) , \tag{22.17}$$

where the principal value is introduced to handle the singularity at $\mu = \nu$ and λ is to be determined. It is readily verified that the solution (22.17) solves (22.11) by direct calculation and recalling the identity $x\delta(x) = 0$. To find λ, we integrate (22.17) and use (22.10). We thus obtain

$$\frac{a\nu}{2}P\int_{-1}^{1}\frac{d\mu}{\nu - \mu} + \lambda(\nu) = 1 . \tag{22.18}$$

Carrying out the above integration and comparing to (22.13), we find that $\lambda(\nu) = \Lambda(\nu)$. Thus we see that there is a continuous spectrum of eigenvalues $\nu \in [-1, 1]$, with corresponding eigenfunctions ϕ_ν given by

$$\phi_\nu(\mu) = \frac{a\nu}{2}P\frac{1}{\nu - \mu} + \Lambda(\nu)\delta(\mu - \nu) . \tag{22.19}$$

It is important to note that ϕ_ν is singular and is not an L^2-function; ϕ_ν is thus referred to as a singular eigenfunction.

The eigenfunctions ϕ satisfying (22.7) are orthogonal. To see this, consider two eigenfunctions ϕ, ϕ' with corresponding eigenvalues ν, ν'. We then have

$$\langle\phi', A\phi\rangle = \nu\langle\phi', B\phi\rangle , \tag{22.20}$$

$$\langle\phi', A\phi'\rangle = \nu'\langle\phi, B\phi'\rangle , \tag{22.21}$$

where the inner product is defined by

$$\langle\phi, \phi'\rangle = \int_{-1}^{1}\phi(\mu)\phi'(\mu)d\mu . \tag{22.22}$$

Since A and B are self-adjoint, upon subtraction of the above relations, we obtain

$$\left(\frac{1}{\nu} - \frac{1}{\nu'}\right) \langle A\phi, \phi' \rangle = 0 . \tag{22.23}$$

Using the fact that A corresponds to multiplication by μ, it follows that eigenfunctions corresponding to distinct eigenvalues are orthogonal under the weighted L^2 inner product

$$\left(\phi, \phi'\right) = \int_{-1}^{1} \phi(\mu)\phi'(\mu)\mu d\mu . \tag{22.24}$$

It is easily seen that the discrete eigenfunctions $\phi_\pm$ are normalizable according to

$$(\phi_\pm, \phi_\pm) = N_\pm , \tag{22.25}$$

where

$$N_\pm = \frac{a\nu_\pm^3}{2}\left(\frac{a}{\nu_\pm^2 - 1} - \frac{1}{\nu_\pm^2}\right) . \tag{22.26}$$

The singular eigenfunctions are not normalizable. However, by making use of the Poincaré-Bertrand formula

$$P\frac{1}{\nu-\mu}P\frac{1}{\nu'-\mu} = \frac{1}{\nu-\nu'}\left(P\frac{1}{\nu'-\mu} - P\frac{1}{\nu-\mu}\right) + \pi^2\delta(\mu-\nu)\delta(\mu-\nu') , \tag{22.27}$$

the singular eigenfunctions can be seen to be normalizable in a formal sense. That is,

$$(\phi_\nu, \phi_\nu) = N(\nu) , \tag{22.28}$$

where

$$N(\nu) = \nu\left[\left(\frac{\pi a \nu}{2}\right)^2 + \left(1 - a\nu \tanh^{-1}\nu\right)^2\right] . \tag{22.29}$$

In summary, we have shown that the eigenfunctions obey the orthogonality condition

$$\left(\phi_\nu, \phi'_\nu\right) = \mathcal{N}(\nu)\delta_{\nu\nu'} , \tag{22.30}$$

where

$$\mathcal{N}(\nu) = \begin{cases} N_\pm & \text{if } \nu = \pm , \\ N(\nu) & \text{if } \nu \in [-1, 1] . \end{cases} \tag{22.31}$$

Note that here $\delta_{\nu\nu'}$ denotes the Kronecker delta when ν and ν' belong to the discrete spectrum and the Dirac delta when ν and ν' belong to the continuous spectrum.

We can now expand an arbitrary function $f(\mu)$ in discrete and continuous eigenfunctions of the form

$$f(\mu) = A_+\phi_+(\mu) + A_-\phi_-(\mu) + \int_{-1}^{1} A(\nu)\phi_\nu(\mu)d\nu \ . \tag{22.32}$$

Here the coefficients $A_\pm$ and $A(\nu)$, which can be obtained from the orthogonality condition (22.30), are given by

$$A_\pm = \frac{1}{N_\pm}(\phi_\pm, f) \ , \quad A(\nu) = \frac{1}{N(\nu)}(\phi_\nu, f) \ . \tag{22.33}$$

It is important to note that the expansion (22.32) must be viewed as formal. That is, we have not established a completeness relation for the ϕ_ν; nor have we specified the function space in which such an expansion can be justified.

22.2 Green's Function

In this section, we apply the theory of singular eigenfunctions to deduce the Green function for the RTE (22.5). The Green function $G(x, \mu; x', \mu')$ obeys the equation

$$\mu\frac{\partial}{\partial x}G(x, \mu; x', \mu') + G(x, \mu; x', \mu') = \frac{a}{2}\int_{-1}^{1} G(x, \mu; x', \mu')d\mu' + \delta(x-x')\delta(\mu-\mu') \ . \tag{22.34}$$

It also obeys the boundary conditions

$$\lim_{|x|\to\infty} G(x, \mu; x', \mu') = 0 \ , \quad \lim_{|x'|\to\infty} G(x, \mu; x', \mu') = 0 \tag{22.35}$$

and the jump condition

$$\mu G(x' + \epsilon, \mu; x', \mu') - \mu G(x' - \epsilon, \mu; x', \mu') = \delta(\mu - \mu') \ , \tag{22.36}$$

which is obtained by integrating (22.5) with respect to x over the interval $[x' - \epsilon, x'+\epsilon]$, where ϵ is a positive infinitesimal. The Green function can now be written as an expansion into modes of the form (22.6), obeying the boundary conditions (22.35). For $x > x'$, we have

$$G(x, \mu; x', \mu') = A_+ \exp(-x/\nu_+)\phi_+(\mu) + \int_0^1 A(\nu) \exp(-x/\nu)\phi_\nu(\mu)d\nu \ . \tag{22.37}$$

Likewise, for $x < x'$, we have

$$G(x, \mu; x', \mu') = -A_- \exp(-x/\nu_-)\phi_-(\mu) - \int_{-1}^{0} A(\nu) \exp(-x/\nu)\phi_\nu(\mu)d\nu \ . \tag{22.38}$$

The coefficients $A_\pm$ and $A(\nu)$ are then obtained by applying the jump condition (22.36) in the form

$$A_+ \exp(-x'/\nu_+)\phi_+(\mu) + A_- \exp(-x'/\nu_-)\phi_-(\mu) \tag{22.39}$$

$$+\int_{-1}^{1} A(\nu)\exp(-x'/\nu)\phi_\nu(\mu)d\nu = \frac{1}{\mu}\delta(\mu-\mu') . \tag{22.40}$$

Multiplying both sides of the above relation by $\phi_\nu(\mu)$ and integrating with respect to μ, we find upon applying the orthogonality condition (22.30) that

$$A_\pm = \frac{1}{N_\pm}\exp(x'/\nu_\pm)\phi_\pm(\mu') , \quad A(\nu) = \frac{1}{N(\nu)}\exp(x'/\nu)\phi_\nu(\mu') . \tag{22.41}$$

Putting everything together, we see that the Green's function is given by

$$G(x,\mu;x',\mu') = \frac{1}{N_0}\exp(-|x-x'|/\nu_0)\phi_0(\mu)\phi_0(\mu')$$
$$+\int_0^1 \frac{1}{N(\nu)}\exp(-|x-x'|/\nu)\phi_\nu(\mu)\phi_\nu(\mu')d\nu , \tag{22.42}$$

where

$$\phi_0(\mu) = \frac{a\nu_0}{2}\frac{1}{\nu_0-\mu} , \tag{22.43}$$

$$N_0 = \frac{a\nu_0^3}{2}\left(\frac{a}{\nu_0^2-1}-\frac{1}{\nu_0^2}\right) . \tag{22.44}$$

22.3 Diffusion Approximation

Consider an isotropic point source at the origin. Then the energy density u is given by

$$u(x) = \int G(x,\mu;0,\mu')d\mu d\mu' . \tag{22.45}$$

If $|x| \gg 1$, it follows from (22.42) and (22.10) that

$$u(x) \simeq \frac{1}{N_0}\exp(-|x|/\nu_0) , \tag{22.46}$$

which corresponds to the diffusion approximation discussed in Chapter 24. Thus, far from the source, the contribution from the continuous spectrum can be neglected and the diffusion approximation is obtained solely from the discrete mode.

References and Additional Reading

The RTE in planar and spherical geometries is discussed in detail in:
S. Chandrasekhar, *Radiative Transfer* (Dover, New York, 1960).

A comprehensive presentation of singular eigenfunctions can be found in:
K.M. Case and P.F. Zweifel, *Linear Transport Theory* (Addison-Wesley, Reading, 1967).

The singular eigenfunctions approach is also introduced in:
J.J. Duderstadt and W.R. Martin, *Transport Theory* (Wiley, New York, 1979).

Illuminating papers on singular eigenfunctions include:
K.M. Case, Annals Phys. **9**, 1 (1960).
D.H. Sattinger, J. Math. Analysis Applications **15**, 497 (1966).
N.J. McCormack and I. Kuscer in *Advances in Nuclear Science and Technology*, edited by E. Henley and J. Lewins (Academic Press, New York, 1973).
B.D. Ganapol, Nucl. Sci. Eng. **137**, 400 (2001).

The following papers present the construction of Green's functions and the extension of the singular eigenfunction method to three-dimensional problems:
G. Panasyuk, J. Schotland and V. Markel, J. Phys. A **39**, 115 (2006)
M. Machida, G. Panasyuk, J. Schotland and V. Markel, J. Phys. A **43**, 65402 (2010).
M. Machida, J. Opt. Soc. Am. A **31**, 67 (2014).
M. Machida, J. Computat. Theoretical Transport **45**, 594 (2016).
M. Machida, J. Phys. A **49**, 175001 (2016).

The singular eigenfunction approach can be extended to time-dependent problems in order to introduce the diffusion approximation. See:
R. Pierrat, J.-J. Greffet and R. Carminati, J. Opt. Soc. Am. A **23**, 1106 (2006).

23

Scattering Theory for the Radiative Transport Equation

In this chapter, we develop a scattering theory appropriate to the radiative transport equation (RTE). The theory is illustrated for the case of a pair of point absorbers.[1]

23.1 Integral Equations

In chapter 21, the collision expansion for the RTE was introduced. This expansion accounts for successively higher orders of scattering, beginning with the unscattered (ballistic) solution to the RTE. Alternatively, it is possible to construct a perturbative expansion, beginning with the solution to the RTE in a homogeneous medium. To proceed, we recall that in a volume V the RTE takes the form

$$\hat{\mathbf{k}} \cdot \nabla I + \mu_a I - LI = S, \tag{23.1}$$

where the scattering operator L is defined by

$$LI(\mathbf{r}, \hat{\mathbf{k}}) = \mu_s \int p(\hat{\mathbf{k}}, \hat{\mathbf{k}}') I(\mathbf{r}, \hat{\mathbf{k}}') d\hat{\mathbf{k}}' - \mu_s I \,. \tag{23.2}$$

We also impose the boundary condition $I = 0$ for $\hat{\mathbf{k}} \cdot \hat{\mathbf{n}} < 0$, where $\hat{\mathbf{n}}$ is the outward unit normal on the surface ∂V enclosing the volume V. Thus no light enters V except due to the source S. The solution to (23.1) is given by

$$I(\mathbf{r}, \hat{\mathbf{k}}) = \int d^3r' d\hat{\mathbf{k}}' G(\mathbf{r}, \hat{\mathbf{k}}; \mathbf{r}', \hat{\mathbf{k}}') S(\mathbf{r}', \hat{\mathbf{k}}') \,. \tag{23.3}$$

Here the Green's function G satisfies the equation

$$\hat{\mathbf{k}} \cdot \nabla_{\mathbf{r}} G(\mathbf{r}, \hat{\mathbf{k}}; \mathbf{r}', \hat{\mathbf{k}}') + \mu_a G(\mathbf{r}, \hat{\mathbf{k}}; \mathbf{r}', \hat{\mathbf{k}}') - LG(\mathbf{r}, \hat{\mathbf{k}}; \mathbf{r}', \hat{\mathbf{k}}') = \delta(\mathbf{r} - \mathbf{r}')\delta(\hat{\mathbf{k}} - \hat{\mathbf{k}}') \tag{23.4}$$

and obeys homogeneous boundary conditions.

Suppose that the absorption coefficient μ_a is spatially varying; the more general situation in which the scattering coefficient μ_s also varies is readily handled. It is

[1] Here we follow the approach in A. Kim and J.C. Schotland, J. Opt. Soc. Am. A **23**, 596 (2006).

convenient to decompose μ_a into a constant part μ_{a0} and a spatially varying part $\delta\mu_a$:

$$\mu_a(\mathbf{r}) = \mu_{a0} + \delta\mu_a(\mathbf{r}) . \tag{23.5}$$

Then, (23.1) can be rewritten in the form

$$\hat{\mathbf{k}} \cdot \nabla I + \mu_{a0} I - LI = -\delta\mu_a I + S . \tag{23.6}$$

According to (23.3), the solution to (23.6) is given by

$$I(\mathbf{r}, \hat{\mathbf{k}}) = I_0(\mathbf{r}, \hat{\mathbf{k}}) - \int d^3r' d\hat{\mathbf{k}}' G_0(\mathbf{r}, \hat{\mathbf{k}}; \mathbf{r}', \hat{\mathbf{k}}') \delta\mu_a(\mathbf{r}') I(\mathbf{r}', \hat{\mathbf{k}}') , \tag{23.7}$$

where G_0 is the Green's function for a homogeneous medium with absorption μ_{a0} and

$$I_0(\mathbf{r}, \hat{\mathbf{k}}) = \int d^3r' d\hat{\mathbf{k}}' G_0(\mathbf{r}, \hat{\mathbf{k}}; \mathbf{r}', \hat{\mathbf{k}}') S(\mathbf{r}', \hat{\mathbf{k}}') \tag{23.8}$$

denotes the corresponding specific intensity.

The integral equation (23.7) is analogous to the Lippmann-Schwinger equation of scattering theory [see (10.8)]. A similar integral equation for the Green's function G may be obtained by comparing (23.3) and (23.7) and then using the relation (23.8). The result is of the form

$$G(\mathbf{r}_1, \hat{\mathbf{k}}_1; \mathbf{r}_2, \hat{\mathbf{k}}_2) = G_0(\mathbf{r}_1, \hat{\mathbf{k}}_1; \mathbf{r}_2, \hat{\mathbf{k}}_2) - \int d^3r d\hat{\mathbf{k}} G_0(\mathbf{r}_1, \hat{\mathbf{k}}_1; \mathbf{r}, \hat{\mathbf{k}}) \delta\mu_a(\mathbf{r}) G(\mathbf{r}, \hat{\mathbf{k}}; \mathbf{r}_2, \hat{\mathbf{k}}_2). \tag{23.9}$$

Equation (23.9) may be written in operator form as

$$G = G_0 - G_0 \delta\mu_a G , \tag{23.10}$$

which, upon iteration, yields the Born series

$$G = G_0 - G_0 \delta\mu_a G_0 + G_0 \delta\mu_a G_0 \delta\mu_a G_0 + \cdots . \tag{23.11}$$

The term of order n in $\delta\mu_a$ corresponds to n successive absorption events. If only the first term in (23.11) is retained, we refer to this as the first Born approximation to G.

We now introduce the T-matrix for the RTE. The T-matrix is defined in terms of the Green's functions G and G_0 by the equation

$$G = G_0 + G_0 T G_0 . \tag{23.12}$$

Using (23.11), it is readily seen that the T-matrix may be expressed as the series

$$T = -\delta\mu_a + \delta\mu_a G_0 \delta\mu_a - \delta\mu_a G_0 \delta\mu_a G_0 \delta\mu_a + \cdots , \tag{23.13}$$

or

$$T = -\delta\mu_a + \delta\mu_a G_0 T . \tag{23.14}$$

Using this result, we can rewrite the Lippmann-Schwinger equation (23.7) as

$$I = I_0 - G_0 T I_0 . \tag{23.15}$$

Thus the T-matrix may be used to relate the incident intensity to the scattered intensity by means of the identity $I_s = I - I_0$.

23.2 Point Absorbers

We consider a very small absorber with absorption cross section σ that is localized at the point $\mathbf{r}_0$. In this case, the absorption coefficient is given by $\delta\mu_a(\mathbf{r}) = \sigma\delta(\mathbf{r} - \mathbf{r}_0)$ and (23.7) becomes

$$I(\mathbf{r}, \hat{\mathbf{k}}) = I_0(\mathbf{r}, \hat{\mathbf{k}}) - \sigma \int d\hat{\mathbf{k}}' G_0(\mathbf{r}, \hat{\mathbf{k}}; \mathbf{r}_0, \hat{\mathbf{k}}') I(\mathbf{r}_0, \hat{\mathbf{k}}') . \tag{23.16}$$

For fixed $\mathbf{r}_0$, (23.16) is an equation for $I(\mathbf{r}_0, \hat{\mathbf{k}})$. It can be solved by expanding $I(\mathbf{r}_0, \hat{\mathbf{k}})$ into spherical harmonics of the form

$$I(\mathbf{r}_0, \hat{\mathbf{k}}) = \sum_{l,m} I_{lm}(\mathbf{r}_0) Y_{lm}(\hat{\mathbf{k}}) , \tag{23.17}$$

where I_{lm} are suitable coefficients. Upon the substitution of (23.17) into (23.16) and making the replacement $\mathbf{r} \to \mathbf{r}_0$, we find that I_{lm} satisfies the system of linear algebraic equations

$$\sum_{l',m'} \left(\delta_{ll'}\delta_{mm'} + A_{lm}^{l'm'} \right) I_{l'm'} = I_{lm}^{(0)} . \tag{23.18}$$

Here

$$A_{lm}^{l'm'} = \sigma \int d\hat{\mathbf{k}} d\hat{\mathbf{k}}' Y_{lm}^*(\hat{\mathbf{k}}) Y_{l'm'}(\hat{\mathbf{k}}') G_0(\mathbf{r}_0, \hat{\mathbf{k}}; \mathbf{r}_0, \hat{\mathbf{k}}') , \tag{23.19}$$

$$I_{lm}^{(0)} = \int d\hat{\mathbf{k}} Y_{lm}^*(\hat{\mathbf{k}}) I_0(\mathbf{r}_0, \hat{\mathbf{k}}) . \tag{23.20}$$

Upon solving (23.18), we can determine I from (23.17) and (23.16).

Next, we consider a collection of point absorbers with $\delta\mu_a(\mathbf{r}) = \sum_j \sigma_j \delta(\mathbf{r} - \mathbf{r}_j)$. Here $\mathbf{r}_j$ and σ_j are the position and absorption cross section of the jth absorber, respectively. Equation (23.7) thus becomes

$$I(\mathbf{r}, \hat{\mathbf{k}}) = I_0(\mathbf{r}, \hat{\mathbf{k}}) - \sum_j \sigma_j \int d\hat{\mathbf{k}}' G_0(\mathbf{r}, \hat{\mathbf{k}}; \mathbf{r}_j, \hat{\mathbf{k}}') I(\mathbf{r}_j, \hat{\mathbf{k}}') . \tag{23.21}$$

Now, (23.21) can be solved for $I(\mathbf{r}_j, \hat{\mathbf{k}})$ as before. We thus obtain the system of linear equations

$$\sum_k \sum_{l',m'} \left(\delta_{ll'}\delta_{mm'}\delta_{jk} + A_{lm}^{l'm'}(\mathbf{r}_j, \mathbf{r}_k) \right) I_{l'm'}(\mathbf{r}_k) = I_{lm}^{(0)}(\mathbf{r}_j) , \tag{23.22}$$

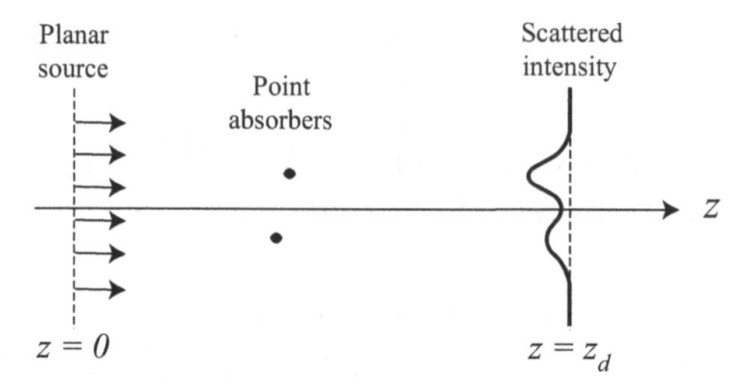

Fig. 23.1 Illustrating the setup. Two point absorbers are located between a planar source at $z = 0$ and the detector plane at $z = z_d$.

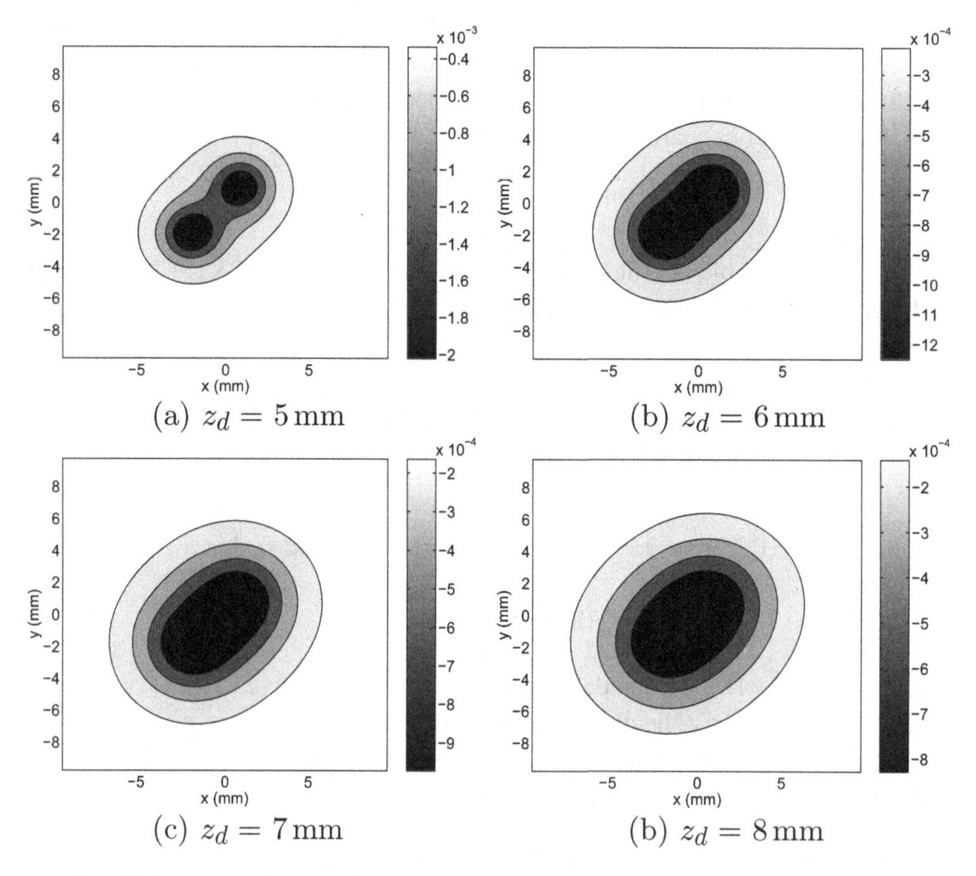

Fig. 23.2 Energy density of the scattered field due to two point absorbers. The optical properties of the medium are $\mu_{a0} = 0.034\,\mathrm{mm}^{-1}$, $\mu_{s0} = 6.11\,\mathrm{mm}^{-1}$ and $g = 0.70$ for a Henyey-Greenstein phase function. The positions of the two point absorbers are $\mathbf{r}_1 = (1.0, 1.0, 4.0)\,\mathrm{mm}$ and $\mathbf{r}_2 = (-2.0, -2.0, 4.1)\,\mathrm{mm}$. Both point absorbers have absorption cross section $\sigma = 0.25\,\mathrm{mm}^2$. The detector planes are located at (a) $z_d = 5\,\mathrm{mm}$, (b) $z_d = 6\,\mathrm{mm}$, (c) $z_d = 7\,\mathrm{mm}$ and (d) $z_d = 8\,\mathrm{mm}$.

where the dependence of $I_{lm}^{(0)}$ on the positions of the absorbers has been made explicit and

$$A_{lm}^{l'm'}(\mathbf{r}_j, \mathbf{r}_k) = \sigma_k \int d\hat{\mathbf{k}} d\hat{\mathbf{k}}' Y_{lm}^*(\hat{\mathbf{k}}) Y_{l'm'}(\hat{\mathbf{k}}') G_0(\mathbf{r}_j, \hat{\mathbf{k}}; \mathbf{r}_k, \hat{\mathbf{k}}') . \tag{23.23}$$

Solving the linear system of equations (23.22) then determines I.

Rather than expanding the specific intensity into spherical harmonics, we may equivalently employ a basis of discrete ordinates. This approach is illustrated for the case of two absorbers in Fig. 23.1. In Fig. 23.2, we plot the energy density in several planes parallel to the source. The absorbers are readily distinguished in the plane closest to the source and become indistinct in more distant planes. That is, resolution is lost due to multiple scattering at large distances from the source.

References and Additional Reading

This paper presents the scattering theory for the RTE in a mathematically rigorous fashion:
G. Bal, Inverse Problems **25**, 053001 (2009).

These papers develop the Green's functions for the RTE:
A.D. Kim, J. Opt. Soc. Am. A **21**, 820 (2004).
A.D. Kim, Waves in Random and Complex Media **15**, 17 (2005).

These papers discuss scattering theory for the RTE:
A. Kim and J.C. Schotland, J. Opt. Soc. Am. A **23**, 596 (2006).
V. Markel and J.C. Schotland, J. Inv. Prob. Imag. **1**, 181 (2006).

24
Diffusion Approximation

As we have seen, exact solutions to the radiative transport equation (RTE) are known in only a small number of cases. However, under certain conditions, the solution to the RTE can be approximated by a solution to a corresponding diffusion equation. This so-called diffusion approximation (DA) is widely used in applications. In this chapter, we derive the DA in two different ways: first as an expansion in angular moments of the specific intensity, and second as an asymptotic expansion in a suitable small parameter. We also discuss the DA from the standpoint of the Bethe–Salpeter equation.

24.1 Angular Moments

We will take as our starting point the time-dependent RTE (19.28)

$$\frac{1}{v_E}\frac{\partial I}{\partial t} + \hat{\mathbf{k}} \cdot \nabla I + \mu_e I - LI = S\,, \tag{24.1}$$

which holds for nonresonant scattering or resonant scattering by point scatterers (see Chapter 19). Here the scattering operator L is defined by

$$LI(\mathbf{r}, \hat{\mathbf{k}}) = \mu_s \int p(\hat{\mathbf{k}}, \hat{\mathbf{k}}')I(\mathbf{r}, \hat{\mathbf{k}}')d\hat{\mathbf{k}}' \tag{24.2}$$

and we have introduced a source term S. The energy velocity v_E coincides with the phase velocity in the medium for nonresonant scattering. We note that (24.1) must also be supplemented by suitable initial and boundary conditions.

We recall from the study of the collision expansion in Chapter 21 that the specific intensity can be decomposed into a sum of terms of the form

$$I = I_0 + I_1 + I_d\,. \tag{24.3}$$

Here the ballistic term I_0 is highly singular and obeys

$$\frac{1}{v_E}\frac{\partial I_0}{\partial t} + \hat{\mathbf{k}} \cdot \nabla I_0 + \mu_e I_0 = S \,. \tag{24.4}$$

The term I_1, which is one order less singular, corresponds to single scattering of the incident field and obeys

$$\frac{1}{v_E}\frac{\partial I_1}{\partial t} + \hat{\mathbf{k}} \cdot \nabla I_1 + \mu_e I_1 = L I_0 \,. \tag{24.5}$$

Finally, the diffuse intensity I_d, which accounts for all orders of scattering beyond the first, is smooth and obeys

$$\frac{1}{v_E}\frac{\partial I_d}{\partial t} + \hat{\mathbf{k}} \cdot \nabla I_d + \mu_e I_d - L I_d = L I_1 \,. \tag{24.6}$$

The idea of the DA is to treat the singular parts of the specific intensity exactly and approximate the smooth part. To proceed, we rewrite (24.6) as

$$\frac{1}{v_E}\frac{\partial I_d}{\partial t} + \hat{\mathbf{k}} \cdot \nabla I_d + \mu_e I_d - L I_d = Q \,, \tag{24.7}$$

where the source term Q is given in terms of I_0 by $Q = LG_0LI_0$, with G_0 the ballistic Green's function defined by (21.3). We note that since G_0 is exponentially small, Q can be neglected at distances from the source larger than ℓ_e. It will prove useful to introduce the first two angular moments of I_d:

$$u = \frac{1}{v_E}\int I_d \, d\hat{\mathbf{k}} \,, \tag{24.8}$$

$$\mathbf{J} = \int \hat{\mathbf{k}} I_d \, d\hat{\mathbf{k}} \,. \tag{24.9}$$

We immediately recognize that u and $\mathbf{J}$ are the diffuse energy density and current, respectively. The specific intensity can then be expanded in its angular moments according to

$$I_d = \frac{v_E}{4\pi} u + \frac{3}{4\pi} \mathbf{J} \cdot \hat{\mathbf{k}} + \cdots \,. \tag{24.10}$$

We now obtain the equations obeyed by the angular moments. Upon integration of (24.7) with respect to $\hat{\mathbf{k}}$, we obtain

$$\frac{\partial u}{\partial t} + \nabla \cdot \mathbf{J} + v_E \mu_a u = S_d \,, \tag{24.11}$$

where $S_d = \int Q d\hat{\mathbf{k}}$. Here we have used the fact that $\int p(\hat{\mathbf{k}}, \hat{\mathbf{k}}')d\hat{\mathbf{k}}' = 1$. Next, we multiply (24.7) by $\hat{\mathbf{k}}$ and then integrate with respect to $\hat{\mathbf{k}}$. We find that

$$\frac{1}{v_E}\frac{\partial \mathbf{J}}{\partial t} + \nabla \cdot \int \hat{\mathbf{k}} \otimes \hat{\mathbf{k}} \, I_d \, d\hat{\mathbf{k}} + \left[\mu_a + (1-g)\mu_s\right]\mathbf{J} = \mathcal{Q} \,, \tag{24.12}$$

where $\mathcal{Q} = \int \hat{\mathbf{k}} Q d\hat{\mathbf{k}}$ and $\mu_a = \mu_e - \mu_s$. The anisotropy factor g is defined by

$$g = \int \hat{\mathbf{k}} \cdot \hat{\mathbf{k}}' p(\hat{\mathbf{k}}, \hat{\mathbf{k}}') d\hat{\mathbf{k}} . \tag{24.13}$$

Here we have assumed that p depends only on the angle between $\hat{\mathbf{k}}$ and $\hat{\mathbf{k}}'$, corresponding to a statistically homogeneous and isotropic random medium. We recall from Chapter 20 that $-1 \leq g \leq 1$ and that $g = 0$ corresponds to isotropic scattering, $g = 1$ to forward scattering, and $g = -1$ to backward scattering. Next, we substitute (24.10) into (24.12) and carry out the indicated integrations using the relations

$$\int \hat{\mathbf{k}} d\hat{\mathbf{k}} = 0 , \quad \int \hat{\mathbf{k}} \otimes \hat{\mathbf{k}} d\hat{\mathbf{k}} = \frac{4\pi}{3}\mathbf{I} , \quad \int \hat{\mathbf{k}} \otimes \hat{\mathbf{k}} \otimes \hat{\mathbf{k}} d\hat{\mathbf{k}} = 0 , \tag{24.14}$$

with $\mathbf{I}$ the unit second-rank tensor. We find that

$$\mathbf{J} + \tau \frac{\partial \mathbf{J}}{\partial t} = -D\nabla u + \ell_t' \mathcal{Q} , \tag{24.15}$$

where $\tau = \ell_t'/v_E$. Here the diffusion coefficient D and ℓ_t' are defined by

$$D = \frac{1}{3} v_E \ell_t' , \quad \ell_t' = \frac{1}{\mu_a + (1-g)\mu_s} . \tag{24.16}$$

We can now derive the diffusion equation obeyed by the energy density u by substituting the expression (24.15) into (24.11). At long times ($t \gg \tau$), the second term on the left-hand side of (24.15) can be neglected, and we find that

$$\frac{\partial u}{\partial t} - \nabla \cdot D\nabla u + v_E \mu_a u = S_d - \ell_t' \nabla \cdot \mathcal{Q} . \tag{24.17}$$

Using (24.10) and (24.15), we see that the DA to the diffuse intensity is given by

$$I_d = \frac{v_E}{4\pi} \left(u - \ell_t' \hat{\mathbf{k}} \cdot \nabla u \right) , \tag{24.18}$$

where the second term on the right-hand side of (24.15), can be neglected far from the source. Evidently, the DA is accurate if $\ell_t' |\nabla u| \ll u$, which means that u cannot vary rapidly on the scale of ℓ_t'. This condition breaks down when the system size is of order ℓ_t, the medium is weakly scattering or strongly absorbing, or the point at which the field is measured is a distance of order ℓ_t from the boundary. Note that within the accuracy of the DA, the second term on the right-hand side of (24.17) can be neglected, and thus the energy density obeys the diffusion equation

$$\frac{\partial u}{\partial t} - \nabla \cdot D\nabla u + \alpha u = S_d , \tag{24.19}$$

where the absorption coefficient $\alpha = v_E \mu_a$.

24.2 Asymptotic Analysis

In this section, we derive the DA using asymptotic methods. The advantage of this approach is that it is mathematically economical and treats corrections to the DA in a simple manner. To proceed, we note that the DA holds for small absorption, large distances, long times and far from the source. Accordingly, we perform the rescaling

$$\mu_a \to \epsilon^2 \mu_a \,, \quad \mathbf{r} \to \epsilon \mathbf{r} \,, \quad t \to \epsilon^2 t \,, \quad Q \to \epsilon^2 Q \,, \tag{24.20}$$

where $\epsilon \ll 1$. Thus the RTE (24.7) becomes

$$\epsilon^2 \frac{1}{v_E} \frac{\partial I_d}{\partial t} + \epsilon \hat{\mathbf{k}} \cdot \nabla I_d + \epsilon^2 \mu_a I_d + \mu_s I_d - L I_d = \epsilon^2 Q \,. \tag{24.21}$$

We then introduce the asymptotic expansion for the specific intensity

$$I_d = I_0 + \epsilon I_1 + \epsilon^2 I_2 + \cdots \,, \tag{24.22}$$

which we substitute into Eq. (24.21). Upon collecting terms of $O(1)$, $O(\epsilon)$ and $O(\epsilon^2)$, we have

$$L I_0 = I_0 \,, \tag{24.23}$$

$$L I_1 - \mu_s I_1 = \hat{\mathbf{k}} \cdot \nabla I_0 \,, \tag{24.24}$$

$$\frac{1}{v_E} \frac{\partial I_1}{\partial t} + \hat{\mathbf{k}} \cdot \nabla I_1 + \mu_a I_0 + \mu_s I_2 - L I_2 = Q \,. \tag{24.25}$$

Making use of the normalization condition $\int p(\hat{\mathbf{k}}, \hat{\mathbf{k}}') d\hat{\mathbf{k}}' = 1$, we see from (24.23) that I_0 can depend only upon the spatial coordinate $\mathbf{r}$. To solve equation (24.24), we use the identity

$$\int \hat{\mathbf{k}}' p(\hat{\mathbf{k}}, \hat{\mathbf{k}}') d\hat{\mathbf{k}}' = g \hat{\mathbf{k}} \,, \tag{24.26}$$

which holds when $p(\hat{\mathbf{k}}, \hat{\mathbf{k}}')$ depends only on the angle between $\hat{\mathbf{k}}$ and $\hat{\mathbf{k}}'$. Then, if we make the ansatz $I_1 = C \hat{\mathbf{k}} \cdot \nabla I_0$, where C is constant, we see that (24.24) is satisfied if $C = -1/[\mu_s(1-g)]$. That is,

$$I_1 = -\frac{1}{(1-g)\mu_s} \hat{\mathbf{k}} \cdot \nabla I_0 \,. \tag{24.27}$$

Finally, we insert the above expression for I_1 into (24.25) and integrate over $\hat{\mathbf{k}}$ using (24.14). We thus obtain the diffusion equation

$$\frac{\partial u}{\partial t} - \nabla \cdot D \nabla u + \alpha u = S_d \,, \tag{24.28}$$

where the energy density u is defined by the relation $I_0 = v_E u/4\pi$. The diffusion coefficient D is given by

$$D = \frac{1}{3} v_E \ell_t \, , \quad \ell_t = \frac{1}{(1-g)\mu_s} \, , \tag{24.29}$$

where ℓ_t is the transport mean free path. Note that $\ell_t' \simeq \ell_t$ when μ_a is sufficiently small. Indeed, the two expressions are asymptotically equivalent since $\mu_a/\mu_s = O(\epsilon^2)$. Finally, it follows from (24.22) and (24.27) that we recover the asymptotic formula (24.18) for the specific intensity.

24.3 Bethe–Salpeter to Diffusion

So far, we have derived the DA from the RTE. Here we show how to obtain the DA directly from the Beth–Salpeter equation. We begin by recalling from Chapter 16 that the Bethe–Salpeter equation is of the form

$$\begin{aligned}&\langle G(\mathbf{r}_1, \mathbf{r}_2) G^*(\mathbf{r}_1', \mathbf{r}_2')\rangle = \langle G(\mathbf{r}_1, \mathbf{r}_2)\rangle \langle G^*(\mathbf{r}_1', \mathbf{r}_2')\rangle \\ &+ k_0^4 \int d^3R d^3R' \langle G(\mathbf{r}_1, \mathbf{R})\rangle \langle G^*(\mathbf{r}_1', \mathbf{R})\rangle C(\mathbf{R}-\mathbf{R}') \langle G(\mathbf{R}', \mathbf{r}_2) G^*(\mathbf{R}', \mathbf{r}_2')\rangle . \end{aligned} \tag{24.30}$$

For simplicity, we consider the case of Gaussian disorder with white-noise correlations, where $C(\mathbf{r}-\mathbf{r}') = A\delta(\mathbf{r}-\mathbf{r}')$. It follows that the second-order correlation function of the field obeys the integral equation

$$\langle U(\mathbf{r}) U^*(\mathbf{r}')\rangle = \langle U(\mathbf{r})\rangle \langle U^*(\mathbf{r}')\rangle + k_0^4 A \int d^3R \langle G(\mathbf{r}, \mathbf{R})\rangle \langle G^*(\mathbf{r}', \mathbf{R})\rangle \langle U(\mathbf{R}) U^*(\mathbf{R})\rangle . \tag{24.31}$$

We now introduce the average intensity $I(\mathbf{r}) = \langle U(\mathbf{r}) U^*(\mathbf{r})\rangle$. We find that I obeys the integral equation

$$I(\mathbf{r}) = I_0(\mathbf{r}) + \frac{4\pi}{\ell_s} \int d^3r' |\langle G(\mathbf{r}, \mathbf{r}')\rangle|^2 I(\mathbf{r}') \, , \tag{24.32}$$

where $I_0(\mathbf{r}) = |\langle U(\mathbf{r})\rangle|^2$ and we have used the relation $\ell_s = 4\pi/(k_0^4 A)$, which follows from (15.49). The integral equation (24.32) is identical to (21.29), which arises in the discussion of isotropic scattering for the RTE. Making use of (15.38), we obtain

$$|\langle G(\mathbf{r}, \mathbf{r}')\rangle|^2 = \frac{1}{(4\pi)^2} \frac{\exp(-|\mathbf{r}-\mathbf{r}'|/\ell_s)}{|\mathbf{r}-\mathbf{r}'|^2} \, , \tag{24.33}$$

and we thus find that I obeys the diffusion equation

$$-\frac{1}{3}\ell_s^2 \nabla^2 I = I_0 \, , \tag{24.34}$$

which agrees with (24.28) for isotropic scattering in the absence of absorption. Further connections between the Bethe–Salpeter equation and the diffusion approximation is established in Chapter 31.

References and Additional Reading

A comprehensive treatment of transport equations can be found in:

S. Chapman and T.G. Cowling, *The Mathematical Theory of Non-uniform Gases* (Cambridge University Press, Cambridge, 1991).

J.J. Duderstadt and W.R. Martin, *Transport Theory* (Wiley, New York, 1979).

The diffusion equation has been extensively studied in the context of conductive heat transfer. See:

H.S. Carslaw and J.C. Jaeger, *Heat Conduction in Solids*, 2nd edition (Oxford University Press, Oxford, 1959).

The moment method for the derivation of the diffusion equation is presented in:

A. Ishimaru, *Wave Propagation and Scattering in Random Media* (IEEE Press, Piscataway, 1997), chap. 9.

The asymptotics for the derivation of the diffusion approximation is introduced in:

E.W. Larsen and J.B. Keller, J. Math. Phys. **15**, 75 (1974).

The diffusion approximation is derived from the singular eigenfunctions approach in:

K.M. Case and P.F. Zweifel, *Linear Transport Theory* (Addison-Wesley, Reading, 1967), chap. 5 and 6.

The singular eigenfunction approach is used in the following paper to derive the expression of the diffusion coefficient in the presence of absorption:

R. Pierrat, J.-J. Greffet and R. Carminati, J. Opt. Soc. Am. A **23**, 1106 (2006).

Derivations of the diffusion approximation from the Bethe–Salpeter equation can be found in:

P. Sheng, *Introduction to Wave Scattering, Localization, and Mesoscopic Phenomena* (Academic Press, San Diego, 1995).

A. Lagendijk and B.A. van Tiggelen, Phys. Rep. **270**, 143 (1996).

E. Akkermans and G. Montambaux, *Mesoscopic Physics of Electrons and Photons* (Cambridge University Press, Cambridge, 2007).

M.C.W. van Rossum and Th.M. Nieuwenhuizen, Rev. Mod. Phys. **71**, 313 (1999).

The diffusion approximation has been widely used to model forward and inverse problems in biomedical optics. For a review, see for example:

S.R. Arridge, Inverse Problems **15**, R41 (1999).

S.R. Arridge and J.C. Schotland, Inverse Problems **25**, 123010 (2009).

The diffusion approximation can be refined, in order to improve its accuracy. See for example:

M. Machida, G.Y. Panasyuk, J.C. Schotland and V.A. Markel, J. Opt. Soc. Am. A **26**, 1291 (2009).

U. Tricoli, C.M. MacDonald, A. Da Silva and V.A. Markel, J. Opt. Soc. Am. A **35**, 356 (2018).

These papers study the transition from ballistic to diffusive transport for light in scattering media:

Z.Q. Zhang, I.P. Jones, H.P. Schriemer, J.H. Page, D.A. Waitz and P. Sheng, Phys. Rev. E **60**, 4843 (1999).
R. Elaloufi, R. Carminati and J.-J. Greffet, J. Opt. Soc. Am. A **21**, 1430 (2004).

These papers address the validity of the diffusion approximation at short length scales and its breakdown:

I. Freund, M. Kaveh and M. Rosenbluh, Phys. Rev. Lett. **60**, 1130 (1988).
K.M. Yoo, F. Liu and R.R. Alfano, Phys. Rev. Lett. **64**, 2647 (1990).
R.H.J. Kop, P. de Vries, R. Sprik and A. Lagendijk, Phys. Rev. Lett. **79**, 4369 (1997).
K.K. Bizheva, A.M. Siegel and D.A. Boas, Phys. Rev. E **58**, 7664 (1998).
A.D. Kim and A. Ishimaru, Appl. Opt. **37**, 5313 (1998).

25

Diffuse Light

In this chapter, we study the propagation of multiply-scattered light in the diffusion approximation to the radiative transport equation. We begin by formulating the proper boundary conditions for the diffusion equation. We then consider the propagation of diffuse light in simple geometries, including the half-space and slab.

25.1 Boundary Conditions

We recall from Chapter 24 that the energy density u obeys the diffusion equation

$$-\nabla \cdot D\nabla u + \alpha u = S_d \quad \text{in} \quad V , \tag{25.1}$$

where $\alpha = v_E \mu_a$ is the absorption coefficient. We also recall that the diffuse intensity is given by

$$I_d = \frac{v_E}{4\pi}\left(u - \ell_t \hat{\mathbf{k}} \cdot \nabla u\right) . \tag{25.2}$$

We now consider the problem of specifying appropriate boundary conditions for the diffusion equation (25.1). There are two cases to consider: diffuse-diffuse and diffuse-nondiffuse interfaces. A diffuse-diffuse interface is a surface dividing two homogeneous diffuse media. In medium 1 the energy density u_1 is taken to obey

$$-D_1 \nabla^2 u_1 + \alpha_1 u_1 = S_1 , \tag{25.3}$$

and in medium 2 the energy density u_2 obeys

$$-D_2 \nabla^2 u_2 + \alpha_2 u_2 = 0 . \tag{25.4}$$

Here α_1, D_1 and α_2, D_2 denote the pairs of absorption and diffusion coefficients in each medium, and the source is assumed to be located in medium 1. The boundary conditions that are satisfied by the energy density across the interface are given by

$$u_1 = u_2 \tag{25.5}$$

$$D_1 \frac{\partial u_1}{\partial n} = D_2 \frac{\partial u_2}{\partial n} , \tag{25.6}$$

where $\partial/\partial n = \hat{\mathbf{n}} \cdot \nabla$, with $\hat{\mathbf{n}}$ the unit normal to the boundary. Equations (25.5) and (25.6) express the continuity of the energy density and the current across the interface.

A diffuse-nondiffuse interface is a surface dividing a diffuse medium and vacuum. Assuming that the source is contained in the diffuse medium, the most general boundary condition compatible with the diffusion equation (25.1) is of the form

$$u + \ell \frac{\partial u}{\partial n} = f \quad \text{on} \quad \partial V , \tag{25.7}$$

where f is a boundary source, $\hat{\mathbf{n}}$ is the outward unit normal to the boundary and ℓ is called the extrapolation length. We note that if ℓ is nonnegative, the solution to the diffusion equation (25.1) obeying the boundary condition (25.7) is unique. There are two important cases to consider. If $\ell = 0$, the boundary is said to be absorbing. If $\ell \to \infty$, the boundary is reflecting. The physical meaning of ℓ may be understood as follows. If ℓ is sufficiently small, we have from (25.7) that $u(\mathbf{r} + \ell\hat{\mathbf{n}}) = 0$. Thus u obeys absorbing boundary conditions on the extrapolated boundary consisting of all points of the form $\mathbf{r} + \ell\hat{\mathbf{n}}$, where $\mathbf{r} \in \partial V$. It follows from (25.2) that the diffuse intensity in the outward normal direction is given by

$$I_d = \frac{v_E}{4\pi}\left(1 + \frac{\ell_t}{\ell}\right) u . \tag{25.8}$$

Thus the energy density determines the observable intensity.

It is possible to justify the boundary condition (25.7) from radiative transport theory. To proceed, we recall that according to (20.11), the boundary condition obeyed by the diffuse intensity I_d is of the form $I_d(\mathbf{r}, \hat{\mathbf{k}}) = 0$ for $\hat{\mathbf{n}} \cdot \hat{\mathbf{k}} < 0$ and $\mathbf{r} \in \partial V$. Evidently, the energy density u cannot satisfy this condition. Instead, we impose a condition of the form

$$\int_{\hat{\mathbf{k}}\cdot\hat{\mathbf{n}}<0} \hat{\mathbf{k}} \cdot \hat{\mathbf{n}}\, I_d(\mathbf{r}, \hat{\mathbf{k}}) d\hat{\mathbf{k}} = 0 \quad \text{for} \quad \mathbf{r} \in \partial V , \tag{25.9}$$

which is consistent with the boundary condition obeyed by I_d. We note that (25.9) says that the inward current of diffuse light vanishes on the boundary. We can evaluate the above integral by making use of the diffusion approximation

$$I_d = \frac{v_E}{4\pi} u + \frac{3}{4\pi} \mathbf{J} \cdot \hat{\mathbf{k}} \tag{25.10}$$

and the relation

$$\mathbf{J} = -D\nabla u + \ell_t \mathcal{Q} . \tag{25.11}$$

We thus obtain

$$u + \frac{2}{3}\ell_t \frac{\partial u}{\partial n} = 2\frac{\ell_t}{v_E}\mathcal{Q}\cdot\hat{\mathbf{n}} \quad \text{on} \quad \partial V , \tag{25.12}$$

where we have used the integrals

$$\int_{\hat{\mathbf{k}}\cdot\hat{\mathbf{n}}<0} \hat{\mathbf{k}}\cdot\hat{\mathbf{n}}\, d\hat{\mathbf{k}} = -\pi , \quad \int_{\hat{\mathbf{k}}\cdot\hat{\mathbf{n}}<0} \hat{\mathbf{k}}\cdot\hat{\mathbf{n}}\,\mathbf{J}\cdot\hat{\mathbf{k}}\, d\hat{\mathbf{k}} = \frac{2\pi}{3}\mathbf{J}\cdot\hat{\mathbf{n}} . \tag{25.13}$$

We observe that (25.12) takes the form of the boundary condition (25.7), with the extrapolation length given by $\ell = 2/3\ell_t$, a result that is often used in applications.

As noted above, the boundary conditions for the diffusion equation and the radiative transport equation are not mathematically consistent. The proper treatment of this problem is to obtain the boundary conditions for the diffusion equation from an asymptotic analysis. The asymptotic approach of Section 24.2 can be modified to include a boundary layer that accounts for the crossover from radiative transport near the boundary to diffusion in the interior of the medium.

25.2 Homogeneous Media

We now consider the propagation of diffuse light in a homogeneous medium. Inhomogeneous media will be taken up in Chapter 27. If the diffusion and absorption coefficients are uniform, then the diffusion equation (25.1) may be written in the form

$$-\nabla^2 u + k_0^2 u = \frac{1}{D}S_d , \tag{25.14}$$

where $k_0 = \sqrt{\alpha/D}$ is known as the diffuse wavenumber. We note an important correspondence between the diffusion equation (25.14) and the wave equation for scalar fields (5.1): the two are related by the transformation of the wavenumber $k_0 \to ik_0$. Thus the theory of diffuse light can be developed in parallel with the theory of scalar waves as discussed in Part I.

Following the exposition of Green's functions and the Kirchoff integral formula in Chapter 5, we find that in a volume V the solution to the diffusion equation (25.14) obeying the boundary condition (25.7) is given by

$$u(\mathbf{r}) = \frac{1}{D}\int_V G(\mathbf{r},\mathbf{r}')S_d(\mathbf{r}')d^3r' + \frac{1}{\ell}\int_{\partial V} G(\mathbf{r},\mathbf{r}')f(\mathbf{r}')d^2r' . \tag{25.15}$$

Here the Green's function G obeys the equation

$$-\nabla^2 G(\mathbf{r},\mathbf{r}') + k_0^2 G(\mathbf{r},\mathbf{r}') = \delta(\mathbf{r}-\mathbf{r}') , \tag{25.16}$$

along with the homogeneous boundary condition

$$G(\mathbf{r},\mathbf{r}') + \ell\hat{\mathbf{n}}\cdot\nabla G(\mathbf{r},\mathbf{r}') = 0 . \tag{25.17}$$

The Green's function G_0 for an infinite medium is given by

$$G_0(\mathbf{r}, \mathbf{r}') = \frac{\exp(-k_0|\mathbf{r} - \mathbf{r}'|)}{4\pi|\mathbf{r} - \mathbf{r}'|}, \tag{25.18}$$

which is an immediate consequence of (5.16). The Green's function G can then be expressed in the form

$$G(\mathbf{r}, \mathbf{r}') = G_0(\mathbf{r}, \mathbf{r}') + G_1(\mathbf{r}, \mathbf{r}'), \tag{25.19}$$

where it follows from (25.16) that G_1 obeys the homogeneous diffusion equation

$$-\nabla^2 G_1 + k_0^2 G_1 = 0 \quad \text{in} \quad V \tag{25.20}$$

together with the inhomogeneous boundary condition

$$G_1 + \ell \frac{\partial G_1}{\partial n} = -G_0 - \ell \frac{\partial G_0}{\partial n} \quad \text{on} \quad \partial V. \tag{25.21}$$

Once G is determined by solving (25.20), the solution to the diffusion equation may be obtained from the formula (25.15). We will illustrate this procedure in Section 25.4 for the case of a semi-infinite medium.

We now consider several special cases of particular interest. We begin with a point source in an infinite medium of the form $S_d(\mathbf{r}) = S_0\delta(\mathbf{r} - \mathbf{r}_0)$, where S_0 is the source power and $\mathbf{r}_0$ is the position of the source. It follows that the energy density is given by

$$u(\mathbf{r}) = \frac{S_0}{4\pi D} \frac{\exp(-k_0|\mathbf{r} - \mathbf{r}_0|)}{|\mathbf{r} - \mathbf{r}_0|}. \tag{25.22}$$

Evidently, the energy density propagates as a spherical excitation that decays exponentially on the length scale $1/k_0$. This is a characteristic feature of diffuse light. Next, we consider a source S_d contained in a finite volume. In this case, the energy density is given by

$$u(\mathbf{r}) = \frac{1}{D} \int_V G(\mathbf{r}, \mathbf{r}') S_d(\mathbf{r}') d^3r'. \tag{25.23}$$

Finally, we consider a source f on the surface of a finite volume. We then have

$$u(\mathbf{r}) = \frac{1}{\ell} \int_{\partial V} G(\mathbf{r}, \mathbf{r}') f(\mathbf{r}') d^2r'. \tag{25.24}$$

This case arises in optical imaging, where diffuse light is used to probe a medium of interest and measurements are carried out on the boundary.

25.3 Plane-Wave Expansions

We consider the propagation of diffuse light in a homogeneous medium in the absence of sources. The energy density then obeys the equation

$$-\nabla^2 u + k_0^2 u = 0 . \tag{25.25}$$

In strict analogy to the development in Section 6.1, we find that the general solution of (25.25) takes the form of a decomposition into plane-wave modes of the form

$$u(\mathbf{r}) = \int \frac{d^2 q}{(2\pi)^2} \left[A(\mathbf{q}) \exp\left(i\mathbf{q}\cdot\boldsymbol{\rho} + Q(q)z\right) + B(\mathbf{q}) \exp\left(i\mathbf{q}\cdot\boldsymbol{\rho} - Q(q)z\right) \right] , \tag{25.26}$$

where $\mathbf{r} = (\boldsymbol{\rho}, z)$, $A(\mathbf{q})$ and $B(\mathbf{q})$ are arbitrary coefficients, and

$$Q(q) = \sqrt{q^2 + k_0^2} . \tag{25.27}$$

The first term on the right-hand side of (25.26) corresponds to propagation in the negative z-direction, while the second term corresponds to propagation in the positive z-direction. We note that each term consists of a superposition of evanescent plane-wave modes. In contrast to the corresponding plane-wave expansion (6.8) for scalar waves, there are no propagating modes present in (25.26).

The Green's function G_0 for an infinite medium can be expanded into plane-wave modes. To see this, we write G_0 as the Fourier integral

$$G_0(\mathbf{r}, \mathbf{r}') = \int \frac{d^3 k}{(2\pi)^3} \frac{\exp(i\mathbf{k}\cdot(\mathbf{r} - \mathbf{r}'))}{k^2 + k_0^2} . \tag{25.28}$$

Next, we use the identity

$$\int_{-\infty}^{\infty} dk_z \frac{\exp(ik_z z)}{k_z^2 + q^2 + k_0^2} = \frac{\pi}{Q(q)} \exp(-Q(q)|z|) \tag{25.29}$$

to carry out the integration over k_z. We find that

$$G_0(\mathbf{r}, \mathbf{r}') = \frac{1}{2(2\pi)^2} \int \frac{d^2 q}{Q(q)} \exp\left[i\mathbf{q}\cdot(\boldsymbol{\rho} - \boldsymbol{\rho}') - Q(q)|z - z'|\right] , \tag{25.30}$$

which is the desired plane-wave expansion.

25.4 Half-Space Geometry

In this section, we discuss the propagation of diffuse light in the half-space geometry, which provides a useful application of the method of plane-wave expansions. We consider a system in which a homogeneous diffuse medium occupies the half-space $z \geq 0$ and the half-space $z < 0$ is vacuum. Thus, the plane $z = 0$ is a diffuse-nondiffuse interface. The half-space Green's function G can be obtained from (25.19), where G_0 is expanded into plane waves according to (25.30) and G_1 consists of plane-wave modes that propagate into the $z \geq 0$ half-space. Thus

$$G_1(\mathbf{r}, \mathbf{r}') = \int \frac{d^2q}{(2\pi)^2} A(\mathbf{q}) \exp\left(i\mathbf{q} \cdot \rho - Q(q)z\right) , \tag{25.31}$$

where the dependence of A on the coordinate $\mathbf{r}'$ is not indicated. Applying the boundary condition (25.17) with $\hat{\mathbf{n}} = -\hat{\mathbf{z}}$, we find that A is given by

$$A(\mathbf{q}) = -\frac{1}{2Q(q)} \frac{1 - Q(q)\ell}{1 + Q(q)\ell} \exp\left(-i\mathbf{q} \cdot \rho' - Q(q)z'\right) . \tag{25.32}$$

Putting everything together, we see that the Green's function G can be written in the form

$$G(\mathbf{r}, \mathbf{r}') = \int \frac{d^2q}{(2\pi)^2} \exp(i\mathbf{q} \cdot (\rho - \rho'))g(z, z'; q) , \tag{25.33}$$

where

$$g(z, z'; q) = \frac{1}{2Q(q)} \left[\exp(-Q(q)|z - z'|) - \frac{1 - Q(q)\ell}{1 + Q(q)\ell} \exp(-Q(q)|z + z'|) \right] . \tag{25.34}$$

Two special cases of the above result should be noted. First, when $\ell = 0$, which corresponds to absorbing boundary conditions, it is easily seen that $G_1(\mathbf{r}, \mathbf{r}') = -G_0(\mathbf{r}, \tilde{\mathbf{r}}')$. Here $\tilde{\mathbf{r}} = (\rho, -z)$ is the reflection of the point $\mathbf{r}$ about the $z = 0$ plane. Thus G becomes

$$G(\mathbf{r}, \mathbf{r}') = G_0(\mathbf{r}, \mathbf{r}') - G_0(\mathbf{r}, \tilde{\mathbf{r}}') , \tag{25.35}$$

a result that can be obtained from the method of images. Second, when $\ell \to \infty$, which corresponds to reflecting boundary conditions, we find that $G_1(\mathbf{r}, \mathbf{r}') = G_0(\mathbf{r}, \tilde{\mathbf{r}})$. We thus obtain

$$G(\mathbf{r}, \mathbf{r}') = G_0(\mathbf{r}, \mathbf{r}') + G_0(\mathbf{r}, \tilde{\mathbf{r}}') . \tag{25.36}$$

In Fig. 25.1 we plot the energy density from a point source at the origin in a half-space with absorbing boundary conditions.

25.5 Slab Geometry

We consider the case of a homogeneous non-absorbing diffuse medium located between the planes $z = 0$ and $z = L$, where the regions $z < 0$ and $z > L$ are taken

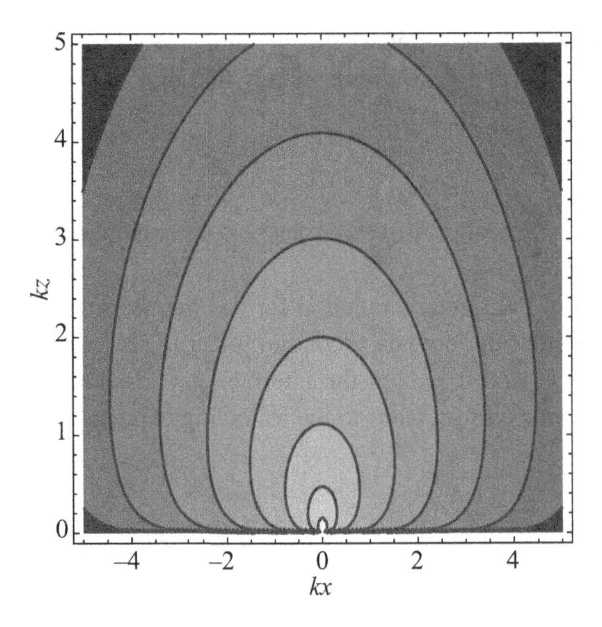

Fig. 25.1 Energy density of diffuse light from a point source at the origin in a half-space with absorbing boundary conditions.

to be vacuum. If the plane $z = 0$ is uniformly illuminated and light is detected on the plane $z = L$, the energy density obeys the equation

$$\frac{d^2u}{dz^2} = 0 \quad \text{for} \quad 0 \leq z \leq L \,, \tag{25.37}$$

$$u - \ell\frac{\partial u}{\partial z} = u_0 \quad \text{for} \quad z = 0 \,, \tag{25.38}$$

$$u + \ell\frac{\partial u}{\partial z} = 0 \quad \text{for} \quad z = L \,, \tag{25.39}$$

where ℓ is the extrapolation length and u_0 is a source that is proportional to the incident flux. By solving the above boundary value problem, it is easily seen that

$$u(z) = \frac{u_0}{L + 2\ell}(L + \ell - z) \ . \tag{25.40}$$

Next, we define the optical conductance per unit area as

$$\Sigma = \frac{\phi}{u(0) - u(L)} \,, \tag{25.41}$$

where $\phi = -D\partial u/\partial z$ is the outgoing flux per unit area. Making use of Eqs. (25.39) and (25.40) with $\ell = (2/3)\ell_t$, we immediately find that

$$\Sigma = \frac{D}{L} = \frac{v_E}{3}\frac{\ell_t}{L}. \tag{25.42}$$

The above result is typical of diffusive transport, as exemplified by Ohm's law for electrical conduction.

The derivation of the Green's function for the slab is similar to the case of the half-space, except that the boundary condition (25.17) is applied with $\hat{\mathbf{n}} = -\hat{\mathbf{z}}$ at the interface $z = 0$, and $\hat{\mathbf{n}} = \hat{\mathbf{z}}$ at the interface $z = L$. The detailed derivation, which is given as an exercise, leads to the following expression for g:

$$g(z, z'; q) = \frac{[1 + (Q\ell)^2]\cosh[Q(L - |z - z'|)] - [1 - (Q\ell)^2]\cosh[Q(L - |z + z'|)] + 2Q\ell\,\sinh[Q(L - |z - z'|)]}{2Q[\sinh(QL) + 2Q\ell\,\cosh(QL) + (Q\ell)^2\sinh(QL)]} \tag{25.43}$$

where $Q = \sqrt{q^2 + k_0^2}$. In the particular case where the Green's function is calculated for points coinciding with the slab interfaces, we can take $z' = 0$ and $z = L$ (or equivalently $z = 0$ and $z' = L$), which leads to

$$g(0, L; q) = \frac{Q\ell^2}{\sinh(QL) + 2Q\ell\,\cosh(QL) + (Q\ell)^2\sinh(QL)}. \tag{25.44}$$

25.6 Time-Dependent Diffusion

We now consider the time-dependent diffusion equation (24.28) in a homogeneous medium:

$$\frac{\partial u}{\partial t} - D\nabla^2 u + \alpha u = S_d \quad \text{in} \quad V. \tag{25.45}$$

The above equation can be used to study the propagation of diffuse light from a pulsed source of the form $S_d(\mathbf{r}, t) = S_d(\mathbf{r})\delta(t)$. The energy density u is also taken to satisfy the initial and boundary conditions

$$u|_{t=0} = 0 \quad \text{in} \quad V, \tag{25.46}$$

$$u + \ell\frac{\partial u}{\partial n} = f \quad \text{on} \quad \partial V, \tag{25.47}$$

where ℓ is the extrapolation length and f is a boundary source. Thus no light enters the medium except from the sources.

The solution to (25.45) is readily obtained by Laplace transforms. To proceed, we define the Laplace transform of u by

$$\hat{u}(\mathbf{r}, \Omega) = \int_0^\infty \exp(-\Omega t) u(\mathbf{r}, t) dt \ , \tag{25.48}$$

where Ω is complex-valued and $\mathrm{Re}(\Omega) > 0$. We find that $\hat{u}$ obeys

$$-D\nabla^2 \hat{u} + (\alpha + \Omega)\hat{u} = \hat{S}_d \quad \text{in} \quad V \ , \tag{25.49}$$

$$\hat{u} + \ell \frac{\partial \hat{u}}{\partial n} = \hat{f} \quad \text{on} \quad \partial V \ . \tag{25.50}$$

Equation (25.49) has the form of the time-independent diffusion equation (25.14). It follows from (25.15) that

$$\hat{u}(\mathbf{r}, \Omega) = \frac{1}{D} \int_V \hat{G}(\mathbf{r}, \mathbf{r}'; \Omega) \hat{S}_d(\mathbf{r}') d^3 r' + \frac{1}{\ell} \int_{\partial V} \hat{G}(\mathbf{r}, \mathbf{r}'; \Omega) \hat{f}(\mathbf{r}', \Omega) d^2 r' \ . \tag{25.51}$$

Here the Green's function $\hat{G}$ obeys the equation

$$-\nabla^2 \hat{G}(\mathbf{r}, \mathbf{r}'; \Omega) + \frac{\alpha + \Omega}{D} \hat{G}(\mathbf{r}, \mathbf{r}'; \Omega) = \delta(\mathbf{r} - \mathbf{r}') \tag{25.52}$$

along with the homogeneous boundary condition

$$\hat{G}(\mathbf{r}, \mathbf{r}'; \Omega) + \ell \frac{\partial}{\partial n} \hat{G}(\mathbf{r}, \mathbf{r}'; \Omega) = 0 \ . \tag{25.53}$$

Upon inversion of the Laplace transforms in (25.51), we find that

$$\begin{aligned} u(\mathbf{r}, t) = {} & \frac{1}{D} \int_0^t dt' \int_V G(\mathbf{r}, \mathbf{r}'; t - t') S_d(\mathbf{r}', t') d^3 r' \\ & + \frac{1}{\ell} \int_0^t dt' \int_{\partial V} G(\mathbf{r}, \mathbf{r}'; t - t') f(\mathbf{r}', t') d^2 r' \ . \end{aligned} \tag{25.54}$$

We will refer to G as the diffusion propagator, which is easily seen to obey

$$\frac{\partial}{\partial t} G(\mathbf{r}, \mathbf{r}'; t) = D\nabla^2 G(\mathbf{r}, \mathbf{r}'; t) - \alpha G(\mathbf{r}, \mathbf{r}'; t) \quad \text{in} \quad V \ , \tag{25.55}$$

$$G(\mathbf{r}, \mathbf{r}'; t) + \ell \frac{\partial}{\partial n} G(\mathbf{r}, \mathbf{r}'; t) = 0 \quad \text{on} \quad \partial V \ , \tag{25.56}$$

$$\lim_{t \to 0} G(\mathbf{r}, \mathbf{r}'; t) = \delta(\mathbf{r} - \mathbf{r}') \ . \tag{25.57}$$

We now discuss some cases of particular physical interest. In an infinite medium, it follows from (25.18) that $\hat{G}$ is given by

$$\hat{G}(\mathbf{r}, \mathbf{r}'; \Omega) = \frac{\exp\left[-\left((\alpha + \Omega)/D\right)^{1/2} |\mathbf{r} - \mathbf{r}'|\right]}{4\pi |\mathbf{r} - \mathbf{r}'|} . \tag{25.58}$$

Inversion of the Laplace transform yields the following expression for the propagator:

$$G(\mathbf{r}, \mathbf{r}'; t) = \frac{D}{(4\pi Dt)^{3/2}} \exp\left[-\frac{|\mathbf{r} - \mathbf{r}'|^2}{4Dt} - \alpha t\right] . \tag{25.59}$$

For a point source of the form $S_d(\mathbf{r}, t) = S_0 \delta(\mathbf{r} - \mathbf{r}_0)\delta(t)$, where S_0 is the source power and $\mathbf{r}_0$ is the position of the source, we see that u is of the form

$$u(\mathbf{r}, t) = \frac{S_0}{(4\pi Dt)^{3/2}} \exp\left[-\frac{|\mathbf{r} - \mathbf{r}_0|^2}{4Dt} - \alpha t\right] . \tag{25.60}$$

Evidently, for a fixed separation $L = |\mathbf{r} - \mathbf{r}_0|$ of the source and detector, u decays on the scale of the diffusion time $\tau = L^2/D$. The half-space and slab geometries described in Sections 25.4 and 25.5 can be treated along similar lines. For a point source in a half-space with an absorbing boundary, it follows from (25.35) that u is given by

$$u(\mathbf{r}, t) = \frac{S_0}{(4\pi Dt)^{3/2}} \left(\exp\left[-\frac{|\mathbf{r} - \mathbf{r}_0|^2}{4Dt} - \alpha t\right] - \exp\left[-\frac{|\mathbf{r} - \tilde{\mathbf{r}}_0|^2}{4Dt} - \alpha t\right]\right) . \tag{25.61}$$

Likewise, for a point source in a slab of width L with absorbing boundaries, we find that

$$u(\mathbf{r}, t) = \frac{S_0}{(4\pi Dt)^{3/2}} \sum_{n=-\infty}^{\infty} \left(\exp\left[-\frac{|\mathbf{r} - \mathbf{r}_n|^2}{4Dt} - \alpha t\right] - \exp\left[-\frac{|\mathbf{r} - \tilde{\mathbf{r}}_n|^2}{4Dt} - \alpha t\right]\right) , \tag{25.62}$$

where $\mathbf{r}_n = \mathbf{r}_0 + 2nL\hat{\mathbf{z}}$. This result can be rewritten by making use of the Poisson summation formula

$$\sum_{n=-\infty}^{\infty} \delta(z - 2\pi n) = \frac{1}{2\pi} \sum_{n=-\infty}^{\infty} \exp(inz) . \tag{25.63}$$

We find that

$$\begin{aligned} u(\mathbf{r}, t) &= \frac{2S_0}{4\pi DtL} \exp\left[-\frac{|\boldsymbol{\rho} - \boldsymbol{\rho}_0|^2}{4Dt} - \alpha t\right] \sum_{n=-\infty}^{\infty} \sin\left(\frac{n\pi z}{L}\right) \sin\left(\frac{n\pi z_0}{L}\right) \\ &\times \exp\left(-\frac{\pi^2 n^2 Dt}{L^2}\right) , \end{aligned} \tag{25.64}$$

where $\mathbf{r} = (\boldsymbol{\rho}, z)$ and $\mathbf{r}_0 = (\boldsymbol{\rho}_0, z_0)$. Note that at long times, the time dependence of the energy density is dominated by the contribution of the lowest-order Fourier mode and

$$u(\mathbf{r}, t) \sim \exp\left(-\frac{\pi^2 D t}{L^2}\right) . \tag{25.65}$$

The energy density decays exponentially on the diffusion time τ, which is typical of diffusive transport.

References and Additional Reading

The diffusion equation, Green's functions and general solutions are analyzed in:

P. M. Morse and H. Feshbach, *Methods of Theoretical Physics* (McGraw-Hill, New York, 1953), vol. 2, chap. 12.

Boundary conditions for the diffusion equation are discussed in:

E.W. Larsen and J.B. Keller, J. Math. Phys. **15**, 75 (1974).
R. Aronson, J. Opt. Soc. Am. A **12**, 2532 (1995).
A. Ishimaru, *Wave Propagation and Scattering in Random Media* (IEEE Press, Piscataway, 1997), chap. 9.
A.D. Kim, J. Opt. Soc. Am. A **28**,1007 (2011).

The introduction of the boundary conditions in this chapter follows the approach presented in:

V.A. Markel and J.C. Schotland, J. Opt. Soc. Am. A **19**, 558 (2002).
This paper also contains a derivation of the Green's functions for half-space and slab geometries.

Boundary conditions for the diffusion equation at the interface between index mismatched media are derived in:

J.X. Zhu, D.J. Pine and D.A. Weitz, Phys. Rev. A **44**, 3948 (1991).
J. Ripoll and M. Nieto-Vesperinas, J. Opt. Soc. Am. A **16**, 1947 (1999).
J. Ripoll Lorenzo, *Principles of Diffuse Light Propagation* (World Scientific, Singapore, 2012).

26

Diffuse Optics

In this chapter, we study the optics of diffuse waves. We focus on the basic properties of such waves, including interference, refraction and diffraction, and contrast our findings with the corresponding phenomena for scalar waves. We note that there are applications of diffuse light to many fields, including biomedical optics and soft-matter physics.

26.1 Diffuse Waves

We will take as our starting point the time-dependent diffusion equation (24.28) in a homogeneous medium:

$$\frac{\partial u}{\partial t} - D\nabla^2 u + \alpha u = S_d \tag{26.1}$$

We consider an amplitude-modulated continuous wave source of the form

$$S_d(\mathbf{r}, t) = (1 + A\exp(-i\omega t))S_d(\mathbf{r}) , \tag{26.2}$$

where ω is the modulation frequency and the amplitude $A < 1$. If $u(\mathbf{r}, t)$ is decomposed into a zero-frequency component $u_0(\mathbf{r})$ and a frequency-dependent component $u(\mathbf{r})$ according to

$$u(\mathbf{r}, t) = u_0(\mathbf{r}) + A\exp(-i\omega t)u(\mathbf{r}) , \tag{26.3}$$

then u obeys the equation

$$-\nabla \cdot D\nabla u + (\alpha - i\omega)u = S_d . \tag{26.4}$$

It is important to note that solutions to (26.4) can be constructed following the methods in Chapter 25. The only modification that is needed is to redefine the diffuse wavenumber k as

$$k^2 = \frac{\alpha - i\omega}{D} . \tag{26.5}$$

26.2 Wave Properties

Modulation has important effects on the propagation of diffuse light. To understand this point, we begin by considering a point source in a homogeneous medium. It follows immediately from (25.22) that the energy density is given by

$$u(\mathbf{r}) = \frac{S_0}{4\pi D} \frac{\exp(-k|\mathbf{r} - \mathbf{r}_0|)}{|\mathbf{r} - \mathbf{r}_0|} , \tag{26.6}$$

where the source has power S_0 and position $\mathbf{r}_0$. Here the wavenumber k is complex-valued and we write $k = k' - ik''$, where

$$k' = \frac{1}{\sqrt{2D}} \left[\left(\omega^2 + \alpha^2\right)^{1/2} + \alpha \right]^{1/2} , \tag{26.7}$$

$$k'' = \frac{1}{\sqrt{2D}} \left[\left(\omega^2 + \alpha^2\right)^{1/2} - \alpha \right]^{1/2} . \tag{26.8}$$

Thus (26.6) becomes

$$u(\mathbf{r}) = \frac{S_0}{4\pi D} \frac{\exp(ik''|\mathbf{r} - \mathbf{r}_0|)}{|\mathbf{r} - \mathbf{r}_0|} \exp(-k'|\mathbf{r} - \mathbf{r}_0|) . \tag{26.9}$$

Equation (26.9) corresponds to an outgoing spherical wave with wavelength $\lambda_d = 2\pi/k''$ that decays exponentially on the scale of the attenuation length $L_d = 1/k'$. The phase velocity of the wave is $v_p = \omega/|k''|$. In the high-frequency limit $\omega \gg \alpha$, we obtain $v_p \sim \sqrt{D\omega}$. We also find that the group velocity $v_g = |d\omega/dk''| \sim \sqrt{D\omega}$. Thus the diffuse wave is dispersive. We note that the divergence of v_g at high frequencies is unphysical and is an artifact of the diffusion approximation. In many applications, $\omega \ll \alpha$, which implies that $k'' \ll k'$. Thus $\lambda_d \gg L_d$ and the wave propagates several wavelengths before it is fully attenuated.

We emphasize that diffuse optics shares many features of the optics of scalar waves including, as we will see, interference, refraction and diffraction. However, the attenuation of diffuse waves, which is akin to the exponential decay of evanescent waves in near-field optics, is responsible for a variety of novel effects, as will be illustrated later in this chapter.

26.3 Interference

We consider a pair of point sources with power density

$$S_d(\mathbf{r}) = S_0\delta(\mathbf{r} - \mathbf{r}_0 + \mathbf{a}/2) + \exp(i\varphi)S_0\delta(\mathbf{r} - \mathbf{r}_0 - \mathbf{a}/2) . \tag{26.10}$$

Here $\mathbf{r}_0$ is the center of the source, $\mathbf{a}$ is the separation between the point sources and φ is the phase difference. The corresponding energy density is given by

$$u(\mathbf{r}) = \frac{S_0}{4\pi D}\left(\frac{\exp(-k|\mathbf{r}-\mathbf{r}_0+\mathbf{a}/2|)}{|\mathbf{r}-\mathbf{r}_0+\mathbf{a}/2|} + \exp(i\varphi)\frac{\exp(-k|\mathbf{r}-\mathbf{r}_0-\mathbf{a}/2|)}{|\mathbf{r}-\mathbf{r}_0-\mathbf{a}/2|}\right). \tag{26.11}$$

If $\varphi = 0$, there is constructive interference, and if $\varphi = \pi$, there is destructive interference. Evidently, if $a \gg 2\pi/k'$, there is no interference and the sources act independently. If $\varphi = \pi$, we refer to the pair as a dipole source and

$$u(\mathbf{r}) = \frac{S_0}{4\pi D}\left(\frac{\exp(-k|\mathbf{r}-\mathbf{r}_0+\mathbf{a}/2|)}{|\mathbf{r}-\mathbf{r}_0+\mathbf{a}/2|} - \frac{\exp(-k|\mathbf{r}-\mathbf{r}_0-\mathbf{a}/2|)}{|\mathbf{r}-\mathbf{r}_0-\mathbf{a}/2|}\right). \tag{26.12}$$

We note that there is a null plane with normal $\hat{\mathbf{a}}$ containing the point $\mathbf{r} = \mathbf{r}_0$, which corresponds to the center of the dipole. The presence of a small absorber perturbs the null plane, an effect that can be used to localize the absorber. Finally, we note that if the point of observation is very far from the center of the dipole, then

$$u(\mathbf{r}) = \frac{1}{4\pi D}\mathbf{p}\cdot\nabla\left(\frac{\exp(-k|\mathbf{r}-\mathbf{r}_0|)}{|\mathbf{r}-\mathbf{r}_0|}\right), \tag{26.13}$$

where the dipole moment $\mathbf{p} = S_0\mathbf{a}$.

26.4 Refraction

Consider a planar interface separating two semi-infinite homogeneous diffuse media. Medium 1 consists of the half-space $z < 0$. It contains a point source of unit strength at the position $\mathbf{r}_0 = (0, 0, z_0)$, has optical properties (α_1, D_1) and wavenumber k_1. Medium 2 consists of the half-space $z \geq 0$. It has optical properties (α_2, D_2) and wavenumber k_2. In each half-space, the energy densities obey

$$-\nabla^2 u_1 + k_1^2 u_1 = \frac{1}{D_1}\delta(\mathbf{r}-\mathbf{r}_0), \tag{26.14}$$

$$-\nabla^2 u_2 + k_2^2 u_2 = 0. \tag{26.15}$$

The plane $z = 0$ is a diffuse-diffuse interface on which the energy densities obey the boundary conditions

$$u_1 = u_2, \tag{26.16}$$

$$D_1\frac{\partial u_1}{\partial z} = D_2\frac{\partial u_2}{\partial z}. \tag{26.17}$$

To proceed, we expand u_1 and u_2 into plane waves of the form

$$u_1(\mathbf{r}) = u_0(\mathbf{r}) + \int \frac{d^2q}{(2\pi)^2} A_1(\mathbf{q}) \exp\left(i\mathbf{q}\cdot\rho + Q_1(q)z\right), \tag{26.18}$$

$$u_2(\mathbf{r}) = \int \frac{d^2q}{(2\pi)^2} A_2(\mathbf{q}) \exp\left(i\mathbf{q}\cdot\boldsymbol{\rho} - Q_2(q)z\right) , \tag{26.19}$$

where $\mathbf{r} = (\boldsymbol{\rho}, z)$ and $Q_i(q) = \sqrt{q^2 + k_i^2}$ for $i = 1, 2$. In addition, u_0 is the energy density of a point source in an infinite medium, which according to (25.30) is given by

$$u_0(\mathbf{r}) = \frac{1}{2D_1(2\pi)^2} \int \frac{d^2q}{Q_1(q)} \exp\left(i\mathbf{q}\cdot\boldsymbol{\rho} - Q_1(q)|z - z_0|\right) . \tag{26.20}$$

Applying the above boundary conditions, we find that the coefficients A_1 and A_2 obey the linear equations

$$A_1(q) - A_2(q) = -\frac{1}{2DQ_1(q)} \exp(Q_1(q)z_0), \tag{26.21}$$

$$D_1Q_1(q)A_1(q) + D_2Q_2(q)A_2(q) = \frac{1}{2} \exp(Q_1(q)z_0). \tag{26.22}$$

Next, we define the reflection coefficient $R = A_1/A_0$ and transmission coefficient $T = A_2/A_0$, where $A_0 = 1/(2D_1Q_1)\exp(Q_1z_0)$. Upon solving (26.21) and (26.22), we obtain

$$R(q) = \frac{D_1Q_1(q) - D_2Q_2(q)}{D_1Q_1(q) + D_2Q_2(q)} \tag{26.23}$$

$$T(q) = \frac{2D_1Q_1(q)}{D_1Q_1(q) + D_2Q_2(q)} . \tag{26.24}$$

Note that $T = 1 + R$. Putting everything together, we find that

$$\begin{aligned} u_1(\mathbf{r}) &= \frac{1}{2D_1(2\pi)^2} \int \frac{d^2q}{Q_1(q)} \exp(i\mathbf{q}\cdot\boldsymbol{\rho}) \\ &\quad \left[\exp((-Q_1(q)|z - z_0|) + R(q)\exp(-Q_1(q)|z + z_0|)\right] , \end{aligned} \tag{26.25}$$

$$u_2(\mathbf{r}) = \frac{1}{2D_1(2\pi)^2} \int \frac{d^2q}{Q_1(q)} T(q) \exp\left(i\mathbf{q}\cdot\boldsymbol{\rho} - Q_2(q)z + Q_1(q)z_0\right) . \tag{26.26}$$

Evidently, (26.25) and (26.26) determine the energy density everywhere in space. See Fig. 28.1, in which the refraction of a diffuse wave at a flat interface is shown.

We now examine the behavior of the above plane-wave modes. Suppose that a unit-amplitude plane wave is incident from medium 1. The energy density in medium 1 is a superposition of the incident wave and reflected wave:

$$u_1(\mathbf{r}) = \exp(i\mathbf{q}_1\cdot\boldsymbol{\rho} + Q_1(q_1)z) + R(q)\exp(i\mathbf{q}_1'\cdot\boldsymbol{\rho} + Q_1(q_1')z) . \tag{26.27}$$

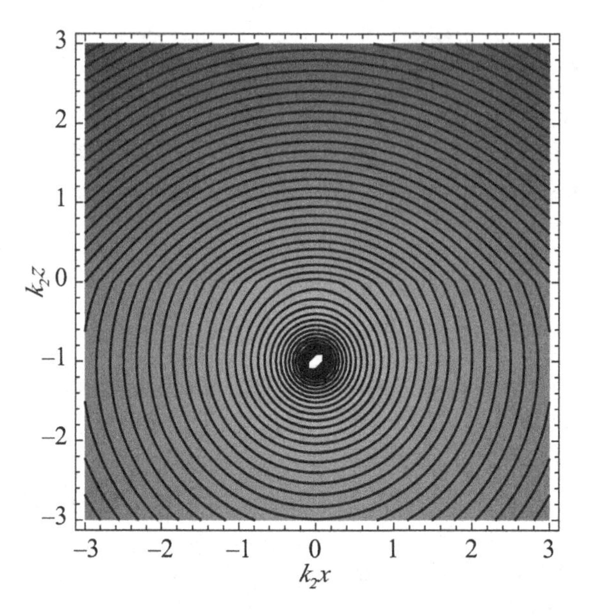

Fig. 26.1 Refraction of diffuse light emitted by a point source. Here $k_2z_0 = -1$ and $k_2/k_1 = 2$.

The energy density in medium 2 is given by

$$u_2(\mathbf{r}) = T(q_2)\exp(i\mathbf{q}_2 \cdot \boldsymbol{\rho} - Q_2(q_2)z) , \tag{26.28}$$

which corresponds to the transmitted wave. Here $\mathbf{q}_1$, $\mathbf{q}_1'$ and $\mathbf{q}_2$ are the transverse wave vectors of the incident, reflected and transmitted waves, respectively. We note that the law of reflection and Snell's law hold in the forms

$$\theta_1 = \theta_1' , \tag{26.29}$$

$$k_1 \sin\theta_1 = k_2 \sin\theta_2 . \tag{26.30}$$

Here $\theta_1, \theta_1', \theta_2$ are the angles between the transverse wave vectors $\mathbf{q}_1, \mathbf{q}_1', \mathbf{q}_2$ and the normal to the interface, respectively. We also note that diffuse waves do not undergo total internal reflection since the transmission coefficient $T(q)$ is non-vanishing for all q. However, there are analogs of Brewster modes, where the reflection coefficient $R(q)$ vanishes. It is easily seen that $R(q_B) = 0$, where

$$q_B = \sqrt{\frac{D_2\alpha_2 - D_1\alpha_1}{D_1^2 - D_2^2}} \tag{26.31}$$

and, for simplicity, we have assumed that $\omega = 0$.

26.5 Diffraction

We now discuss the theory of diffraction for diffuse light. We consider an infinite homogeneous medium with a source present in the half-space $z < 0$. We suppose that the energy density is known on the plane $z = 0$ and wish to determine the energy density in the $z \geq 0$ half-space. This problem may be solved by making use of Rayleigh–Sommerfeld theory, suitably adapted to diffuse waves. The energy density in the $z \geq 0$ half-space obeys

$$-\nabla^2 u + k^2 u = 0 . \tag{26.32}$$

Following the development in Section 7.1, we find that energy density is given by

$$u(\boldsymbol{\rho}, z) = \int d^2\rho K(\boldsymbol{\rho}, \boldsymbol{\rho}'; z) u(\boldsymbol{\rho}, 0) . \tag{26.33}$$

Here the propagator K can be represented as the Fourier integral

$$K(\boldsymbol{\rho}, \boldsymbol{\rho}'; z) = \int \frac{d^2q}{(2\pi)^2} \exp\left[i\mathbf{q} \cdot (\boldsymbol{\rho} - \boldsymbol{\rho}') - Q(q)z\right] . \tag{26.34}$$

As may be expected,

$$\lim_{z \to 0} K(\boldsymbol{\rho}, \boldsymbol{\rho}'; z) = \delta(\boldsymbol{\rho} - \boldsymbol{\rho}') , \tag{26.35}$$

so that the energy density is reproduced on the $z = 0$ plane. Note that there is exponential decay of all frequencies on propagation. In contrast, frequency components of scalar waves below the band-limit $|\mathbf{q}| \leq k$ survive propagation, which gives rise to the diffraction-limited resolution of optical instruments.

The above results can be used to analyze the spatial resolution of diffuse light imaging. Suppose that the energy density $u(\boldsymbol{\rho}, 0) = \delta(\boldsymbol{\rho})$, which corresponds to an incident field that has passed through a small aperture in the $z = 0$ plane. It follows from (26.33) that the energy density in the $z > 0$ half-space is given by $u(\boldsymbol{\rho}, z) = K(\boldsymbol{\rho}, 0; z)$. We interpret the quantity $h(\boldsymbol{\rho}) = u(\boldsymbol{\rho}, z_0)$ as the image of the aperture in the plane $z = z_0$; it is also known as the point spread function. In Fig. 26.2, we plot h for different values of the distance of propagation. We see that the maximum of h occurs at the origin and that as z_0 increases, h becomes broader, indicating a loss of resolution. In a more precise manner, we define the spatial resolution ΔR to be the full width at half maximum of h. That is, $\Delta R = 2\rho$, where $\boldsymbol{\rho}$ solves the equation

$$u(\boldsymbol{\rho}, z) = \frac{1}{2} u(0, z) . \tag{26.36}$$

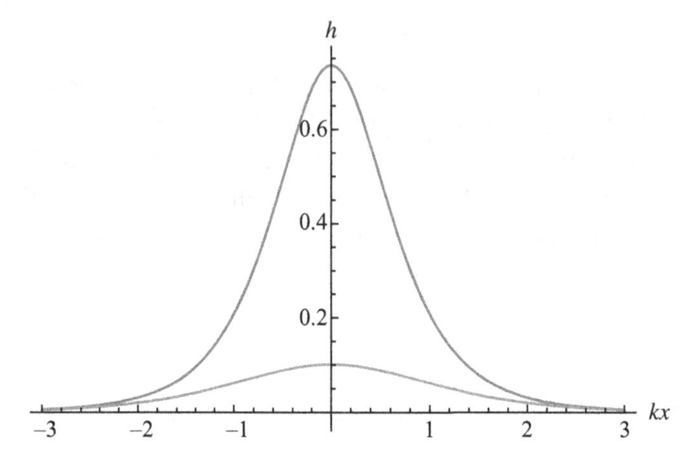

Fig. 26.2 Dependence of the point spread function $h(x, 0, z)$ on the distance of propagation z. The top curve corresponds to $kz_0 = 1$ and the bottom curve to $kz_0 = 2$.

To proceed, we use the facts that

$$u(0, z) = \frac{1}{2\pi z^2} e^{-kz}(1 + kz) , \tag{26.37}$$

and for $kz \ll 1$,

$$u(\rho, z) = \frac{1}{2\pi \left(1 + \dfrac{\rho^2}{z^2}\right)^{3/2}} \left(\frac{1}{z^2} - \frac{k^2}{2}\right) . \tag{26.38}$$

Both of the above results follow from evaluating the integral in (26.34). We then find that the resolution is given by

$$\Delta R = z\left[a - b(kz)^3 + O\left((kz)^4\right)\right] , \tag{26.39}$$

where

$$a = 2\sqrt{2^{2/3} - 1} \simeq 1.53 , \quad b = \frac{2\sqrt{2^{2/3} - 1}}{3(2 - 2^{1/3})} \simeq 0.69 . \tag{26.40}$$

In Fig. 26.3, we illustrate the dependence of ΔR on the distance of propagation. We note that in the absence of absorption $\Delta R \propto z$. This differs from the case of scalar waves, in which the far-field resolution reaches the diffraction limit of $\lambda/2$. We also note that in the presence of absorption, ΔR decreases. This finding may be explained by the exponential loss of long diffusion paths.

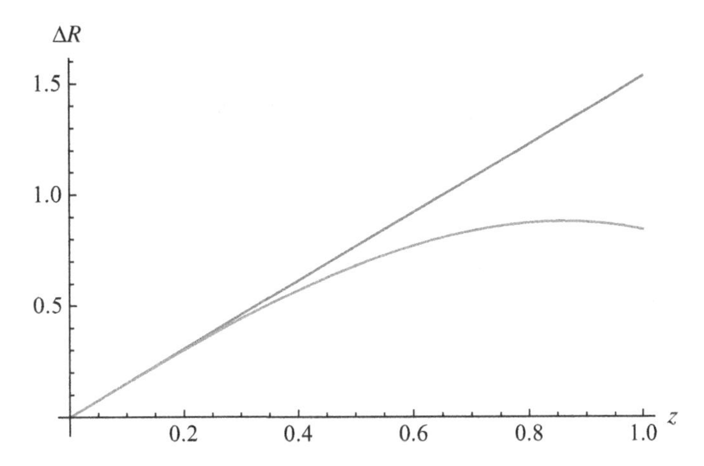

Fig. 26.3 Dependence of the resolution ΔR on the distance of propagation z. The top curve corresponds to $k = 0$ and the bottom curve to $k = 1\,\text{cm}^{-1}$. All lengths are measured in units of centimeters.

References and Additional Reading

The concept of diffuse waves to image inside scattering media was introduced in:

J.B. Fishkin, E. Gratton, J. Opt. Soc. Am. A **10**, 27 (1993).

The use of diffuse waves to image strongly scattering media, including biological tissues, was highlighted in the following paper:

A.G. Yodh and B. Chance, Phys. Today **48**, 34 (1995).

A measurement of refraction of diffuse waves was reported in:

M.A. O'Leary, D.A. Boas, B. Chance and A.G. Yodh, Phys. Rev. Lett. **69**, 2658 (1992).

The following paper gives an interesting overview of diffuse waves, beyond optics:

A. Mandelis, Phys. Today **53**, 29 (2000).

The spatial resolution of diffuse waves is analyzed in the following paper, using concepts from diffraction theory:

J. Ripoll, M. Nieto-Vesperinas, and R. Carminati, J. Opt. Soc. Am. A **16**, 1466 (1999).

A comprehensive treatment of diffuse waves and their use in biomedical optics can be found in the following textbook:

J. Ripoll Lorenzo, *Principles of Diffuse Light Propagation* (World Scientific, Singapore, 2012).

27

Scattering of Diffuse Waves

In this chapter, we study the propagation of diffuse light in inhomogeneous media. We begin by developing the integral equation formulation of scattering theory with diffuse waves. We then apply this theory to the case of small inhomogeneities. Finally, we discuss surface integral equation methods and the extinction theorem.

27.1 Integral Equations

We consider the propagation of a diffuse wave in an inhomogeneous medium contained in the volume V. The energy density satisfies the equation

$$-\nabla \cdot D(\mathbf{r})\nabla u(\mathbf{r}) + [\alpha(\mathbf{r}) - i\omega]u(\mathbf{r}) = S_d(\mathbf{r}) \qquad (27.1)$$

and obeys the boundary condition

$$u + \ell\frac{\partial u}{\partial n} = 0\,. \qquad (27.2)$$

To proceed, we decompose α and D into constant and spatially varying parts:

$$\alpha(\mathbf{r}) = \alpha_0 + \delta\alpha(\mathbf{r})\,, \qquad D(\mathbf{r}) = D_0 + \delta D(\mathbf{r})\,, \qquad (27.3)$$

where α_0 and D_0 are the absorption and diffusion coefficients in a homogeneous reference medium. Equation (27.1) thus becomes

$$-D_0\nabla^2 u(\mathbf{r}) + (\alpha_0 - i\omega)u(\mathbf{r}) = \nabla \cdot \delta D(\mathbf{r})\nabla u(\mathbf{r}) - \delta\alpha(\mathbf{r})u(\mathbf{r}) + S_d(\mathbf{r}). \qquad (27.4)$$

The solution to (27.4) is given by

$$u(\mathbf{r}) = u_0(\mathbf{r}) - \int d^3r' \left[G(\mathbf{r},\mathbf{r}')u(\mathbf{r}')\delta\alpha(\mathbf{r}') + \nabla_{\mathbf{r}}G(\mathbf{r},\mathbf{r}')\cdot\nabla u(\mathbf{r}')\delta D(\mathbf{r}')\right]\,, \qquad (27.5)$$

where we have integrated by parts and have assumed that δD vanishes on the boundary. Here the Green's function G obeys

$$-D_0\nabla_{\mathbf{r}}^2 G(\mathbf{r},\mathbf{r}') + (\alpha_0 - i\omega)G(\mathbf{r},\mathbf{r}') = \delta(\mathbf{r}-\mathbf{r}'), \tag{27.6}$$

along with homogeneous boundary conditions, and u_0, which is the energy density in the reference medium produced by the source S_d, is defined by

$$u_0(\mathbf{r}) = \int G(\mathbf{r},\mathbf{r}')S_d(\mathbf{r}')d^3r' . \tag{27.7}$$

The integral equation (27.5) is the analog of the Lippmann–Schwinger equation for diffuse waves [see (10.8)]. It can be iterated to produce a Born series of the form

$$u(\mathbf{r}) = \int d^3r_1 K_i^{(1)}(\mathbf{r};\mathbf{r}_1)\eta_i(\mathbf{r}_1) + \int d^3r_1 d^3r_2 K_{ij}^{(2)}(\mathbf{r};\mathbf{r}_1,\mathbf{r}_2)\eta_i(\mathbf{r}_1)\eta_j(\mathbf{r}_2) + \cdots , \tag{27.8}$$

where

$$\eta(\mathbf{r}) = \begin{pmatrix} \eta_1(\mathbf{r}) \\ \eta_2(\mathbf{r}) \end{pmatrix} = \begin{pmatrix} \delta\alpha(\mathbf{r}) \\ \delta D(\mathbf{r}) \end{pmatrix} , \tag{27.9}$$

and summation over repeated indices is implied with $i, j = 1, 2$ (the Born series and the Born approximation are introduced in Chapter 10). The components of the operators $K^{(1)}$ and $K^{(2)}$ are defined by

$$K_1^{(1)}(\mathbf{r};\mathbf{r}_1) = G(\mathbf{r},\mathbf{r}_1)u_0(\mathbf{r}_1) , \tag{27.10}$$
$$K_2^{(1)}(\mathbf{r};\mathbf{r}_1) = \nabla_{\mathbf{r}}G(\mathbf{r},\mathbf{r}_1)\cdot\nabla u_0(\mathbf{r}_1) , \tag{27.11}$$
$$K_{11}^{(2)}(\mathbf{r};\mathbf{r}_1,\mathbf{r}_2) = -G(\mathbf{r},\mathbf{r}_1)G(\mathbf{r}_1,\mathbf{r}_2)u_0(\mathbf{r}_2) , \tag{27.12}$$
$$K_{12}^{(2)}(\mathbf{r};\mathbf{r}_1,\mathbf{r}_2) = -G(\mathbf{r},\mathbf{r}_1)\nabla_{\mathbf{r}_1}G(\mathbf{r}_1,\mathbf{r}_2)\cdot\nabla u_0(\mathbf{r}_2) , \tag{27.13}$$
$$K_{21}^{(2)}(\mathbf{r};\mathbf{r}_1,\mathbf{r}_2) = -G(\mathbf{r},\mathbf{r}_1)\nabla_{\mathbf{r}_2}G(\mathbf{r}_1,\mathbf{r}_2)\cdot\nabla u_0(\mathbf{r}_2) , \tag{27.14}$$
$$K_{22}^{(2)}(\mathbf{r};\mathbf{r}_1,\mathbf{r}_2) = -\nabla_{\mathbf{r}}G(\mathbf{r},\mathbf{r}_1)\cdot\nabla_{\mathbf{r}_1}\left[\nabla_{\mathbf{r}_2}G(\mathbf{r}_1,\mathbf{r}_2)\cdot\nabla u_0(\mathbf{r}_2)\right] . \tag{27.15}$$

The Born approximation corresponds to truncating the Born series at first order in η. Equation (27.8) thus becomes

$$u(\mathbf{r}) = u_0(\mathbf{r}) - \int d^3r' \left[G(\mathbf{r},\mathbf{r}')u_0(\mathbf{r}')\delta\alpha(\mathbf{r}') + \nabla_{\mathbf{r}}G(\mathbf{r},\mathbf{r}')\cdot\nabla u_0(\mathbf{r}')\delta D(\mathbf{r}')\right] . \tag{27.16}$$

The Born approximation is accurate for small weak inhomogeneities. Higher-order terms in the Born series correspond to multiple scattering of the incident diffuse wave.

27.2 Small Inhomogeneities

One of the very few cases in which it is possible to analytically solve the integral equations of scattering theory corresponds to the physical setting of small inhomogeneities. For simplicity, we consider an infinite medium and begin with the case of a single point absorber. The absorption coefficient is then given by

$$\alpha(\mathbf{r}) = \alpha_0 + \delta\alpha_0 V \delta(\mathbf{r} - \mathbf{r}_0) , \tag{27.17}$$

where $\mathbf{r}_0$, $\delta\alpha_0$ and V are the position, strength and volume of the absorber, respectively. Substituting (27.17) into (27.5), we obtain

$$u(\mathbf{r}) = u_0(\mathbf{r}) - \delta\alpha_0 V G(\mathbf{r}, \mathbf{r}_0) u(\mathbf{r}_0) . \tag{27.18}$$

Here the Green's function G is given by

$$G(\mathbf{r}, \mathbf{r}') = \frac{1}{4\pi D_0} \frac{\exp(-k|\mathbf{r} - \mathbf{r}'|)}{|\mathbf{r} - \mathbf{r}'|} , \tag{27.19}$$

where the wavenumber k is defined as

$$k^2 = \frac{\alpha_0 - i\omega}{D_0} . \tag{27.20}$$

Equation (27.18) is an algebraic equation for u which has the solution

$$u(\mathbf{r}) = u_0(\mathbf{r}) - \delta\alpha V G(\mathbf{r}, \mathbf{r}_0) u_0(\mathbf{r}_0) , \tag{27.21}$$

where the renormalized absorption $\delta\alpha$ is defined by

$$\delta\alpha = \frac{\delta\alpha_0}{1 + \delta\alpha_0 V G(\mathbf{r}_0, \mathbf{r}_0)} . \tag{27.22}$$

The physical interpretation of the renormalized absorption is somewhat subtle. The quantity $G(\mathbf{r}_0, \mathbf{r}_0)$ is divergent, and thus $\delta\alpha$ vanishes. In order to rectify the situation, it is necessary to regularize the divergence. To do so, we examine the behavior of $G(\mathbf{r}, \mathbf{r}')$ for small $|\mathbf{r} - \mathbf{r}'|$:

$$G(\mathbf{r}, \mathbf{r}') = \frac{1}{4\pi D_0 |\mathbf{r} - \mathbf{r}'|} - \frac{k}{4\pi D_0} + O(|\mathbf{r} - \mathbf{r}'|) . \tag{27.23}$$

It can be seen that the singular part of G is isolated in the first term above. Now, consider the Fourier integral representation

$$\frac{1}{|\mathbf{r} - \mathbf{r}'|} = \int \frac{d^3 q}{(2\pi)^3} \frac{\exp(i\mathbf{q} \cdot (\mathbf{r} - \mathbf{r}'))}{q^2} . \tag{27.24}$$

We introduce a high-frequency cutoff on the wave vector integration to regularize the divergence:

$$G(\mathbf{r}_0, \mathbf{r}_0) = \frac{1}{4\pi D_0} \int_{|\mathbf{q}| \le \frac{2\pi}{\Lambda}} \frac{d^3q}{(2\pi)^3} \frac{1}{q^2} - \frac{k}{4\pi D_0}$$
$$= \frac{1}{4\pi D_0} \left(\frac{1}{\pi \Lambda} - k \right) , \tag{27.25}$$

where Λ defines the cutoff. We may identify Λ with the linear size of the absorber so that $k\Lambda \ll 1$. Note that $G(\mathbf{r}_0, \mathbf{r}_0)$ does not depend upon $\mathbf{r}_0$, as may be expected from translational invariance. An alternative to the above regularization procedure is to assign a finite size to the absorber and take the limit $V \to 0$, as is done for the problem of small scatterers in Chapter 12.

We now generalize the above results to the case of a collection of point absorbers. We consider a homogeneous medium containing N point absorbers with positions $\mathbf{R}_1, \ldots, \mathbf{R}_N$, absorptions $\delta\alpha_1, \ldots, \delta\alpha_N$, and volumes $V_1, \ldots, V_N$. The total absorption of the medium is given by

$$\alpha(\mathbf{r}) = \alpha_0 + \sum_i \delta\alpha_i V_i \delta(\mathbf{r} - \mathbf{R}_i) , \tag{27.26}$$

where α_0 is the background absorption. For convenience, each absorber is assumed to have a distinct volume, even though only the product of $\delta\alpha_i$ and V_i arises in (27.26). Inserting the above expression for α into the integral equation (27.5), we find that energy density obeys the equation

$$u(\mathbf{r}) = u_0(\mathbf{r}) - \sum_i G(\mathbf{r}, \mathbf{R}_i) \delta\alpha_i V_i u(\mathbf{R}_i) . \tag{27.27}$$

Evidently, (27.27) determines u self-consistently. This observation leads to a system of algebraic equations for u of the form

$$\sum_j M_{ij} u(\mathbf{R}_j) = u_0(\mathbf{R}_i) , \tag{27.28}$$

where

$$M_{ij} = \delta_{ij} + G(\mathbf{R}_i, \mathbf{R}_j) \delta\alpha_j V_j . \tag{27.29}$$

Solving (27.28) for $u(\mathbf{R}_i)$, we obtain

$$u(\mathbf{r}) = u_0(\mathbf{r}) - \sum_{i,j} G(\mathbf{r}, \mathbf{R}_i) T_{ij} u_0(\mathbf{R}_j) , \tag{27.30}$$

where $T_{ij} = \delta\alpha_i V_i M_{ij}^{-1}$ is the analog of the renormalized absorption. Note that as in (27.25), the diagonal elements of M need to be properly regularized. If the absorbers are well separated, such that $k|\mathbf{R}_i - \mathbf{R}_j| \gg 1$, then

$$M_{ij} = \delta_{ij}(1 + \delta\alpha_i V_i G(0)(\mathbf{R}_i, \mathbf{R}_i)) \tag{27.31}$$

and

$$u(\mathbf{r}) = u_0(\mathbf{r}) - \sum_i G(\mathbf{r}, \mathbf{R}_i) \frac{\delta\alpha_i V_i}{1 + \delta\alpha_i V_i G(\mathbf{R}_i, \mathbf{R}_i)} u_0(\mathbf{R}_i) . \tag{27.32}$$

In this case, u corresponds to the superposition of the energy density for N isolated point absorbers. However, if the absorbers are sufficiently close so as to interact, then M depends in a nontrivial way upon the positions of all the absorbers in the system. In Fig. 27.1, we plot the energy density in the plane of two identical interacting point absorbers.

27.3 Extinction Theorem

We now consider the propagation of diffuse light in piecewise homogeneous media. For simplicity, we consider a volume V_1, bounded by a surface S_1, that contains a

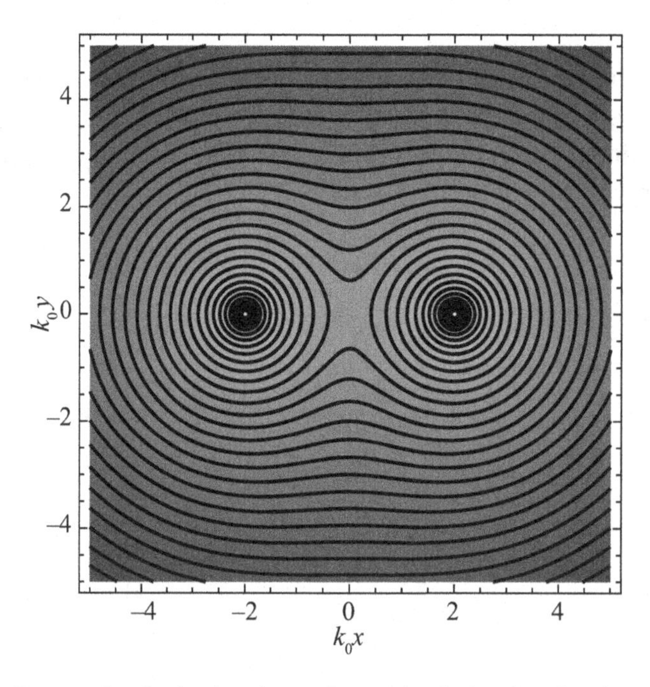

Fig. 27.1 Energy density in the plane of two identical point absorbers. A point source of unit amplitude is placed at the origin. Here $k_0^3 V = 0.005$ and $\delta\alpha_0/\alpha_0 = 0.1$.

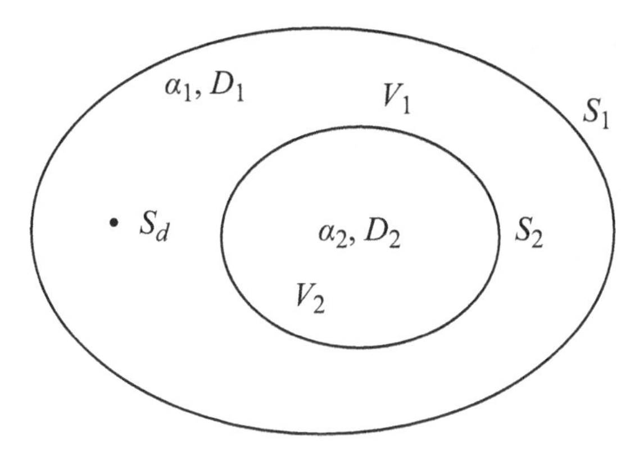

Fig. 27.2 Illustrating the geometry in the extinction theorem.

volume V_2 bounded by a surface S_2, as illustrated in Fig. 27.2. We assume that V_1 and V_2 are homogeneous with optical properties (α_1, D_1) and (α_2, D_2), respectively. Thus u_1 and u_2, which are the energy densities in V_1 and V_2, respectively, obey

$$- D_1\nabla^2 u_1 + \alpha_1 u_1 = S_d \quad \text{in} \quad V_1 , \tag{27.33}$$

$$-D_2\nabla^2 u_2 + \alpha_2 u_2 = 0 \quad \text{in} \quad V_2 , \tag{27.34}$$

where we have assumed that the source is contained in V_1 and we have not yet specified the boundary conditions satisfied by u_1 and u_2.

We now derive the extinction theorem. To proceed, we introduce the Green's functions G_1 and G_2, which satisfy

$$- D_1\nabla^2 G_1(\mathbf{r}, \mathbf{r}') + \alpha_1 G_1(\mathbf{r}, \mathbf{r}') = \delta(\mathbf{r} - \mathbf{r}') \quad \text{in} \quad V_1 , \tag{27.35}$$

$$-D_2\nabla^2 G_2(\mathbf{r}, \mathbf{r}') + \alpha_2 G_2(\mathbf{r}, \mathbf{r}') = \delta(\mathbf{r} - \mathbf{r}') \quad \text{in} \quad V_2 . \tag{27.36}$$

Once again, we have not yet specified the boundary conditions obeyed by G_1 and G_2. Following the development in Section 5.1, we find that

$$\int_{V_1} \delta(\mathbf{r} - \mathbf{r}')u_1(\mathbf{r}')d^3r' = u_0(\mathbf{r}) + D_1 \int_{S_1\cup S_2} d^2r' \left[G_1(\mathbf{r}, \mathbf{r}')\frac{\partial u_1(\mathbf{r}')}{\partial n} - u_1(\mathbf{r}')\frac{\partial G_1(\mathbf{r}, \mathbf{r}')}{\partial n} \right], \tag{27.37}$$

$$\int_{V_2} \delta(\mathbf{r} - \mathbf{r}')u_2(\mathbf{r}')d^2r' = D_2 \int_{S_2} d^2r' \left[G_2(\mathbf{r}, \mathbf{r}')\frac{\partial u_2(\mathbf{r}')}{\partial n} - u_2(\mathbf{r}')\frac{\partial G_2(\mathbf{r}, \mathbf{r}')}{\partial n} \right], \tag{27.38}$$

which is a direct consequence of the above definitions and integration by parts. Here the incident field u_0 is defined by

$$u_0(\mathbf{r}) = \int_{V_1} G_1(\mathbf{r}, \mathbf{r}')S_d(\mathbf{r}')d^3r' \, . \tag{27.39}$$

It follows that if $\mathbf{r} \in V_1$,

$$u_1(\mathbf{r}) = u_0(\mathbf{r}) + D_1 \int_{S_1 \cup S_2} d^2r' \left[G_1(\mathbf{r}, \mathbf{r}') \frac{\partial u_1(\mathbf{r}')}{\partial n} - u_1(\mathbf{r}') \frac{\partial G_1(\mathbf{r}, \mathbf{r}')}{\partial n} \right] , \tag{27.40}$$

$$\int_{S_2} d^2r' \left[G_2(\mathbf{r}, \mathbf{r}') \frac{\partial u_2(\mathbf{r}')}{\partial n} - u_2(\mathbf{r}') \frac{\partial G_2(\mathbf{r}, \mathbf{r}')}{\partial n} \right] = 0 \, . \tag{27.41}$$

Likewise, if $\mathbf{r} \in V_2$,

$$u_2(\mathbf{r}) = D_2 \int_{S_2} d^2r' \left[G_2(\mathbf{r}, \mathbf{r}') \frac{\partial u_2(\mathbf{r}')}{\partial n} - u_2(\mathbf{r}') \frac{\partial G_2(\mathbf{r}, \mathbf{r}')}{\partial n} \right] , \tag{27.42}$$

$$u_0(\mathbf{r}) + D_1 \int_{S_1 \cup S_2} d^2r' \left[G_1(\mathbf{r}, \mathbf{r}') \frac{\partial u_1(\mathbf{r}')}{\partial n} - u_1(\mathbf{r}') \frac{\partial G_1(\mathbf{r}, \mathbf{r}')}{\partial n} \right] = 0. \tag{27.43}$$

Equation (27.43) is known as the extinction theorem. The name stems from the idea that the scattered field extinguishes the incident field inside the scatterer.

27.4 Surface Integral Equations

We now apply (27.40) and (27.42) to the solution of boundary value and scattering problems. We begin by considering the boundary value problem

$$-D_1 \nabla^2 u_1 + \alpha_1 u_1 = S_d \quad \text{in} \quad V_1 \, , \tag{27.44}$$

$$-D_2 \nabla^2 u_2 + \alpha_2 u_2 = 0 \quad \text{in} \quad V_2 \, , \tag{27.45}$$

$$u_1 = 0 \quad \text{on} \quad S_1 \cup S_2 \, , \tag{27.46}$$

$$u_2 = 0 \quad \text{on} \quad S_2 \, . \tag{27.47}$$

Here the inner and outer boundaries are absorbing and the problem is to find u_1 everywhere in V_1. Applying the above boundary conditions, we find that (27.40) becomes

$$u_1(\mathbf{r}) = u_0(\mathbf{r}) + D_1 \int_{S_1 \cup S_2} G_1(\mathbf{r}, \mathbf{r}') \frac{\partial u_1(\mathbf{r}')}{\partial n} d^2r' \, . \tag{27.48}$$

We must now specify the Green's function G_1. Note that we may choose any convenient solution of (27.35), since boundary conditions on G_1 were not used in the derivation of (27.40). In this case, we will choose

$$G_1(\mathbf{r}, \mathbf{r}') = \frac{1}{4\pi D_1} \frac{\exp(-k_1|\mathbf{r} - \mathbf{r}'|)}{|\mathbf{r} - \mathbf{r}'|} , \tag{27.49}$$

where

$$k_1^2 = \frac{\alpha_1 - i\omega}{D_1} . \tag{27.50}$$

Applying the boundary conditions on u_1 once again, we find that (27.48) becomes

$$-u_0(\mathbf{r}) = D_1 \int_{S_1 \cup S_2} G_1(\mathbf{r}, \mathbf{r}') \frac{\partial u_1(\mathbf{r}')}{\partial n} d^2 r' , \quad \mathbf{r} \in S_1 \cup S_2 . \tag{27.51}$$

Equation (27.51) is a so-called surface integral equation that may be solved for the normal derivative $\sigma = \partial u_1/\partial n|_{S_1 \cup S_2}$. We do not consider here the numerical methods for solving surface integral equations, except to note that singularities must be handled with care. It follows that once σ is known, the energy density u_1 can then be calculated from the formula

$$u_1(\mathbf{r}) = u_0(\mathbf{r}) + D_1 \int_{S_1 \cup S_2} G_1(\mathbf{r}, \mathbf{r}')\sigma(\mathbf{r}') d^2 r' . \tag{27.52}$$

We note that the above approach can be adapted to other boundary conditions.

Next, we consider the scattering problem in which a diffuse wave in V_1 scatters from the inhomogeneity V_2:

$$-D_1 \nabla^2 u_1 + \alpha_1 u_1 = S_d \quad \text{in} \quad V_1 , \tag{27.53}$$

$$-D_2 \nabla^2 u_2 + \alpha_2 u_2 = 0 \quad \text{in} \quad V_2 , \tag{27.54}$$

$$u_1 = 0 \quad \text{on} \quad S_1 , \tag{27.55}$$

$$u_1 = u_2 \quad \text{on} \quad S_2 , \tag{27.56}$$

$$D_1 \frac{\partial u_1}{\partial n} = D_2 \frac{\partial u_2}{\partial n} \quad \text{on} \quad S_2 . \tag{27.57}$$

Here the inner boundary is a diffuse-diffuse interface and the outer boundary is absorbing. We take the Green's function G_1 to obey absorbing boundary conditions on S_1 and

$$G_2(\mathbf{r}, \mathbf{r}') = \frac{1}{4\pi D_2} \frac{\exp(-k_2|\mathbf{r} - \mathbf{r}'|)}{|\mathbf{r} - \mathbf{r}'|} , \tag{27.58}$$

where

$$k_2^2 = \frac{\alpha_2 - i\omega}{D_2} . \tag{27.59}$$

Note that G_1 may be obtained by solving a boundary value problem using a surface integral equation method. Applying the above boundary conditions, we find that

(27.40) and (27.42) become

$$u_1(\mathbf{r}) = u_0(\mathbf{r}) + \int_{S_2} d^2r' \left[G_1(\mathbf{r},\mathbf{r}')\psi(\mathbf{r}') - D_1 \frac{\partial G_1(\mathbf{r},\mathbf{r}')}{\partial n}\phi(\mathbf{r}') \right] , \quad (27.60)$$

$$u_2(\mathbf{r}) = \int_{S_2} d^2r' \left[G_2(\mathbf{r},\mathbf{r}')\psi(\mathbf{r}') - D_2 \frac{\partial G_2(\mathbf{r},\mathbf{r}')}{\partial n}\phi(\mathbf{r}') \right] , \quad (27.61)$$

where $\phi = u_1|_{S_2} = u_2|_{S_2}$ and $\psi = \partial u_1/\partial n|_{S_2} = \partial u_2/\partial n|_{S_2}$. Making use of the boundary condition (27.57) once again, we find that ϕ and ψ obey the surface integral equations

$$\frac{1}{2}\phi(\mathbf{r}) = u_0(\mathbf{r}) + \int_{S_2} d^2r' G_1(\mathbf{r},\mathbf{r}')\psi(\mathbf{r}') - D_1 P \int_{S_2} d^2r' \frac{\partial G_1(\mathbf{r},\mathbf{r}')}{\partial n}\phi(\mathbf{r}') , \quad \mathbf{r} \in S_2, \quad (27.62)$$

$$\frac{1}{2}\phi(\mathbf{r}) = \int_{S_2} d^2r' G_2(\mathbf{r},\mathbf{r}')\psi(\mathbf{r}') - D_2 P \int_{S_2} d^2r' \frac{\partial G_2(\mathbf{r},\mathbf{r}')}{\partial n}\phi(\mathbf{r}') , \quad \mathbf{r} \in S_2. \quad (27.63)$$

Note that the factor of one half in the above formulas arises from putting $\mathbf{r} \in S_2$ in the Dirac delta functions appearing in (27.37) and (27.38). The principal value is present to regularize the singularity in the normal derivative of the Green's function G_2. Equations (27.62) and (27.63) are surface integral equations that may be solved for ϕ and ψ. Once ϕ and ψ are determined, the energy densities u_1 and u_2 can then be calculated everywhere in space from (27.60) and (27.61).

References and Additional Reading

Early experimental studies of scattering of diffuse waves by localized objects were reported in the following papers:

P.N. den Outer, Th.M. Nieuwenhuizen and A. Lagendijk, J. Opt. Soc. Am. A **10**, 1209 (1993).

D.A. Boas, M.A. O'Leary, B. Chance and A.G. Yodh, Proc. Natl. Acad. Sci. (USA) **19**, 4887 (1994).

S.A. Walker, D.A. Boas, and E. Gratton, Appl. Opt. 37, 1935–1944 (1998).

This paper describes a perturbation theory to solve the problem of scattering of a diffuse wave by a small inhomogeneity:

S. Feng, F. Zeng and B. Chance, Appl. Opt. **35**, 3826 (1995).

The following papers describe theoretical approaches to the problem of scattering of a diffuse wave by an object below an interface, using integral equations:

J. Ripoll and M. Nieto-Vesperinas, J. Opt. Soc. Am. A **16**, 1453 (1999).

J.-B. Thibaud, R. Carminati and J.-J. Greffet, J. Appl. Phys. **87**, 7638 (2000).

The following papers discuss the scattering of diffuse waves in the context of optical tomography:

J. Schotland and V. Markel, J. Opt. Soc. Am. A **18**, 2767 (2001).
J. Schotland and V. Markel, J. Opt. Soc. Am. A **19**, 558 (2002).
J. Schotland and V. Markel, Phys. Rev. E **70**, 056616 (2004).

Exercises

IV.1 For near-infrared light, biological tissues are characterized by an effective refractive index $n = 1.4$, scattering mean free path $\ell_s \simeq 100\ \mu$m, anisotropy factor $g \simeq 0.9$ and absorption mean free path $\ell_a > 1$ cm.

a. How much ballistic light is transmitted through a slice of biological tissues with thickness $d = 1$ cm?

b. Is the diffusion approximation accurate for modeling light transport in this medium?

IV.2 A white paint is characterized by a transport mean free path $\ell_t = 50\ \mu$m at $\lambda = 500$ nm and a refractive index $n = 1.5$. Consider a layer of this paint with thickness $L = 0.5$ mm, illuminated with an extended beam of visible light assumed to be a plane wave. What is the order of magnitude of the fraction of the incident power that is reflected?

IV.3 A pulse of light, with duration $\tau_p = 100$ fs, is incident on a layer of scattering material with thickness $L = 2$ mm. The scattering mean free path of the material is $\ell_s = 10\ \mu$m, the transport mean free path is $\ell_t = 100\ \mu$m and the effective refractive index is $n = 1.5$. The total transmitted diffuse intensity $T(t)$ (integrated over all output angles) is measured.

a. Justify the use of the diffusion approximation to describe the long-time behavior of $T(t)$?

b. What is the characteristic time scale of the diffusion process?

c. What is the expected long time behavior of $T(t)$?

IV.4 A strongly scattering material is characterized by a transport mean free path $\ell_t = 5\ \mu$m at $\lambda = 650$ nm and an effective refractive index $n = 1.5$. Consider a layer of this material with thickness $L = 0.5$ mm, illuminated by an extended beam assumed to be a plane wave. What is the typical path length of the diffuse transmitted light? Compare this to the path length for ballistic light.

IV.5 A biological tissue is characterized by an effective refractive index $n = 1.4$, a transport mean free path $\ell_t \simeq 0.9$ mm and an absorption mean free path $\ell_a \simeq 20$ mm, for an illumination at a wavelength $\lambda = 850$ nm. A diffuse wave is created by illumination with a light beam modulated in intensity at a frequency ω. What modulation frequency should be used to produce a diffuse wave with an attenuation length of 1 cm?

IV.6 A cold atomic cloud (an ensemble of atoms assumed to be at rest) is illuminated on resonance by a laser with wavelength $\lambda = 600$ nm. The polarizability of the atoms is $\alpha(\omega) = i\,6\pi c^3/\omega^3$ with $\omega = 2\pi c/\lambda$, c being the speed of light in vacuum.

(a) Derive a formula for the scattering cross section σ_s and compute its numerical value.

(b) Consider an extinction measurement through an atomic cloud with thickness $d = 1$ cm and $d/\ell_s = 100$, with ℓ_s as the scattering mean free path. What is the number density ρ of the cloud?

(c) By performing a time-resolved measurement of the diffuse transmission, an optical diffusion constant $D = 0.5\ \text{m}^2\ \text{s}^{-1}$ is obtained. Is there a difference between the transport mean free path ℓ_t and the scattering mean free path ℓ_s for atoms? Deduce the energy velocity v_E and discuss the result.

(d) A strong reduction of the energy velocity results from the "trapping" of light in the atoms, due to their internal resonance. The trapping time is the lifetime τ of the excited state. In the limit $\tau \gg \ell_s/c$, find the energy velocity v_E. Deduce the order of magnitude of τ. Hint: Use a random walk picture to describe the diffusion process.

IV.7 A random laser is made of scatterers embedded in a gain medium. Assume the medium is a sphere with radius R and model the gain using a negative absorption coefficient μ_a, taken to be uniform.

a. Using the diffusion approximation, evaluate the average distance traveled by a photon before exiting the medium as a function of R and the transport mean free path ℓ_t.

b. The lasing threshold is reached when this distance equals the gain length $\ell_g = -1/\mu_a$. Deduce the critical size R_c of the medium to observe a lasing threshold.

IV.8 In Chapter 25, the Green's function for the diffusion equation was calculated for the half-space geometry. Following the development in Section 25.5, derive the expression for the Green's function for a slab of scattering material located between the planes $z = 0$ and $z = L$, as given by (25.43).

IV.9 Consider the propagation of a diffuse wave with energy density u in an infinite homogeneous medium. Find u in the half-space $z \geq 0$ from knowledge of $\partial u/\partial z$ on the plane $z = 0$. That is, in the absence of sources, show that

$$u(\mathbf{r}) = \int G(\boldsymbol{\rho} - \boldsymbol{\rho}'; z) \frac{\partial u(\boldsymbol{\rho}', z)}{\partial z}\bigg|_{z=0} d^2\rho', \tag{27.64}$$

where

$$G(\boldsymbol{\rho} - \boldsymbol{\rho}'; z) = -\frac{1}{2\pi} \frac{\exp(-k_0|\mathbf{r} - \mathbf{r}'|)}{|\mathbf{r} - \mathbf{r}'|}\bigg|_{z'=0}. \tag{27.65}$$

Here $\mathbf{r} = (\boldsymbol{\rho}, z)$ and k_0 is the diffuse wavenumber.

IV.10 Consider the scattering of a diffuse wave from a point absorber in the semi-infinite medium $z \geq 0$. The energy density U obeys

$$-D\nabla^2 U + \alpha_0 \delta(\mathbf{r} - \mathbf{r}_0) U = 0\,,$$

$$U + \ell \frac{\partial U}{\partial z} = \delta(\mathbf{r} - \mathbf{r}_s) \quad \text{on} \quad z = 0\,.$$

Here $\mathbf{r}_0$ is the position of the absorber, the source is taken to have unit power and α_0 is constant. Find the intensity of light at a point on the boundary (i) for a weak absorber (within the Born approximation) and (ii) exactly.

Part V

Speckle and Interference Phenomena

28

Intensity Statistics

The scattering of a coherent wave by a disordered medium, such as a powder, a sheet of paper or a rough surface, leads to a complicated spatial distribution of intensity known as a speckle pattern. An example is shown in Fig. 28.1. The detailed analysis of a particular speckle pattern is often out of reach and possibly even useless. Nevertheless, speckle patterns can be characterized statistically. An interesting feature of speckle patterns is that a wide class of them, known as fully developed speckles, exhibit universal statistical properties. In this chapter, we study the statistical distribution of the intensity measured at one point in a speckle pattern.

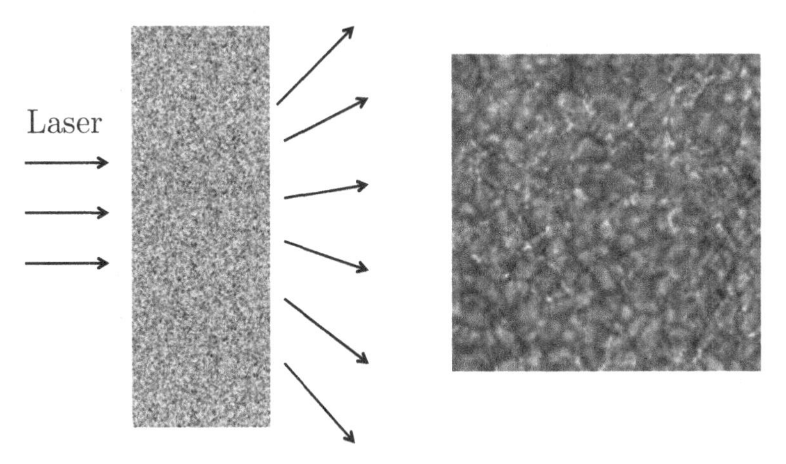

Fig. 28.1 Speckle pattern generated by illuminating a slab of scattering medium by a coherent beam.

28.1 Fully Developed Speckle

Let us denote by $I(\mathbf{r})$ the intensity of the field measured at the point $\mathbf{r}$ in a speckle pattern. We describe the scattering medium using a statistical model of disorder, where $I(\mathbf{r})$ is a random variable. In practice, considering a real medium as one realization of a random medium, the statistical properties of $I(\mathbf{r})$ can be compared to that obtained from the spatial fluctuations of the intensity.

We consider the field to be monochromatic, with complex amplitude $U(\mathbf{r})$. This field can be described as the superposition of scattered waves emerging from all possible scattering sequences inside the medium and can be written in the form

$$U(\mathbf{r}) = \sum_{\mathcal{S}} A_{\mathcal{S}}(\mathbf{r}) \exp[i\phi_{\mathcal{S}}(\mathbf{r})] , \tag{28.1}$$

where $\mathcal{S}$ denotes any scattering sequence that begins on the surface of the medium and ends at the observation point $\mathbf{r}$. This expansion of the field in terms of scattering sequences is justified in Chapter 32. Each term in the sum involves a real amplitude $A_{\mathcal{S}}(\mathbf{r})$ and a phase $\phi_{\mathcal{S}}(\mathbf{r})$ that are both random variables.

The statistical properties of the intensity produced by a field of the form (28.1) can be obtained by making the following assumptions, which define the model of the fully developed speckle:

(i) For two different sequences $\mathcal{S}$ and $\mathcal{S}'$, the complex amplitudes $U_{\mathcal{S}}(\mathbf{r}) = A_{\mathcal{S}}(\mathbf{r}) \exp[i\phi_{\mathcal{S}}(\mathbf{r})]$ and $U_{\mathcal{S}'}(\mathbf{r}) = A_{\mathcal{S}'}(\mathbf{r}) \exp[i\phi_{\mathcal{S}'}(\mathbf{r})]$ are independent random variables.
(ii) For a given sequence $\mathcal{S}$, the amplitude $A_{\mathcal{S}}(\mathbf{r})$ and the phase $\phi_{\mathcal{S}}(\mathbf{r})$ are uncorrelated.
(iii) The phase $\phi_{\mathcal{S}}(\mathbf{r})$ is uniformly distributed on $[-\pi, +\pi]$.

We will see that under these assumptions, the field obeys Gaussian statistics.[1]

28.2 Amplitude Distribution Function

The sum (28.1) contains a large number of terms that are independent and identically distributed random variables. It follows that it is possible to deduce the statistics of the field amplitude and intensity from the central limit theorem. Let $X = \operatorname{Re} U(\mathbf{r})$ and $Y = \operatorname{Im} U(\mathbf{r})$ be the real and imaginary parts of the field, which are given by

$$X = \sum_{\mathcal{S}} A_{\mathcal{S}}(\mathbf{r}) \cos\phi_{\mathcal{S}}(\mathbf{r}) , \tag{28.2}$$

[1] The approach followed in this chapter is similar to that presented in J.W. Goodman, *Statistical Optics* (Wiley, New York, 1985), sec. 2.9.

$$Y = \sum_{\mathcal{S}} A_{\mathcal{S}}(\mathbf{r}) \sin \phi_{\mathcal{S}}(\mathbf{r}) \,. \tag{28.3}$$

The average, variance and correlation of the random variables X and Y can be easily determined using the three aforementioned hypotheses. Since the amplitudes and phases are uncorrelated, and the phases are uniformly distributed on $[-\pi, +\pi]$, we immediately obtain $\langle X \rangle = 0$ and $\langle Y \rangle = 0$. Here $\langle \cdots \rangle$ denotes averaging over the joint probability density of amplitudes and phases. It is also possible to compute the second moment of X as follows:

$$\begin{aligned}
\langle X^2 \rangle &= \sum_{\mathcal{S}} \sum_{\mathcal{S}'} \langle A_{\mathcal{S}}(\mathbf{r}) A_{\mathcal{S}'}(\mathbf{r}) \rangle \, \langle \cos \phi_{\mathcal{S}}(\mathbf{r}) \cos \phi_{\mathcal{S}'}(\mathbf{r}) \rangle \\
&= \sum_{\mathcal{S}} \langle A^2_{\mathcal{S}}(\mathbf{r}) \rangle \, \langle \cos^2 \phi_{\mathcal{S}}(\mathbf{r}) \rangle \\
&= \frac{1}{2} \sum_{\mathcal{S}} \langle A^2_{\mathcal{S}}(\mathbf{r}) \rangle \,,
\end{aligned} \tag{28.4}$$

where we have used hypothesis 1 in the second line. The same result holds for $\langle Y^2 \rangle$, and we will use the notation $\sigma^2 = \langle X^2 \rangle = \langle Y^2 \rangle$ for the variance of X and Y, which also means that $2\sigma^2 = \langle I \rangle$, with $\langle I \rangle = \langle X^2 \rangle + \langle Y^2 \rangle$ the average intensity. Moreover, it is easy to verify that $\langle XY \rangle = 0$, meaning that the real and imaginary parts of the field are uncorrelated.

Since X is the sum of a large number of independent random variables, it follows from the central limit theorem that X is a Gaussian random variable with probability density

$$p(X) = \frac{1}{2\pi\sigma^2} \exp\left(-\frac{X^2}{2\sigma^2}\right) . \tag{28.5}$$

The same holds for Y, and since X and Y are uncorrelated (which, for Gaussian random variables, is equivalent to independence), we find that their joint probability density is

$$p(X, Y) = \frac{1}{2\pi\sigma^2} \exp\left(-\frac{X^2 + Y^2}{2\sigma^2}\right) . \tag{28.6}$$

The statistics of the field amplitude $A = \sqrt{X^2 + Y^2}$ is obtained by a simple change of variables. That is,

$$\begin{aligned}
p(A) &= \frac{A}{\sigma^2} \exp\left(-\frac{A^2}{2\sigma^2}\right) \quad \text{for } A > 0 \,, \\
p(A) &= 0 \quad \text{for } A < 0 \,.
\end{aligned} \tag{28.7}$$

28.3 Intensity Distribution Function

The probability density of the intensity $I(\mathbf{r}) = |U(\mathbf{r})|^2$ follows directly from (28.8), using the relationship

$$p(I) = p(A = \sqrt{I})\,\frac{dA}{dI} = \frac{1}{2\sqrt{I}}\,p(A = \sqrt{I})\,. \tag{28.8}$$

We thus obtain

$$p(I) = \frac{1}{\langle I \rangle} \exp\left(-\frac{I}{\langle I \rangle}\right) \quad \text{for } I > 0\,, \tag{28.9}$$
$$p(I) = 0 \ \text{ for } I < 0\,.$$

This result is known as Rayleigh statistics and is a feature of speckle patterns in the Gaussian approximation. We note that the most likely value of the intensity is $I = 0$.

28.4 Speckle Contrast

In order to characterize the intensity fluctuations in a speckle pattern, we calculate the variance $\mathrm{Var}(I) = \langle I^2 \rangle - \langle I \rangle^2$. The speckle contrast is defined as the normalized standard deviation $\sigma_I / \langle I \rangle$ with $\sigma_I = \sqrt{\mathrm{Var}(I)}$. The second moment of the intensity is readily obtained from the probability density according to

$$\langle I^2 \rangle = \int_0^\infty I^2\, p(I)\, dI = 2\,\langle I \rangle^2 \tag{28.10}$$

We thus find that

$$\mathrm{Var}(I) = \langle I \rangle^2\,, \tag{28.11}$$

which is a distinguishing feature of Rayleigh statistics. In terms of the speckle contrast, this relation is equivalent to

$$\frac{\sigma_I}{\langle I \rangle} = 1\,. \tag{28.12}$$

We see that a speckle pattern exhibits a large contrast, with intensity fluctuations on the same order as the average intensity, consistent with the experiment.

28.5 Intensity Statistics of Unpolarized Electromagnetic Waves

The expression (28.9) describes the intensity statistics for a scalar wave field. In optics, this corresponds to the intensity of a linearly polarized field. Finding the statistical distribution of the full intensity is a more complicated task. Here we show how to determine the statistics in the particular case of unpolarized light observed in the far-field.

In a three-dimensional speckle, such as that observed inside a disordered medium or in the near-field of the output surface, the complex amplitude of the field is the superposition of three-independent complex random variables. Intensity correlations and fluctuations in unpolarized three-dimensional speckles are addressed in Chapter 44. When the speckle is observed in the far-zone, the field is locally a plane wave and only two components, E_α and E_β, need to be accounted for. The statistical distribution of the intensity $I = |E_\alpha|^2 + |E_\beta|^2$ can be deduced from that of $I_\alpha = |E_\alpha|^2$ and $I_\beta = |E_\beta|^2$. Since for unpolarized light the full intensity I is the sum of two independent random components, its probability density $p(I)$ is the convolution of the probability densities of I_α and I_β, both being given by (28.9). Noting that $\langle I_\alpha \rangle = \langle I_\beta \rangle = \langle I \rangle /2$, the convolution takes the form

$$p(I) = \left(\frac{2}{\langle I \rangle} \right)^2 \int_0^I \exp\left[-\frac{2(I-x)}{\langle I \rangle} \right] \exp\left(-\frac{2x}{\langle I \rangle} \right) dx \, , \tag{28.13}$$

and $p(I) = 0$ for $I < 0$. This leads immediately to the final result

$$\begin{aligned} p(I) &= \left(\frac{2}{\langle I \rangle} \right)^2 I \exp\left(-\frac{2I}{\langle I \rangle} \right) \quad \text{for } I > 0 \, , \\ p(I) &= 0 \quad \text{for } I < 0 \, . \end{aligned} \tag{28.14}$$

We note that for this modified Rayleigh statistics, which applies to a two-dimensional unpolarized field, the most likely value of the intensity is not zero. Moreover, this distribution leads to a reduced speckle contrast $\sigma_I / \langle I \rangle = 1/\sqrt{2}$. The calculation of the speckle contrast for unpolarized light is left as an exercise.

References and Additional Reading

The first paper reporting the observation of a speckle pattern produced by laser light seems to be:

R.V. Langmuir, Appl. Phys. Lett. **2**, 29 (1963).

A comprehensive study of speckle statistics is found in:

J.W. Goodman, *Statistical Optics* (Wiley, New York, 1985), sec. 2.9.

J.W. Goodman, *Speckle Phenomena in Optics* (Roberts and Company, Greenwood, 2007).

The following papers contain a study of polarization statistics in speckles:

I. Freund, M. Kaveh, R. Berkovits and M. Rosenbluh, Phys. Rev. B **42**, R2613 (1990).

S.M. Cohen, D. Eliyahu, I. Freund and M. Kaveh, Phys. Rev. A **43**, R5748 (1991).

A. Dogariu and R. Carminati, Phys. Rep. **559**, 1 (2015).

29

Some Properties of Rayleigh Statistics

In Chapter 28, we have shown that in the setting of fully developed speckle, the field obeys Gaussian statistics. In this chapter, we discuss some properties of Rayleigh statistics and make the connection with the diagrammatic approach developed in multiple scattering theory.

29.1 High-Order Moments of the Intensity

As shown in Chapter 28, the intensity obeys Rayleigh statistics with the probability density

$$p(I) = \frac{1}{\langle I \rangle} \exp\left(-\frac{I}{\langle I \rangle}\right) \quad \text{for } I > 0\,, \tag{29.1}$$
$$p(I) = 0 \ \text{ for } I < 0\,.$$

The second moment of the intensity was found to be $\langle I^2 \rangle = 2\langle I \rangle^2$. The higher-order moments are easily seen to be given in terms of the first moment by the formula

$$\langle I^p \rangle = p\,!\,\langle I \rangle^p\,, \tag{29.2}$$

where p is a nonnegative integer.

We note that specifying the above moments is equivalent to knowing the distribution p. That is, the moments of the random variable I determine the probability density $p(I)$ through the relation

$$\tilde{p}(I) = \int_{-\infty}^{+\infty} \sum_{p=0}^{\infty} \frac{(ik)^p}{p\,!} \langle I^p \rangle \exp(-ikI)\, \frac{dk}{2\pi}\,, \tag{29.3}$$

where $\tilde{p}$ is the Fourier transform of p. Inserting (29.2) into (29.3) leads to the Rayleigh distribution $p(I)$.

29.2 Field and Intensity Correlations

Rayleigh statistics leads to an interesting relationship between the intensity correlation function (for example, between two different points) and the field correlation function, which is widely used in the study of speckle correlations.

Let us consider an infinite disordered medium illuminated by a monochromatic point source located at the point $\mathbf{r}_s$. The intensity $I(\mathbf{r}) = |U(\mathbf{r})|^2$ in the medium forms a speckle pattern, which we describe statistically by assuming Gaussian statistics for the field or Rayleigh statistics for the intensity. Spatial correlations are characterized by the intensity correlation function, which is defined as

$$\langle \delta I(\mathbf{r})\, \delta I(\mathbf{r}') \rangle = \langle I(\mathbf{r})\, I(\mathbf{r}') \rangle - \langle I(\mathbf{r}) \rangle \langle I(\mathbf{r}') \rangle, \tag{29.4}$$

where $\mathbf{r}$ and $\mathbf{r}'$ are two points within the medium. As a measure of the degree of correlation, we introduce the normalized correlation function

$$C_I(\mathbf{r}, \mathbf{r}') = \frac{\langle \delta I(\mathbf{r})\, \delta I(\mathbf{r}') \rangle}{\langle I(\mathbf{r}) \rangle \langle I(\mathbf{r}') \rangle} \,. \tag{29.5}$$

In terms of the complex amplitude field, this correlation function becomes

$$C_I(\mathbf{r}, \mathbf{r}') = \frac{\langle U(\mathbf{r})\, U^*(\mathbf{r})\, U(\mathbf{r}')\, U^*(\mathbf{r}') \rangle}{\langle |U(\mathbf{r})|^2 \rangle \langle |U(\mathbf{r}')|^2 \rangle} - 1 \,, \tag{29.6}$$

which is a fourth-order correlation function.

29.2.1 *Factorization of the Intensity Correlation Function*

A Gaussian random variable is fully characterized by its first and second moments. As a consequence, high-order correlation functions always reduce to a sum of all possible products of second-order correlation functions. For the complex field amplitude, the only non-vanishing terms in the factorization are

$$\begin{aligned} \langle U(\mathbf{r})\, U^*(\mathbf{r})\, U(\mathbf{r}')\, U^*(\mathbf{r}') \rangle &= \langle U(\mathbf{r})\, U^*(\mathbf{r}) \rangle \langle U(\mathbf{r}')\, U^*(\mathbf{r}') \rangle \\ &\quad + \langle U(\mathbf{r})\, U^*(\mathbf{r}') \rangle \langle U^*(\mathbf{r})\, U(\mathbf{r}') \rangle \,. \end{aligned} \tag{29.7}$$

Here the term $\langle U(\mathbf{r}) U(\mathbf{r}') \rangle \langle U^*(\mathbf{r}) U^*(\mathbf{r}') \rangle$ vanishes and does not contribute since the real and imaginary parts of U are independent and identically distributed random variables, as discussed in Chapter 28. We immediately see that the above result can be rewritten as

$$\langle U(\mathbf{r})\, U^*(\mathbf{r})\, U(\mathbf{r}')\, U^*(\mathbf{r}') \rangle = \langle |U(\mathbf{r})|^2 \rangle \langle |U(\mathbf{r}')|^2 \rangle + |\langle U(\mathbf{r})\, U^*(\mathbf{r}') \rangle|^2 \,. \tag{29.8}$$

Introducing this expression into (29.6) leads to

$$C_I(\mathbf{r}, \mathbf{r}') = |C_U(\mathbf{r}, \mathbf{r}')|^2 \,, \tag{29.9}$$

where

$$C_U(\mathbf{r}, \mathbf{r}') = \frac{\langle U(\mathbf{r})\, U^*(\mathbf{r}')\rangle}{\sqrt{\langle |U(\mathbf{r})|^2\rangle}\sqrt{\langle |U(\mathbf{r}')|^2\rangle}} \tag{29.10}$$

is the normalized field correlation function.

It is interesting to note that in the particular case where $\mathbf{r} = \mathbf{r}'$, (29.9) leads to

$$\mathrm{Var}[I(\mathbf{r})] = \langle \delta I(\mathbf{r})^2\rangle = \langle I(\mathbf{r})\rangle^2 \,, \tag{29.11}$$

or equivalently

$$\langle I(\mathbf{r})^2\rangle = 2\langle I(\mathbf{r})\rangle^2 \,, \tag{29.12}$$

which is consistent with the assumption of the intensity obeying Rayleigh statistics.

29.2.2 Diagrammatic Representation

The factorization of the intensity correlation function in terms of the field correlation function can also be understood in terms of diagrams. To lowest order in $1/(k_0\ell_s)$, the field correlation function $\langle U(\mathbf{r})\, U^*(\mathbf{r}')\rangle$ can be calculated using the Bethe–Salpeter equation in the ladder approximation (see Chapter 16). Given a point source at $\mathbf{r}_s$, the correlation between the fields at the points $\mathbf{r}$ and $\mathbf{r}'$, both located a large distance from the source (compared to the scattering mean free path ℓ_s), is described by the ladder diagram in Fig. 29.1. At distances $\lesssim \ell_s$, a contribution $\langle U(\mathbf{r})\rangle\langle U^*(\mathbf{r}')\rangle$ from the ballistic fields should be included.

In order to compute the intensity correlation function, we need to describe the propagation of four fields from $\mathbf{r}_s$ to $\mathbf{r}$ and $\mathbf{r}'$. The two possibilities to achieve this in the ladder approximation are sketched in Fig. 29.2. In the left panel, the upper diagram represents $\langle I(\mathbf{r})\rangle$ while the lower diagram represents $\langle I(\mathbf{r}')\rangle$. The contribution of these two independent diagrams to the intensity correlation function is $\langle I(\mathbf{r})\rangle\langle I(\mathbf{r}')\rangle$. In the right panel, fields propagating along two different ladders are interchanged at the last scattering event, thus creating an intensity correlation (the two resulting intensities are no more independent). The contribution of these crossed diagrams to the intensity correlation function is $|\langle U(\mathbf{r})\, U^*(\mathbf{r}')\rangle|^2$. We thus

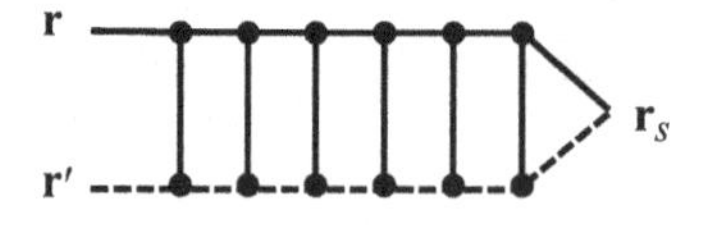

Fig. 29.1 Diagram representing the field correlation function $\langle U(\mathbf{r})U^*(\mathbf{r}')\rangle$ in the ladder approximation. For the sake of clarity, we use a slightly different rule for the diagrams than in Chapter 16. Here the solid (dashed) line represents the retarded (advanced) average Green's function.

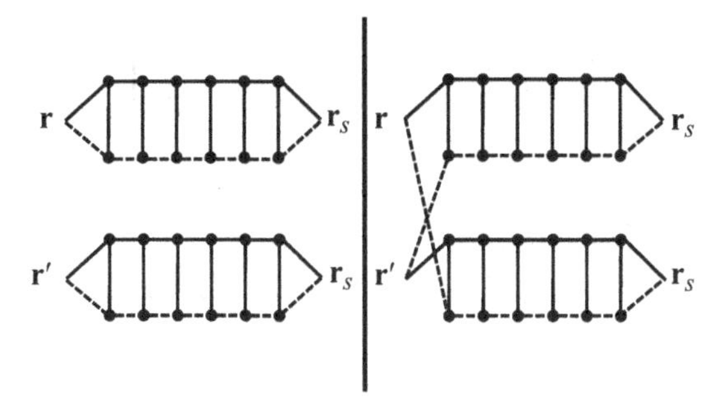

Fig. 29.2 Contributions to the intensity correlation function in the ladder approximation, to leading order in $1/(k_0\ell_s)$. Left panel: Diagrams representing $\langle I(\mathbf{r})\rangle$ and $\langle I(\mathbf{r}')\rangle$. Right panel: Diagrams of the same order in which fields between two different ladders are interchanged at the last scattering event. We recognize the appearance of the product of the diagram in Fig. 29.1 and its complex conjugate.

obtain the result that to lowest order in $1/(k_0\ell_s)$

$$\langle I(\mathbf{r})\, I(\mathbf{r}')\rangle = \langle I(\mathbf{r})\rangle\langle I(\mathbf{r}')\rangle + |\langle U(\mathbf{r})\, U^*(\mathbf{r}')\rangle|^2 , \tag{29.13}$$

which after proper normalization leads to (29.9).

29.3 Diagrammatic View of Rayleigh Statistics

Rayleigh statistics for the intensity, or equivalently Gaussian statistics for the field, can be derived from the hypotheses for fully developed speckle stated in Chapter 28. It is natural to inquire whether Rayleigh statistics can also be derived from multiple scattering theory. In this section, we describe a qualitative argument justifying Rayleigh statistics in the weak-scattering regime $k_0\ell_s \gg 1$.

In the previous section, we have obtained (29.13) by considering the dominant diagrams the in weak-scattering regime, as represented in Fig. 29.2. For $\mathbf{r} = \mathbf{r}'$, Equation (29.13) leads to the relation $\langle I(\mathbf{r})^2\rangle = 2\langle I(\mathbf{r})\rangle^2$. By repeating this process, it is possible to calculate the higher-order moments of the intensity. In Fig. 29.3, we illustrate some of the diagrams contributing to the third-order correlation function of the intensity $\langle I(\mathbf{r})\, I(\mathbf{r}') I(\mathbf{r}'')\rangle$. The right panel shows an example of the pairing of fields resulting from two different ladder diagrams. We note that five possible pairings exist. These five configurations of diagrams, and the diagrams in the left panel, correspond to the following expression:

$$\begin{aligned}\langle I(\mathbf{r})\, I(\mathbf{r}') I(\mathbf{r}'')\rangle = \; & \langle I(\mathbf{r})\rangle\langle I(\mathbf{r}')\rangle\langle I(\mathbf{r}'')\rangle \\ & + |\langle U(\mathbf{r})\, U^*(\mathbf{r}')\rangle|^2\langle I(\mathbf{r}'')\rangle \\ & + |\langle U(\mathbf{r}')\, U^*(\mathbf{r}'')\rangle|^2\langle I(\mathbf{r})\rangle\end{aligned}$$

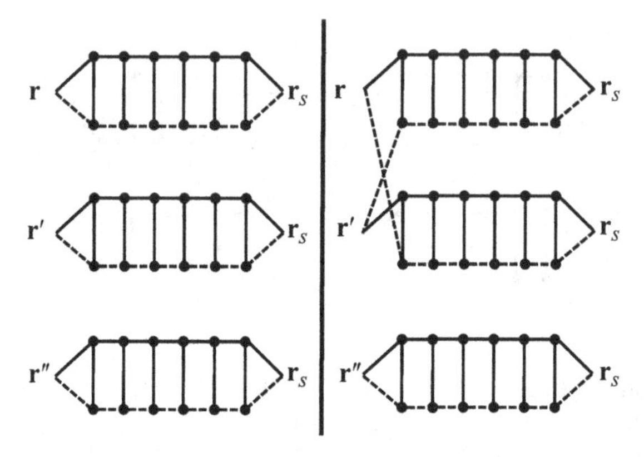

Fig. 29.3 Contributions to the third-order intensity correlation function in the ladder approximation. Left panel: Diagrams representing the contribution $\langle I(\mathbf{r})\rangle\langle I(\mathbf{r}')\rangle\langle I(\mathbf{r}'')\rangle$. Right panel: Diagrams of the same order in which fields between two different ladders are interchanged in the final scattering event. Only one of five possible pairings is shown.

$$
\begin{aligned}
&+ |\langle U(\mathbf{r})\, U^*(\mathbf{r}'')\rangle|^2 \langle I(\mathbf{r}')\rangle \\
&+ \langle U(\mathbf{r})\, U^*(\mathbf{r}')\rangle \langle U^*(\mathbf{r})\, U(\mathbf{r}'')\rangle \langle U(\mathbf{r}')\, U^*(\mathbf{r}'')\rangle \\
&+ \langle U^*(\mathbf{r})\, U(\mathbf{r}')\rangle \langle U(\mathbf{r})\, U^*(\mathbf{r}'')\rangle \langle U^*(\mathbf{r}')\, U(\mathbf{r}'')\rangle\,. \qquad (29.14)
\end{aligned}
$$

Note that when $\mathbf{r} = \mathbf{r}' = \mathbf{r}''$, the above expression becomes

$$\langle I(\mathbf{r})^3\rangle = 6\,\langle I(\mathbf{r})\rangle^3 = 3\,!\langle I(\mathbf{r})\rangle^3\,, \qquad (29.15)$$

which gives the third moment of the intensity. Continuing in the same fashion, we obtain the moments of the Rayleigh distribution (29.1),

$$\langle I(\mathbf{r})^p\rangle = p\,!\langle I(\mathbf{r})\rangle^p\,, \qquad (29.16)$$

thus establishing the equivalence of Rayleigh statistics and the ladder approximation.

References and Additional Reading

A connection between diagrammatic calculations and Rayleigh statistics is made in the following papers:

B. Shapiro, Phys. Rev. Lett. **57**, 2168 (1986).

E. Kogan, M. Kaveh, R. Baumgartner and R. Berkovits, Phys. Rev. B **48**, 9404 (1993).

T.M. Nieuwenhuizen and M.C.W. van Rossum, Phys. Rev. Lett. **74**, 2674 (1995).

30

Bulk Speckle Correlations

In this chapter, we study the spatial correlations of the intensity in a speckle pattern in an infinite medium. The analysis is carried out within the ladder approximation, which holds in the weak-scattering regime.

30.1 Model of Disorder

We consider a disordered medium with a dielectric function of the form $\epsilon(\mathbf{r}) = 1 + \delta\epsilon(\mathbf{r})$. Here the fluctuations $\delta\epsilon(\mathbf{r})$ are taken to be real-valued, corresponding to a nonabsorbing medium, and obey Gaussian white-noise statistics with first and second moments

$$\langle \delta\epsilon(\mathbf{r}) \rangle = 0 \,, \tag{30.1}$$

$$\langle \delta\epsilon(\mathbf{r})\, \delta\epsilon(\mathbf{r}') \rangle = A\, \delta(\mathbf{r} - \mathbf{r}') \,. \tag{30.2}$$

This model is valid provided that the field does not vary on scales on the order of the microscopic structure of the medium. The constant A is determined by calculating the scattering mean-free path ℓ_s to lowest order in $(k_0\ell_s)^{-1}$. From (15.49), we find that

$$A = \frac{4\pi}{k_0^4\, \ell_s} \,. \tag{30.3}$$

Also note that within the model of uncorrelated white-noise disorder, the scattering and transport mean-free paths are equal since the phase function is a constant, as can be seen from (16.47).

30.2 Field Correlation Function in the Ladder Approximation

Consider the field $U(\mathbf{r})$ generated by a point source at $\mathbf{r}_0$ in an infinite medium. Far from the source, with $|\mathbf{r} - \mathbf{r}_0| \gg \ell_s$, the average field is exponentially small and can be neglected. It follows from the Bethe–Salpeter equation (16.7) that the field correlation function obeys

$$\langle U(\mathbf{r})\,U^*(\mathbf{r}')\rangle = \int \langle G(\mathbf{r},\mathbf{r}_1)\rangle\,\langle G^*(\mathbf{r}',\mathbf{r}_1')\rangle\,\Gamma(\mathbf{r}_1,\mathbf{r}_2;\mathbf{r}_1',\mathbf{r}_2')\\ \times\langle U(\mathbf{r}_2)\,U^*(\mathbf{r}_2')\rangle\,d^3r_1\,d^3r_1'\,d^3r_2\,d^3r_2'\,, \tag{30.4}$$

where Γ is the irreducible vertex. The average Green function $\langle G\rangle$ is given by (15.38) and can be written as

$$\langle G(\mathbf{r},\mathbf{r}')\rangle = \frac{\exp(ik_{\text{eff}}|\mathbf{r}-\mathbf{r}'|)}{4\pi\,|\mathbf{r}-\mathbf{r}'|}\,, \tag{30.5}$$

where k_{eff} is the effective wavevector, defined such that $\text{Im}(k_{\text{eff}}) = 1/(2\ell_s)$ (see Section 15.3). In the ladder approximation, the irreducible vertex is of the form

$$\Gamma(\mathbf{r}_1,\mathbf{r}_2;\mathbf{r}_1',\mathbf{r}_2') \simeq k_0^4\,\langle\delta\epsilon(\mathbf{r}_1)\,\delta\epsilon(\mathbf{r}_1')\rangle\,\delta(\mathbf{r}_1-\mathbf{r}_2)\,\delta(\mathbf{r}_1'-\mathbf{r}_2')\,. \tag{30.6}$$

Using this result, (30.4) becomes

$$\langle U(\mathbf{r})\,U^*(\mathbf{r}')\rangle = k_0^4\int \langle G(\mathbf{r},\mathbf{r}_1)\rangle\,\langle G^*(\mathbf{r}',\mathbf{r}_1')\rangle\\ \times\langle\delta\epsilon(\mathbf{r}_1)\,\delta\epsilon(\mathbf{r}_1')\rangle\,\langle U(\mathbf{r}_1)\,U^*(\mathbf{r}_1')\rangle\,d^3r_1d^3r_1'\,. \tag{30.7}$$

Making use of (30.2) and (30.3), we find that

$$\langle U(\mathbf{r})\,U^*(\mathbf{r}')\rangle = \frac{4\pi}{\ell_s}\int \langle G(\mathbf{r},\mathbf{r}_1)\rangle\,\langle G^*(\mathbf{r}',\mathbf{r}_1)\rangle\,\langle|U(\mathbf{r}_1)|^2\rangle\,d^3r_1\,. \tag{30.8}$$

This result is the starting point for the computation of the field correlation function. Since $\langle|U|^2\rangle$ is slowly varying, it can be taken outside of the integral in (30.8), which thus becomes

$$\langle U(\mathbf{r})\,U^*(\mathbf{r}')\rangle = \frac{4\pi}{\ell_s}\,\langle|U|^2\rangle\int \langle G(\mathbf{r},\mathbf{r}_1)\rangle\,\langle G^*(\mathbf{r}',\mathbf{r}_1)\rangle\,d^3r_1\,. \tag{30.9}$$

Using the above, we can now obtain the normalized correlation function, without calculating $\langle|U|^2\rangle$ explicitly. A full treatment, in which $\langle|U|^2\rangle$ is calculated within the diffusion approximation, is presented in Chapter 31. To proceed further, we make use of the identity

$$k_0^2\,\text{Im}\varepsilon_{\text{eff}}\int \langle G(\mathbf{r},\mathbf{r}_1)\rangle\,\langle G^*(\mathbf{r}',\mathbf{r}_1)\rangle\,d^3r_1 = \text{Im}\langle G(\mathbf{r},\mathbf{r}')\rangle\,, \tag{30.10}$$

where $\varepsilon_{\text{eff}}k_0^2 = k_{\text{eff}}^2$. The derivation of this identity is suggested as an exercise (a vector form of this relation is derived in Section 40.5). Finally, it follows from (30.9) and (30.10) that

$$\langle U(\mathbf{r})\,U^*(\mathbf{r}')\rangle = \frac{4\pi}{k_0}\,\langle|U|^2\rangle\,\text{Im}\langle G(\mathbf{r},\mathbf{r}')\rangle\,. \tag{30.11}$$

We note that (30.11) can be shown to hold in a more general setting than considered here, namely in statistically homogeneous and isotropic media, where measurements of the spatial correlations of the field can be used to deduce the average Green function. This result provides the foundation for a variety of techniques to image complex media.

By combining (30.11) and (30.5), we obtain the expression for the normalized field correlation function

$$\frac{\langle U(\mathbf{r})\,U^*(\mathbf{r}')\rangle}{\langle |U|^2\rangle} = \operatorname{sinc}\left(\frac{2\pi}{\lambda_{\text{eff}}}\rho\right)\exp\left(-\frac{\rho}{2\ell_s}\right), \tag{30.12}$$

where $\rho = |\mathbf{r} - \mathbf{r}'|$ and $\lambda_{\text{eff}} = 2\pi/\text{Re}(k_{\text{eff}})$ is the effective wavelength in the medium.

30.3 Intensity Correlation Function

As shown in Chapter 29, a field obeying Gaussian statistics, or equivalently an intensity satisfying Rayleigh statistics, has an intensity correlation function which is the squared modulus of the field correlation function. It follows immediately from (30.12) that

$$\frac{\langle \delta I(\mathbf{r})\,\delta I(\mathbf{r}')\rangle}{\langle I(\mathbf{r})\rangle\,\langle I(\mathbf{r}')\rangle} = \operatorname{sinc}^2\left(\frac{2\pi}{\lambda_{\text{eff}}}\rho\right)\exp\left(-\frac{\rho}{\ell_s}\right). \tag{30.13}$$

The width of the correlation function is a measure of the typical size of a speckle spot and is on the order of $\lambda_{\text{eff}}/2$. This contribution to the intensity correlation function is referred to as a short-range correlation. Long-range correlations also exist in speckle patterns and emerge from a description beyond the Gaussian approximation.

References and Additional Reading

Calculations of intensity correlations in speckles produced by volume scattering were initially presented in:

B. Shapiro, Phys. Rev. Lett. **57**, 2168 (1986).

The following textbooks contain chapters or sections on speckle correlations:

E. Akkermans and G. Montambaux, *Mesoscopic Physics of Electrons and Photons* (Cambridge University Press, Cambridge, 2007), chap. 12.

P. Sheng, *Introduction to Wave Scattering, Localization, and Mesoscopic Phenomena* (Springer, Berlin, 2010), chap. 11.

The role of field correlations on mesoscopic fluctuations of electromagnetic waves was put forward in the following paper:

P. Sebbah, B. Hu, A.Z. Genack, R. Pnini and B. Shapiro, Phys. Rev. Lett. **88**, 123901 (2002).

A relation similar to Eq. (30.11) is obtained for equilibrium fields, as a direct application of the fluctuation-dissipation theorem:

G.S. Agarwal, Phys. Rev. A **11**, 230 (1975).

L.D. Landau, E.M. Lifshitz and L.P. Pitaevskii, *Statistical Physics*, 3rd ed. (Pergamon Press, Oxford, 1980), part 1, chap. 12, and part 2, chap. 8.

S.M. Rytov, Yu.A. Kravtsov and V.I. Tatarskii, *Principles of Statistical Radiophysics* (Springer-Verlag, Berlin, 1989), vol. 3, chap. 3.

K. Joulain, J.P. Mulet, F. Marquier, R. Carminati and J.-J. Greffet, Surf. Sci. Rep. **57**, 59 (2005).

A general derivation of Eq. (30.11) for statistically homogeneous and isotropic vector fields is given in:

T. Setälä, K. Blomstedt, M. Kaivola and A.T. Friberg, Phys. Rev. E **67**, 026613 (2003).

Speckle correlations of vector fields are studied in Chapter 44.

The following papers highlight the use of spatial field correlation functions for Green's function retrieval:

R.L. Weaver and O.I. Lobkis, Phys. Rev. Lett. **87**, 134301 (2001).

A. Derode, E. Larose, M. Tanter, J. de Rosny, A. Tourin, M. Campillo and M. Fink, J. Acoust. Soc. Am. **113**, 2973 (2003).

B.A. van Tiggelen, Phys. Rev. Lett. **91**, 243904 (2003).

M. Campillo and A. Paul, Science **299**, 547 (2003).

M. Davy, M. Fink and J. de Rosny, Phys. Rev. Lett. **110**, 203901 (2013).

Long-range intensity correlations beyond the ladder (or Gaussian) approximation have been discussed in:

M.J. Stephen and C. Cwilich, Phys. Rev. Lett. **59**, 285 (1987).

S. Feng, C. Kane, P.A. Lee and A.D. Stone, Phys. Rev. Lett. **61**, 834 (1988).

R. Pnini and B. Shapiro, Phys. Rev. B **39**, 6986 (1989).

R. Berkovits and S. Feng, Phys. Rep. **238**, 135 (1994).

31

Two-Frequency Speckle Correlations

We consider the correlation function of a speckle field at two points and two frequencies. We will recover the results of Chapter 30 for single frequencies as a special case.

31.1 Two-Frequency Bethe–Salpeter Equation

Following the development in Chapter 19, we begin with the Bethe–Salpeter equation for the correlation of the field at two points $\mathbf{r}$ and $\mathbf{r}'$, and two frequencies ω and ω':

$$\langle U(\mathbf{r},\omega)U^*(\mathbf{r}',\omega')\rangle = \langle U(\mathbf{r},\omega)\rangle\langle U^*(\mathbf{r}',\omega')\rangle + \int \langle G(\mathbf{r},\mathbf{r}_1,\omega)\rangle\langle G^*(\mathbf{r}',\mathbf{r}_2,\omega')\rangle \times \Gamma(\mathbf{r}_1,\mathbf{r}_3,\omega;\mathbf{r}_2,\mathbf{r}_4,\omega')\langle U(\mathbf{r}_3,\omega)U^*(\mathbf{r}_4,\omega')\rangle\, d^3r_1 d^3r_2 d^3r_3 d^3r_4\,, \tag{31.1}$$

where $\langle G\rangle$ is the average Green function and Γ is the two-frequency irreducible vertex. Equation (31.1) is established along the same lines as the one-frequency Bethe–Salpeter equation derived in Chapter 16. Within the framework of the ladder approximation, and using the model of white-noise disorder [See (30.1)–(30.3)], the Bethe–Salpeter equation becomes

$$\langle U(\mathbf{r},\omega)U^*(\mathbf{r}',\omega')\rangle = \langle U(\mathbf{r},\omega)\rangle\langle U^*(\mathbf{r}',\omega')\rangle + \frac{4\pi}{\ell_s}\int \langle G(\mathbf{r},\mathbf{r}_1,\omega)\rangle\langle G^*(\mathbf{r}',\mathbf{r}_1,\omega')\rangle \times \langle U(\mathbf{r}_1,\omega)U^*(\mathbf{r}_1,\omega')\rangle\, d^3r_1\,, \tag{31.2}$$

where we have assumed that the scattering mean free path ℓ_s takes the same value at the frequencies ω and ω', corresponding to weak dispersion.

Introducing the two-frequency ladder propagator $L(\mathbf{r}_1,\mathbf{r}_2,\omega,\omega')$ defined by

$$L(\mathbf{r}_1,\mathbf{r}_2,\omega,\omega') = \frac{4\pi}{\ell_s}\,\delta(\mathbf{r}_1-\mathbf{r}_2) + \frac{4\pi}{\ell_s}\int \langle G(\mathbf{r}_1,\mathbf{r}_3,\omega)\rangle\langle G(\mathbf{r}_1,\mathbf{r}_3,\omega')\rangle \times L(\mathbf{r}_3,\mathbf{r}_2,\omega,\omega')\, d^3r_3\,, \tag{31.3}$$

we can rewrite (31.2) as

$$\langle U(\mathbf{r},\omega)U^*(\mathbf{r}',\omega')\rangle = \langle U(\mathbf{r},\omega)\rangle\langle U^*(\mathbf{r}',\omega')\rangle + \int \langle G(\mathbf{r},\mathbf{r}_1,\omega)\rangle\langle G^*(\mathbf{r}',\mathbf{r}_1,\omega')\rangle \\ \times L(\mathbf{r}_1,\mathbf{r}_2,\omega,\omega')\langle U(\mathbf{r}_2,\omega)\rangle\langle U^*(\mathbf{r}_2,\omega')\rangle\, d^3r_1 d^3r_2 \,. \quad (31.4)$$

For later convenience, an equivalent form of this result can be written in terms of Green's functions:

$$\langle G(\mathbf{r},\mathbf{R},\omega)G^*(\mathbf{r}',\mathbf{R}',\omega')\rangle = \langle G(\mathbf{r},\mathbf{R},\omega)\rangle\langle G^*(\mathbf{r}',\mathbf{R}',\omega'\rangle \\ + \int \langle G(\mathbf{r},\mathbf{r}_1,\omega)\rangle\langle G^*(\mathbf{r}',\mathbf{r}_1,\omega')\rangle L(\mathbf{r}_1,\mathbf{r}_2,\omega,\omega') \\ \times \langle G(\mathbf{r}_2,\mathbf{R},\omega)\rangle\langle G^*(\mathbf{r}_2,\mathbf{R}',\omega')\rangle\, d^3r_1 d^3r_2 \,. \quad (31.5)$$

From (31.3) and (31.5), it can be seen that the ladder propagator L is the vertex Λ introduced in (16.11), calculated in the ladder approximation.

31.2 Two-Frequency Ladder Propagator

The ladder propagator $L(\mathbf{r}_1,\mathbf{r}_2,\omega,\omega')$ can be calculated explicitly in an infinite medium under the assumptions

$$|\mathbf{r}_1-\mathbf{r}_2| \gg \ell_s \,, \qquad |\Delta\omega| \ll \frac{c}{\ell_s} \,, \quad (31.6)$$

where $\Delta\omega = \omega - \omega'$. In order to simplify the integral in (31.3), we expand the ladder propagator to second order:

$$L(\mathbf{r}_3,\mathbf{r}_2,\omega,\omega') = L(\mathbf{r}_1,\mathbf{r}_2,\omega,\omega') + (\mathbf{r}_3-\mathbf{r}_1)\cdot\nabla_{\mathbf{r}_2}L(\mathbf{r}_1,\mathbf{r}_2,\omega,\omega') \\ + \frac{1}{2}\left[(\mathbf{r}_3-\mathbf{r}_1)\cdot\nabla_{\mathbf{r}_2}\right]^2 L(\mathbf{r}_1,\mathbf{r}_2,\omega,\omega') \,. \quad (31.7)$$

The integral of the first-order term vanishes. For the second-order term, only the diagonal elements in the operator $(\mathbf{r}_3-\mathbf{r}_1)\cdot\nabla_{\mathbf{r}_2}$ yield a non-zero contribution. We thus obtain

$$L(\mathbf{r}_1,\mathbf{r}_2,\omega,\omega') = \frac{4\pi}{\ell_s}\,\delta(\mathbf{r}_1-\mathbf{r}_2) + \frac{4\pi}{\ell_s}L(\mathbf{r}_1,\mathbf{r}_2,\omega,\omega')I_1 + \frac{4\pi}{6\ell_s}\nabla^2_{\mathbf{r}_2}L(\mathbf{r}_1,\mathbf{r}_2,\omega,\omega')I_2 \,, \quad (31.8)$$

where the integrals I_1 and I_2 are given by

$$I_1 = \int \langle G(\mathbf{r}_1,\mathbf{r}_3,\omega)\rangle\langle G^*(\mathbf{r}_1,\mathbf{r}_3,\omega')\rangle\, d^3r_3 \,, \quad (31.9)$$

$$I_2 = \int |\mathbf{r}_3-\mathbf{r}_1|^2\langle G(\mathbf{r}_1,\mathbf{r}_3,\omega)\rangle\langle G^*(\mathbf{r}_1,\mathbf{r}_3,\omega')\rangle\, d^3r_3 \,. \quad (31.10)$$

Here the average Green function is given by

$$\langle G(\mathbf{r}_1, \mathbf{r}_3)\rangle = \frac{\exp(ik_{\text{eff}}|\mathbf{r}_1 - \mathbf{r}_3|)}{4\pi|\mathbf{r}_1 - \mathbf{r}_3|}, \tag{31.11}$$

where $k_{\text{eff}} = k_0 + i/(2\ell_s)$ is the effective wavenumber to lowest order in $(k_0\ell_s)^{-1}$, as defined in Section 15.3. The integrals I_1 and I_2 can be calculated explicitly:

$$I_1 = \frac{\ell_s}{4\pi}\frac{1}{1 - i\ell_s\Delta\omega/c} \simeq \frac{\ell_s}{4\pi}\left(1 + i\ell_s\frac{\Delta\omega}{c}\right) \tag{31.12}$$

and

$$I_2 = \frac{2\ell_s^3}{4\pi}\frac{1}{(1 - i\ell_s\Delta\omega/c)^3} \simeq \frac{2\ell_s^3}{4\pi}\left(1 + i\ell_s\frac{\Delta\omega}{3c}\right). \tag{31.13}$$

Inserting the above into (31.8) leads to

$$\frac{\ell_s^2}{3}\left(1 + i\ell_s\frac{\Delta\omega}{3c}\right)\nabla^2_{\mathbf{r}_2}L(\mathbf{r}_1, \mathbf{r}_2, \omega, \omega') + i\ell_s\frac{\Delta\omega}{c}L(\mathbf{r}_1, \mathbf{r}_2, \omega, \omega') = -\frac{4\pi}{\ell_s}\delta(\mathbf{r}_1 - \mathbf{r}_2). \tag{31.14}$$

Assuming that the ladder propagator varies smoothly on the scale of ℓ_s,

$$\nabla^2_{\mathbf{r}_2}L(\mathbf{r}_1, \mathbf{r}_2, \omega, \omega') \ll \frac{1}{\ell_s^2}L(\mathbf{r}_1, \mathbf{r}_2, \omega, \omega'), \tag{31.15}$$

we find that (31.14) becomes a diffusion equation of the form

$$\nabla^2_{\mathbf{r}_2}L(\mathbf{r}_1, \mathbf{r}_2, \omega, \omega') + i\frac{\Delta\omega}{D}L(\mathbf{r}_1, \mathbf{r}_2, \omega, \omega') = -\frac{12\pi}{\ell_s^3}\delta(\mathbf{r}_1 - \mathbf{r}_2), \tag{31.16}$$

where the diffusion coefficient $D = (1/3)c\ell_s$. Here we recall that $\ell_s = \ell_t$ for the white-noise model. In an infinite medium, the solution to (31.16) is given by

$$L(\mathbf{r}_1, \mathbf{r}_2, \omega, \omega') = \frac{3}{\ell_s^3|\mathbf{r}_1 - \mathbf{r}_2|}\exp(ik|\mathbf{r}_1 - \mathbf{r}_2|), \tag{31.17}$$

where $k = (1 + i)\sqrt{\Delta\omega/(2D)}$.

31.3 Field Correlation Function in an Infinite Medium

The field correlation function can now be calculated using (31.5) and (31.17). The ladder propagator varies smoothly on the scale of ℓ_s, while the average field and the average Green function have an envelope varying on the scale ℓ_s. As a consequence, (31.5) becomes

$$\langle G(\mathbf{r}, \mathbf{R}, \omega)G^*(\mathbf{r}', \mathbf{R}', \omega')\rangle \simeq L(\mathbf{r}, \mathbf{R}, \omega, \omega')\int\langle G(\mathbf{r}, \mathbf{r}_1, \omega)\rangle\langle G^*(\mathbf{r}', \mathbf{r}_1, \omega')\rangle\, d^3r_1$$
$$\times\int\langle G(\mathbf{r}_2, \mathbf{R}, \omega)\rangle\langle G^*(\mathbf{r}_2, \mathbf{R}', \omega')\rangle\, d^3r_2, \tag{31.18}$$

where the term $\langle G(\mathbf{r}, \mathbf{R}, \omega)\rangle\langle G^*(\mathbf{r}', \mathbf{R}', \omega')\rangle$ has been neglected since we assume that observations take place at large distances ($|\mathbf{r} - \mathbf{R}| \gg \ell_s$ and $|\mathbf{r}' - \mathbf{R}'| \gg \ell_s$). In the above integrals, the term scaling as $\exp(i\Delta\omega\ell_s/c)$ has a negligible contribution compared to the term proportional to $\exp(i\sqrt{\Delta\omega/D}|\mathbf{r} - \mathbf{R}])$. Therefore, the integrals can be evaluated by putting $\omega = \omega'$. We can then make use of the identity [see (30.10)]:

$$\begin{aligned}\int \langle G(\mathbf{r}, \mathbf{r}_1)\rangle \, \langle G^*(\mathbf{r}', \mathbf{r}_1)\rangle d^3 r_1 &= \frac{\ell_s}{k_0} \operatorname{Im}\langle G(\mathbf{r}, \mathbf{r}')\rangle \\ &= \frac{\ell_s}{4\pi} \operatorname{sinc}\left(k_0|\mathbf{r} - \mathbf{r}'|\right) \exp\left(-\frac{|\mathbf{r} - \mathbf{r}'|}{2\ell_s}\right) .\end{aligned} \tag{31.19}$$

Introducing the notation $\Delta r = |\mathbf{r} - \mathbf{r}'|$ and $\Delta R = |\mathbf{R} - \mathbf{R}'|$, we obtain a general expression for the correlation function at two different points and at two different frequencies,

$$\langle G(\mathbf{r}, \mathbf{R}, \omega) G^*(\mathbf{r}', \mathbf{R}', \omega')\rangle = \frac{3}{16\pi^2\ell_s} \frac{\exp(ik|\mathbf{r} - \mathbf{R}|)}{|\mathbf{r} - \mathbf{R}|} f(\Delta r)\, f(\Delta R) , \tag{31.20}$$

where $f(X) = \operatorname{sinc}(kX)\exp[-X/(2\ell_s)]$.

We note that in the particular case of illumination by a single monochromatic point source, with $\Delta R = 0$ and $\Delta\omega = 0$, $\langle G(\mathbf{r}, \mathbf{R}, \omega) G^*(\mathbf{r}', \mathbf{R}, \omega)\rangle$ is translational to the correlation function $\langle U(\mathbf{r}) U^*(\mathbf{r}')\rangle$ studied in Chapter 30. After normalization, we recover (30.12).

References and Additional Reading

Frequency correlations in speckles are discussed in:

B. Shapiro, Phys. Rev. Lett. **57**, 2168 (1986).

J.F. de Boer, M.P. van Albada and A. Lagendijk, Phys. Rev. B **45**, 658 (1992).

A. Lagendijk and B.A. van Tiggelen, Phys. Rep. **270**, 143 (1996).

P. Sheng, *Introduction to Wave Scattering, Localization, and Mesoscopic Phenomena* (Springer, Berlin, 2010), chap. 11.

The presentation developed in this chapter was inspired by:

R. Pnini, in *Waves and Imaging through Complex Media*, edited by P. Sebbah (Kluwer Academic, Dordrecht, 2001), p. 391.

Transport equations for two-frequency field correlation function can be found in:

A. Ishimaru, *Wave Propagation and Scattering in Random Media* (IEEE Press, Piscataway, 1997).

32

Amplitude and Intensity Propagators for Multiply-Scattered Fields

In this chapter, we introduce the amplitude propagator, which connects the scattered field to the incident field in the multiple scattering regime. We then justify the well-known representation of the amplitude propagator in terms of scattering sequences. These results will be applied to the theory of speckle correlations in later chapters.

32.1 Amplitude Propagator

In the slab geometry shown in Fig. 32.1, the complex amplitude of the field $U(\mathbf{r}_b)$ at a point $\mathbf{r}_b$ on the output surface $z = L$ is linearly related to the incident field $U_0(\mathbf{r}_a)$ at a point $\mathbf{r}_a$ on the input surface $z = 0$. We define the amplitude propagator $h(\mathbf{r}_b, \mathbf{r}_a)$ for the transmitted field according to

$$U(\mathbf{r}_b) = \int_{z=0} h(\mathbf{r}_b, \mathbf{r}_a)\, U_0(\mathbf{r}_a)\, d^2 r_a \,. \tag{32.1}$$

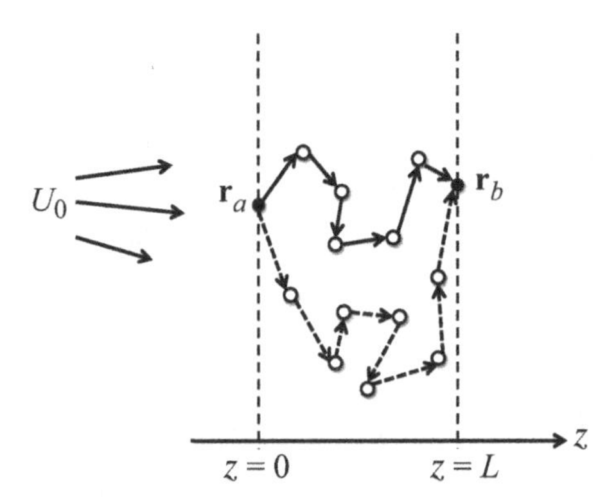

Fig. 32.1 Schematic representation of a scattering sequence. A disordered medium composed of discrete scatterers is confined within a slab of thickness L. White circles are scatterers and black circles represent the input and output points.

In a similar manner, the propagator for the reflected field can be defined by taking both $\mathbf{r}_a$ and $\mathbf{r}_b$ on the surface $z = 0$. This propagator generalizes the Dirichlet propagator introduced in Chapter 7. Note that the input and output surfaces can be chosen arbitrarily.

32.1.1 The Scattering Sequences Picture

The idea of scattering sequences is encapsulated in the following formula for the amplitude propagator:

$$h(\mathbf{r}_b, \mathbf{r}_a) = \sum_{n=0}^{\infty} \sum_{\mathcal{S}_n} A_{\mathcal{S}_n}(\mathbf{r}_b, \mathbf{r}_a) \exp[i\phi_{\mathcal{S}_n}(\mathbf{r}_b, \mathbf{r}_a)] \,. \tag{32.2}$$

Here we characterize a scattering sequence by the number of scattering events n and by the ordered sequence of scattering events $\mathcal{S}_n$. Note that the entry and exit points $\mathbf{r}_a$ and $\mathbf{r}_b$ do not necessarily coincide with scattering events. In the summation, $n = 0$ corresponds to free space propagation of the incident field. We denote by $A_{\mathcal{S}_n}(\mathbf{r}_b, \mathbf{r}_a)$ the real amplitude resulting from scattering along a given sequence $\mathcal{S}_n$ connecting $\mathbf{r}_a$ to $\mathbf{r}_b$, and by $\phi_{\mathcal{S}_n}(\mathbf{r}_b, \mathbf{r}_a)$, the phase shift accumulated along this sequence. The following simplified notation

$$h(\mathbf{r}_b, \mathbf{r}_a) = \sum_{\mathcal{S}_{ab}} A_{\mathcal{S}_{ab}} \exp(i\phi_{\mathcal{S}_{ab}}) \tag{32.3}$$

is often used instead of the double summation in (32.2), where $\mathcal{S}_{ab}$ denotes any scattering sequence connecting $\mathbf{r}_a$ to $\mathbf{r}_b$. Two scattering sequences are illustrated in Fig. 32.1.

32.1.2 Rigorous Definition of a Scattering Sequence

The representation of the scattered field in the form (32.2) can be justified by making use of the Born series. The total field $U = U_0 + U_s$ obeys the Lippman–Schwinger equation, which in operator form is given by

$$U = U_0 + G_0\, V\, U \,, \tag{32.4}$$

where V is the potential and G_0 is the free-space Green function. As usual, the Born series is obtained by iteration and is of the form

$$U = U_0 + G_0\, V\, U_0 + G_0\, V\, G_0\, V\, U_0 + G_0\, V\, G_0\, V\, G_0\, V\, U_0 + \cdots \,. \tag{32.5}$$

Introducing the T-matrix of the medium, we rewrite (32.4) as

$$U = U_0 + G_0\, T\, U_0 \,. \tag{32.6}$$

Assuming that the medium is composed of identical particles, and following the development in Section 18.1, the T-matrix can be shown to be

$$T = \sum_j t_j + \sum_{j,k} t_k\, G_0\, t_j + \sum_{j \neq k, k \neq l} t_l\, G_0\, t_k\, G_0\, t_j + \cdots , \qquad (32.7)$$

where t_j is the T-matrix of an individual scatterer located at the position $\mathbf{r}_j$. Inserting (32.7) into (32.6) leads to

$$U = U_0 + \sum_j G_0\, t_j\, U_0 + \sum_{j,k} G_0\, t_k\, G_0\, t_j\, U_0 + \sum_{j \neq k, k \neq l} G_0\, t_l\, G_0\, t_k\, G_0\, t_j\, U_0 + \cdots , \qquad (32.8)$$

which explicitly describes the scattering process as a series of scattering events from individual particles. It is then possible to identify the field U in (32.8) and (32.1). To proceed, we assume an incident field entering the medium at point $\mathbf{r}_a$, and a total field observed at point $\mathbf{r}_b$, and we write explicitly the second term on the right-hand side of (32.8):

$$\sum_j \int d^3r_1 d^3r_2\, G_0(\mathbf{r}_b - \mathbf{r}_1)\, t_j(\mathbf{r}_1, \mathbf{r}_2)\, U_0(\mathbf{r}_2) =$$

$$\int d^2R_a \left[\sum_j \int G_0(\mathbf{r}_b - \mathbf{r}_1)\, t_j(\mathbf{r}_1, \mathbf{r}_2)\, h_0(\mathbf{r}_2, \mathbf{r}_a) d^3r_1 d^3r_2 \right] U_0(\mathbf{r}_a) . \qquad (32.9)$$

Here h_0 denotes the amplitude propagator in free space. The same idea can be applied to each term of the multiple scattering expansion in (32.8). Comparison with (32.1) leads to

$$h = h_0 + \sum_j G_0\, t_j\, h_0 + \sum_{j,k} G_0\, t_k\, G_0\, t_j\, h_0 + \sum_{j \neq k, k \neq l} G_0\, t_l\, G_0\, t_k\, G_0\, t_j\, h_0 + \cdots \qquad (32.10)$$

In the above, the integrals are implicit, and one has to keep in mind that G_0 connects the last scattering event to the output point $\mathbf{r}_b$. Moreover, in the summations, h_0 connects the input point $\mathbf{r}_a$ to the first scattering event. The expression (32.10) is precisely of the form (32.2). The first summation corresponds to all scattering sequences involving one scatterer [$n = 1$ in (32.2)]. The second summation corresponds to all scattering sequences involving two different scatterers [$n = 2$ in (32.2)] and so on. Note that the summation includes the special case in which the first and third scattering events occur on the same scatterer. We have thus justified and illustrated the concept of scattering sequences, which is an essential tool.

32.2 Correlation Function of the Amplitude Propagator

In the weak-scattering regime $k_0\ell_s \gg 1$, the correlation function of the scattered field obeys the Bethe–Salpeter equation in the ladder approximation (see Chapter 30),

$$\langle U(\mathbf{r})\, U^*(\mathbf{r}')\rangle = \langle U(\mathbf{r})\rangle\langle U^*(\mathbf{r}')\rangle + \frac{4\pi}{\ell_s}\int \langle G(\mathbf{r},\mathbf{r}_1)\rangle\, \langle G^*(\mathbf{r}',\mathbf{r}_1)\rangle\, \langle |U(\mathbf{r}_1)|^2\rangle\, d^3r_1\,, \tag{32.11}$$

where $\langle G\rangle$ is the averaged Green function. Introducing the ladder propagator L defined by

$$L(\mathbf{r}_1,\mathbf{r}_2) = \frac{4\pi}{\ell_s}\,\delta(\mathbf{r}_1-\mathbf{r}_2) + \frac{4\pi}{\ell_s}\int |\langle G(\mathbf{r}_1,\mathbf{r}_3)\rangle|^2\, L(\mathbf{r}_3,\mathbf{r}_2)\, d^3r_3\,, \tag{32.12}$$

we can rewrite (32.11) in the form

$$\langle U(\mathbf{r})\, U^*(\mathbf{r}')\rangle = \frac{4\pi}{\ell_s}\int \langle G(\mathbf{r},\mathbf{r}_1)\rangle\, \langle G^*(\mathbf{r}',\mathbf{r}_1)\rangle\, L(\mathbf{r}_1,\mathbf{r}_2)\, |\langle U(\mathbf{r}_2)\rangle|^2\, d^3r_1 d^3r_2\,. \tag{32.13}$$

The average field $\langle U(\mathbf{r}_2)\rangle$ is related to the incident field by the amplitude propagator, and thus the field correlation function can be rewritten as

$$\begin{aligned}\langle U(\mathbf{r})\, U^*(\mathbf{r}')\rangle = \int &\langle G(\mathbf{r},\mathbf{r}_1)\rangle\, \langle G^*(\mathbf{r}',\mathbf{r}_1)\rangle\, L(\mathbf{r}_1,\mathbf{r}_2)\, \langle h(\mathbf{r}_2,\mathbf{r}_a)\rangle\langle h^*(\mathbf{r}_2,\mathbf{r}'_a)\rangle \\ &\times U_0(\mathbf{r}_a)U_0^*(\mathbf{r}'_a)\, d^3r_1 d^3r_2 d^2r_a d^2r'_a\,.\end{aligned} \tag{32.14}$$

By definition of the amplitude propagator, we can also write

$$\langle U(\mathbf{r})\, U^*(\mathbf{r}')\rangle = \int \langle h(\mathbf{r},\mathbf{r}_a)h^*(\mathbf{r}',\mathbf{r}'_a)\rangle U_0(\mathbf{r}_a)U_0^*(\mathbf{r}'_a)\, d^2r_a d^2r'_a\,. \tag{32.15}$$

Identifying the two preceding expressions, we obtain

$$\begin{aligned}\langle h(\mathbf{r},\mathbf{r}_a)h^*(\mathbf{r}',\mathbf{r}'_a)\rangle = \int &\langle G(\mathbf{r},\mathbf{r}_1)\rangle\, \langle G^*(\mathbf{r}',\mathbf{r}_1)\rangle\, L(\mathbf{r}_1,\mathbf{r}_2) \\ &\times \langle h(\mathbf{r}_2,\mathbf{r}_a)\rangle\langle h^*(\mathbf{r}_2,\mathbf{r}'_a)\rangle\, d^3r_1 d^3r_2\,,\end{aligned} \tag{32.16}$$

which is a general expression of the correlation function of the amplitude propagator.

32.3 Correlation Function in an Infinite Medium

At large scales, such that $|\mathbf{r}_1-\mathbf{r}_2| \gg \ell_s$, we have shown in Chapter 31 that the ladder propagator obeys the diffusion equation [see (31.16)]

$$\nabla^2_{\mathbf{r}_2} L(\mathbf{r}_1,\mathbf{r}_2) = -\frac{12\pi}{\ell_s^3}\,\delta(\mathbf{r}_1-\mathbf{r}_2)\,. \tag{32.17}$$

The ladder propagator L varies smoothly on the scale of the scattering mean free path ℓ_s, while the averaged Green function has an envelope varying on the scale of ℓ_s. Thus (32.16) becomes

$$\langle h(\mathbf{r}, \mathbf{r}_a) h^*(\mathbf{r}', \mathbf{r}'_a) \rangle \simeq L(\mathbf{r}, \mathbf{r}_a) \int \langle G(\mathbf{r}, \mathbf{r}_1) \rangle \langle G^*(\mathbf{r}', \mathbf{r}_1) \rangle \times \langle h(\mathbf{r}_2, \mathbf{r}_a) \rangle \langle h^*(\mathbf{r}_2, \mathbf{r}'_a) \rangle \, d^3 r_1 d^3 r_2 \,. \tag{32.18}$$

The two integrals can be factorized and calculated explicitly. To proceed, we need the expressions for the average Green function and the average amplitude propagator. For simplicity, we consider the case of an infinite medium. (In this case, the input and output surfaces used to define the amplitude propagator can be chosen arbitrarily within the medium, as shown in Fig. 32.2.)

To calculate the first integral, we need the expression for the average outgoing Green function

$$\langle G(\mathbf{r}, \mathbf{r}') \rangle = \frac{\exp(i k_{\text{eff}} |\mathbf{r} - \mathbf{r}'|)}{4\pi |\mathbf{r} - \mathbf{r}'|} \,, \tag{32.19}$$

where the effective wavevector is $k_{\text{eff}} = k_0[1 + i/(2\ell_s)]$ to lowest order in $(k_0 \ell_s)^{-1}$. The calculation is the same as in Chapter 30, and we obtain

$$\begin{aligned} \int \langle G(\mathbf{r}, \mathbf{r}_1) \rangle \langle G^*(\mathbf{r}', \mathbf{r}_1) \rangle d^3 r_1 &= \frac{\ell_s}{k_0} \operatorname{Im} \langle G(\mathbf{r}, \mathbf{r}') \rangle \\ &= \frac{\ell_s}{4\pi} \operatorname{sinc}\left(k_0 |\mathbf{r} - \mathbf{r}'|\right) \exp\left(-\frac{|\mathbf{r} - \mathbf{r}'|}{2\ell_s}\right) . \end{aligned} \tag{32.20}$$

To calculate the second integral, we need the expression of the average amplitude propagator. Since the average Green function obeys a wave equation in an effective homogeneous medium, the connection between the amplitude propagator and the Green function is the same as in free space. Using (7.7), we deduce that

$$\langle h(\mathbf{r}_2, \mathbf{r}_a) \rangle = -2 \frac{\partial}{\partial z} \langle G(\mathbf{r}_2, \mathbf{r}_a) \rangle \,, \tag{32.21}$$

where z is the coordinate along the direction normal to the input and output planes (see Fig. 32.2). In the far-field approximation, the derivative can be simplified according to

$$\langle h(\mathbf{r}_2, \mathbf{r}_a) \rangle \simeq -2 i k_0 \langle G(\mathbf{r}_2, \mathbf{r}_a) \rangle \,. \tag{32.22}$$

The second integral can now be calculated, leading to

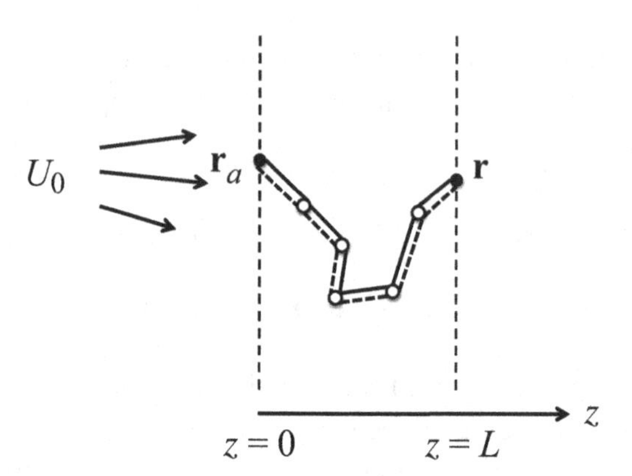

Fig. 32.2 Schematic representation of the correlation function of the amplitude propagator in the ladder approximation.

$$\int \langle h(\mathbf{r}_2, \mathbf{r}_a)\rangle \langle h^*(\mathbf{r}_2, \mathbf{r}'_a)\rangle d^3r_2 = 4k_0\ell_s \,\mathrm{Im}\langle G(\mathbf{r}_a, \mathbf{r}'_a)\rangle$$
$$= \frac{k_0^2\ell_s}{\pi} \,\mathrm{sinc}\left(k_0|\mathbf{r}_a - \mathbf{r}'_a|\right) \exp\left(-\frac{|\mathbf{r}_a - \mathbf{r}'_a|}{2\ell_s}\right) . \tag{32.23}$$

Finally, from (32.18), (32.20) and (32.23), we obtain

$$\langle h(\mathbf{r}, \mathbf{r}_a) h^*(\mathbf{r}', \mathbf{r}'_a)\rangle = \frac{(k_0\ell_s)^2}{4\pi^2} L(\mathbf{r}, \mathbf{r}_a) \,\mathrm{sinc}\left(k_0|\mathbf{r} - \mathbf{r}'|\right) \exp\left(-\frac{|\mathbf{r} - \mathbf{r}'|}{2\ell_s}\right)$$
$$\times \,\mathrm{sinc}\left(k_0|\mathbf{r}_a - \mathbf{r}'_a|\right) \exp\left(-\frac{|\mathbf{r}_a - \mathbf{r}'_a|}{2\ell_s}\right) . \tag{32.24}$$

The above is the formula for the correlation function of the amplitude propagator in an infinite medium, obtained within the ladder approximation.

32.4 Intensity Propagator

The dependences on the quantities $\mathbf{r} - \mathbf{r}'$ and $\mathbf{r}_a - \mathbf{r}'_a$ in (32.24) are the same as obtained in Chapter 30 for the field correlation in bulk speckle patterns [see (30.12)], which describe microscopic field correlations at scales of order ℓ_s (the scale of a speckle spot inside the medium). The computation of field or intensity correlations outside the medium often involves integrations of (32.24) over $\mathbf{r}$ and/or $\mathbf{r}_a$. Since L is a slowly varying function on the scale of ℓ_s, the two functions of $|\mathbf{r}-\mathbf{r}'|$ and $|\mathbf{r}_a-\mathbf{r}'_a|$ behave as Dirac delta functions with respect to such integrations. Therefore, at the macroscopic scale, (32.24) can be simplified by replacing the $\mathrm{sinc}(\cdots)\exp(\cdots)$ functions with delta functions, whose amplitude is found by computing their integrals using

$$\int_0^\infty \sin(aR) \exp(-R/b)\, R\, dR = \frac{2ab^3}{1 + a^2b^2} . \tag{32.25}$$

We find that

$$\langle h(\mathbf{r}, \mathbf{r}_a) h^*(\mathbf{r}', \mathbf{r}'_a) \rangle = P(\mathbf{r}, \mathbf{r}_a)\, \delta(\mathbf{r} - \mathbf{r}')\, \delta(\mathbf{r}_a - \mathbf{r}'_a), \tag{32.26}$$

where

$$P(\mathbf{r}, \mathbf{r}_a) = \frac{64\, \ell_s^4}{k_0^2}\, L(\mathbf{r}, \mathbf{r}_a) \tag{32.27}$$

is the intensity propagator. It describes the transport of intensity from $\mathbf{r}_a$ to $\mathbf{r}$ in the ladder approximation and at scales much larger than the scattering mean free path ℓ_s. The expression (32.26) has a simple meaning that is represented graphically in Fig. 32.2.

References and Additional Reading

The amplitude propagator for propagation in free space is discussed in:

M. Nieto-Vesperinas, *Scattering and Diffraction in Physical Optics*, 2nd edition (World Scientific, Singapore, 2006).

The concept of scattering sequence is widely used in mesoscopic physics, see for example:

E. Akkermans and G. Montambaux, *Mesoscopic Physics of Electrons and Photons* (Cambridge University Press, Cambridge, 2007).

33

Far-Field Angular Speckle Correlations

In this chapter, we study the far-field angular correlation function of the intensity in a speckle pattern produced by transmission of light through a slab of scattering material. We derive the expression for the correlation function, and discuss several of its implications: the memory effect, size of a far-field speckle spot and number of transmission modes.

33.1 Angular Correlation Function

We consider the geometry in Fig. 33.1. We study the speckle pattern produced in transmission through a slab of thickness L, assumed to be infinite along the transverse directions x and y.

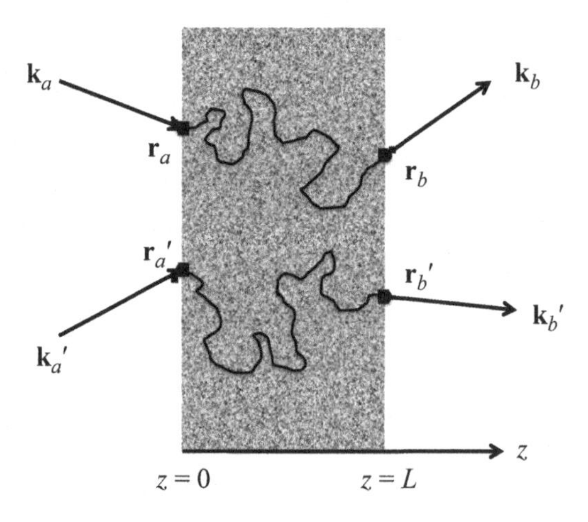

Fig. 33.1 Geometry used for the calculation of the angular correlation function of the intensity in a speckle pattern produced in transmission through a slab of scattering material.

In the analysis of a far-field speckle pattern, a useful quantity is the correlation function between the intensity $I(\mathbf{k}_a, \mathbf{k}_b)$ emerging in direction $\mathbf{k}_b$ when the medium is illuminated by a plane wave with wavevector $\mathbf{k}_a$, and the intensity $I(\mathbf{k}_{a'}, \mathbf{k}_{b'})$ observed in direction $\mathbf{k}_{b'}$ with an illumination from direction $\mathbf{k}_{a'}$. We use the following notation for the projection of vectors along the (x, y) plane, such that $\mathbf{r}_a = (\boldsymbol{\rho}_a, 0)$, $\mathbf{r}_b = (\boldsymbol{\rho}_b, L)$ and $\mathbf{k}_a = (\mathbf{q}_a, k_z(q_a))$.

The angular correlation function of the intensity fluctuations $\delta I = I - \langle I \rangle$ is defined as

$$C^I_{aba'b'} = \frac{\langle \delta I(\mathbf{k}_a, \mathbf{k}_b)\, \delta I(\mathbf{k}_{a'}, \mathbf{k}_{b'}) \rangle}{\langle I(\mathbf{k}_a, \mathbf{k}_b) \rangle \langle I(\mathbf{k}_{a'}, \mathbf{k}_{b'}) \rangle} . \tag{33.1}$$

In the weak-scattering regime $k_0 \ell_s \gg 1$, we can assume that the field obeys Gaussian statistics, or equivalently that transport can be described in the ladder approximation. As a result, the intensity correlation function can be factorized as the square of the field correlation function (see Chapter 29),

$$C^I_{aba'b'} = |C^U_{aba'b'}|^2 , \tag{33.2}$$

and we are left with determining the normalized angular correlation function of the field:

$$C^U_{aba'b'} = \frac{\langle U(\mathbf{k}_a, \mathbf{k}_b)\, U^*(\mathbf{k}_{a'}, \mathbf{k}_{b'}) \rangle}{\sqrt{\langle |U(\mathbf{k}_a, \mathbf{k}_b)|^2 \rangle} \sqrt{\langle |U(\mathbf{k}_{a'}, \mathbf{k}_{b'})|^2 \rangle}} . \tag{33.3}$$

In this expression, $U(\mathbf{k}_a, \mathbf{k}_b)$ denotes the far-field amplitude of the scattered field in the direction $\mathbf{k}_b$, with illumination by a plane wave in the direction of the wavevector $\mathbf{k}_a$.

33.2 Field Angular Correlation Function

The scattered field $U(\mathbf{k}_a, \mathbf{k}_b)$ can be written in terms of the amplitude propagator $h(\mathbf{r}_b, \mathbf{r}_a)$ introduced in Chapter 32. For an incident plane wave $U_0 \exp(i\mathbf{k}_a \cdot \mathbf{r})$, the field emerging at point $\mathbf{r}_b$ on the exit surface $z = L$ is

$$U(\boldsymbol{\rho}_b, L) = \int_{z=0} h(\mathbf{r}_b, \mathbf{r}_a)\, U_0 \exp(i\mathbf{q}_a \cdot \boldsymbol{\rho}_a)\, d^2\rho_a , \tag{33.4}$$

where $\mathbf{r}_a = (\boldsymbol{\rho}_a, 0)$ and $\mathbf{r}_b = (\boldsymbol{\rho}_b, L)$ are the input and output points. The far-field scattered in direction $\mathbf{k}_b$ takes the form

$$U(\mathbf{r}) = \frac{k_z(q_b)}{2i\pi}\, U(\mathbf{q}_b)\, \frac{\exp(ikr)}{r} , \tag{33.5}$$

where $k_z(q) = (k_0^2 - \mathbf{q}^2)^{1/2}$ is the z-component of the wavevector $\mathbf{k}$ and $U(\mathbf{q}_b)$ is the Fourier transform of the scattered field in the plane $z = L$:

$$U(\mathbf{q}_b) = \int_{z=L} U(\boldsymbol{\rho}_b, L) \exp(-i\mathbf{q}_b \cdot \boldsymbol{\rho}_b)\, d^2\rho_b . \tag{33.6}$$

Identifying $U(\mathbf{k}_a, \mathbf{k}_b)$ with $k_z(q_b)U(\mathbf{q}_b)$ leads to

$$U(\mathbf{k}_a, \mathbf{k}_b) = k_z(q_b)\, U_0 \int_{z=0} \int_{z=L} h(\mathbf{r}_b, \mathbf{r}_a)\, \exp(i\mathbf{q}_a \cdot \boldsymbol{\rho}_a - i\mathbf{q}_b \cdot \boldsymbol{\rho}_b)\, d^2\rho_a\, d^2\rho_b\,. \tag{33.7}$$

The above is the expression for the scattered field in the far-zone. In practice, the far-zone conditions are met in the focal plane of a converging lens, or in the Fourier plane of a microscope objective. The angular correlation function of the field is now readily deduced:

$$\begin{aligned}\langle U(\mathbf{k}_a, \mathbf{k}_b)\, U^*(\mathbf{k}_{a'}, \mathbf{k}_{b'})\rangle &= k_z(q_b)k_z(q_{b'})\, |U_0|^2 \int_{z=0} \int_{z=L} \langle h(\mathbf{r}_b, \mathbf{r}_a)h^*(\mathbf{r}_{b'}, \mathbf{r}_{a'})\rangle \\ &\times \exp(i\mathbf{q}_a \cdot \boldsymbol{\rho}_a - i\mathbf{q}_b \cdot \boldsymbol{\rho}_b - i\mathbf{q}_{a'} \cdot \boldsymbol{\rho}_{a'} + i\mathbf{q}_{b'} \cdot \boldsymbol{\rho}_{b'})\, d^2\rho_a d^2\rho_b d^2\rho_{a'} d^2\rho_{b'}\,.\end{aligned} \tag{33.8}$$

The integrals above include contributions from all entry points $\boldsymbol{\rho}_a$ and $\boldsymbol{\rho}'_a$, and exit points $\boldsymbol{\rho}_b$ and $\boldsymbol{\rho}'_b$ (see Fig. 33.1). In the ladder approximation, the correlation function of the amplitude propagator is given by (see Chapter 32)

$$\langle h(\mathbf{r}_b, \mathbf{r}_a)h^*(\mathbf{r}_{b'}, \mathbf{r}_{a'})\rangle \simeq P(\mathbf{r}_b, \mathbf{r}_a)\, \delta(\mathbf{r}_a - \mathbf{r}_{a'})\delta(\mathbf{r}_b - \mathbf{r}_{b'})\,, \tag{33.9}$$

where $P(\mathbf{r}_b, \mathbf{r}_a)$ is the intensity propagator from $\mathbf{r}_a$ to $\mathbf{r}_b$. In the slab geometry, due to translational invariance along the transverse directions, $P(\mathbf{r}_b, \mathbf{r}_a)$ depends only on the difference $\boldsymbol{\rho}_b - \boldsymbol{\rho}_a$, and we can rewrite (33.8) in the form

$$\begin{aligned}\langle U(\mathbf{k}_a, \mathbf{k}_b)\, U^*(\mathbf{k}_{a'}, \mathbf{k}_{b'})\rangle &= k_z(q_b)k_z(q_{b'})\, |U_0|^2 \int_{z=0} \int_{z=L} P(\boldsymbol{\rho}_b - \boldsymbol{\rho}_a) \\ &\times \exp(i\Delta\mathbf{q}_a \cdot \boldsymbol{\rho}_a - i\Delta\mathbf{q}_b \cdot \boldsymbol{\rho}_b)\, d^2\rho_a d^2\rho_b\,.\end{aligned} \tag{33.10}$$

Here we have introduced the notation $\Delta\mathbf{q}_a = \mathbf{q}_a - \mathbf{q}_{a'}$ and $\Delta\mathbf{q}_b = \mathbf{q}_b - \mathbf{q}_{b'}$. In order to evaluate the integrals, we perform the change of variables $\mathbf{X} = \boldsymbol{\rho}_a - \boldsymbol{\rho}_b$ and $\boldsymbol{\rho} = (\boldsymbol{\rho}_a + \boldsymbol{\rho}_b)/2$ (with unit Jacobian), leading to

$$\begin{aligned}\langle U(\mathbf{k}_a, \mathbf{k}_b)\, U^*(\mathbf{k}_{a'}, \mathbf{k}_{b'})\rangle &= k_z(q_b)k_z(q_{b'})\, |U_0|^2 \\ &\int P(\mathbf{X})\, \exp[i(\Delta\mathbf{q}_a + \Delta\mathbf{q}_b) \cdot \mathbf{X}/2]\, d^2X \\ &\times \int \exp[i(\Delta\mathbf{q}_a - \Delta\mathbf{q}_b) \cdot \boldsymbol{\rho}]\, d^2\rho\,.\end{aligned} \tag{33.11}$$

The first integral is the Fourier transform $\widetilde{P}(\mathbf{q})$ of $P(\mathbf{X})$, where $\mathbf{q} = (\Delta\mathbf{q}_a + \Delta\mathbf{q}_b)/2$. The second integral is given by $4\pi^2\, \delta(\Delta\mathbf{q}_a - \Delta\mathbf{q}_b)$. We finally obtain the following simple expression for the field correlation function:

$$\langle U(\mathbf{k}_a, \mathbf{k}_b)\, U^*(\mathbf{k}_{a'}, \mathbf{k}_{b'})\rangle = k_z(q_b)k_z(q_{b'})\, |U_0|^2\, \widetilde{P}(\Delta\mathbf{q}_a)\, \delta(\Delta\mathbf{q}_a - \Delta\mathbf{q}_b). \tag{33.12}$$

This result demonstrates that the angular correlation function of the field is nonvanishing only when $\Delta\mathbf{q}_a = \Delta\mathbf{q}_b$. Moreover, when $\Delta q_a = |\Delta\mathbf{q}_a|$ increases, the range of the correlation is described by the Fourier transform $\widetilde{P}(\Delta\mathbf{q}_a)$ of the intensity propagator.

33.3 Intensity Propagator in the Diffusion Approximation

In Chapter 32, we have shown that the intensity propagator is given by

$$P(\mathbf{r}_b, \mathbf{r}_a) = \frac{64\,\ell_s^4}{k_0^2}\, L(\mathbf{r}_b, \mathbf{r}_a)\,, \tag{33.13}$$

where the ladder propagator L obeys the diffusion equation

$$\nabla_{\mathbf{r}_a}^2 L(\mathbf{r}_b, \mathbf{r}_a) = -\frac{12\pi}{\ell_s^3}\,\delta(\mathbf{r}_b - \mathbf{r}_a)\,, \tag{33.14}$$

along with appropriate boundary conditions. The Green's function of the diffusion equation for a slab geometry is calculated in Chapter 25, using extrapolated boundary conditions. Using (25.44) with $Q = q$ for a non-absorbing medium, we directly obtain

$$\widetilde{P}(\mathbf{q}) = \frac{64 \times 12\pi\,\ell_s}{k_0^2}\;\frac{q\ell^2}{\sinh(qL) + 2q\ell\,\cosh(qL) + (q\ell)^2\sinh(qL)}\,, \tag{33.15}$$

with $\ell = (2/3)\ell_t$. Since $\ell \simeq \ell_t$ and the scale $1/q$ is assumed to be large compared to ℓ_t, we can assume $q\ell \ll 1$, leading to

$$\widetilde{P}(\mathbf{q}) = A\,\frac{q}{\sinh(qL)}\,, \tag{33.16}$$

where $A = (64 \times 12\pi\,\ell_s\,\ell^2)/k_0^2$.

33.4 Intensity Correlation Function and Memory Effect

An explicit expression for the field correlation function can be obtained by inserting (33.16) into (33.12):

$$\langle U(\mathbf{k}_a, \mathbf{k}_b)\,U^*(\mathbf{k}_{a'}, \mathbf{k}_{b'})\rangle = A\,k_z(q_b)k_z(q_{b'})\,|U_0|^2\,\frac{\Delta q_a}{\sinh(\Delta q_a\,L)}\,\delta(\Delta\mathbf{q}_a - \Delta\mathbf{q}_b)\,. \tag{33.17}$$

The normalized angular intensity correlation function can be deduced by using (33.2) and (33.3). We thus obtain

$$C^I_{aba'b'} = \left|\frac{\Delta q_a\,L}{\sinh(\Delta q_a\,L)}\right|^2 \delta_{\Delta\mathbf{q}_a, \Delta\mathbf{q}_b}, \tag{33.18}$$

where $\delta_{\Delta \mathbf{q}_a, \Delta \mathbf{q}_b}$ is a Kronecker delta.

The above angular correlation function describes the so-called "memory effect." We see that by changing the angle of incidence from $\mathbf{q}_a$ to $\mathbf{q}_{a'} = \mathbf{q}_a + \Delta \mathbf{q}_a$, the speckle pattern observed in the direction $\mathbf{q}_{b'} = \mathbf{q}_b + \Delta \mathbf{q}_a$ remains correlated with the initial speckle pattern observed in direction $\mathbf{q}_b$ (the speckle pattern moves as a whole). This effect remains visible as long as the amplitude of the correlation does not vanish when Δq_a increases. When the condition $\Delta q_a L \gg 1$ is satisfied, $C^I_{aba'b'} \sim \exp(-2\Delta q_a L)$, which shows that the angular intensity correlation function describes short-range correlations.

33.5 Size of a Speckle Spot

In a speckle pattern, a complex distribution of bright and dark spots is observed (see, for example, Fig. 28.1 in Chapter 28). The intensity correlation function can be used to characterize the typical size of a speckle spot. To see this, we consider the transmission geometry in Fig. 33.1, but with a beam of finite transverse size. In the paraxial approximation, the complex amplitude of the field produced by such a beam in the plane $z = 0$ is of the form $U_0(\boldsymbol{\rho}_a) \exp(i\mathbf{q}_a \cdot \boldsymbol{\rho}_a)$. In this situation, (33.10) becomes

$$\langle U(\mathbf{k}_a, \mathbf{k}_b)\, U^*(\mathbf{k}_{a'}, \mathbf{k}_{b'})\rangle = k_z(q_b)k_z(q_{b'}) \int_{z=0} \int_{z=L} I(\boldsymbol{\rho}_a)\, P(\boldsymbol{\rho}_b - \boldsymbol{\rho}_a) \times \exp(i\Delta \mathbf{q}_a \cdot \boldsymbol{\rho}_a - i\Delta \mathbf{q}_b \cdot \boldsymbol{\rho}_b)\, d^2\rho_a d^2\rho_b\,, \tag{33.19}$$

where $I(\boldsymbol{\rho}_a) = |U_0(\boldsymbol{\rho}_a)|^2$ is the intensity of the beam in the input plane $z = 0$. Once again using the same change of variables that leads to (33.11) and performing the Fourier transforms yields

$$\langle U(\mathbf{k}_a, \mathbf{k}_b)\, U^*(\mathbf{k}_{a'}, \mathbf{k}_{b'})\rangle = k_z(q_b)k_z(q_{b'})\, \widetilde{P}(\Delta \mathbf{q}_b)\, \tilde{H}(\Delta \mathbf{q}_b - \Delta \mathbf{q}_a)\,. \tag{33.20}$$

A measure of the angular size of a speckle spot is the width of the correlation function (33.20), considered as a function of the observation direction $\mathbf{q}_b$, for a fixed direction of incidence $\mathbf{q}_a$. (In practice, one usually measures the width of the intensity correlation function, which is essentially the square modulus of the field correlation function.) We therefore need to evaluate the width of $\langle U(\mathbf{k}_a, \mathbf{k}_b)\, U^*(\mathbf{k}_{a'}, \mathbf{k}_{b'})\rangle$, considered as a function of $\Delta \mathbf{q}_b$ with $\Delta \mathbf{q}_a = 0$. The result depends on the respective widths of the two functions on the right-hand side of (33.20).

For a slab with thickness $L \gg \ell^*$, the solution of the diffusion equation reveals that $P(\boldsymbol{\rho}_a - \boldsymbol{\rho}_b)$ leads to a spatial distribution of diffuse intensity of size L in the output plane $z = L$. The function $I(\boldsymbol{\rho}_a)$ has a width W, corresponding to the beam size in the plane $z = 0$. Two different situations need to be considered.

Extended Beam $(W \gg L)$

In this case, the angular width of the correlation function is driven by the function $\tilde{I}(\Delta\mathbf{q}_b)$. This is given by $\Delta q_b \sim 2\pi/W$. If one observes the speckle pattern in the focal plane of an imaging system with focal length f, the size of the speckle spot is $\Delta R \sim f\lambda/W$.

Focused Illumination $(W \ll L)$

In this case, the angular width of the correlation function is controlled by the function $\widetilde{P}(\Delta\mathbf{q}_b)$. As discussed in the previous section, the width of $\widetilde{P}$ is $\Delta q_b \sim 2\pi/L$. This gives a speckle spot size $\Delta R \sim f\lambda/L$ in the focal plane.

33.6 Number of Transmission Modes

In the case of illumination with a beam of finite transverse size, the angular size of a speckle spot can be associated with the size of a transmitted mode. This assertion is based on the intuitive picture that two transmitted wavevectors will be independent (and will describe two different modes) when their angular separation is larger than the angular range of the intensity correlation function.

Let us denote by θ the angle between two transmitted wavevectors $\mathbf{k}_b$ and $\mathbf{k}_{b'}$. We have $|\Delta\mathbf{q}_b|^2 \simeq |\Delta\mathbf{k}_b|^2 = 2k_0^2(1 - \cos\theta)$. For a beam of width W satisfying $W \gg L$, we have $|\Delta\mathbf{q}_b| \simeq 2\pi/W$. Therefore, the angle θ defines the angular extent of a mode when

$$2k_0^2(1 - \cos\theta) \simeq \frac{4\pi^2}{W^2} . \tag{33.21}$$

The angle θ also corresponds to a solid angle through the relation $\Delta\Omega = 2\pi(1 - \cos\theta)$, so that a transmission mode corresponds to a solid angle $\Delta\Omega \simeq \pi\lambda^2/W^2$. The number of transmission modes is defined as

$$N = \frac{2\pi}{\Delta\Omega} \simeq \frac{2\,W^2}{\lambda^2} . \tag{33.22}$$

The number of modes is analogous to the number of transverse modes that is used in the transport of waves through waveguides, as in mesoscopic electronic transport. It is also related to the number of degrees of freedom that are available in light transmission through a disordered medium in the multiple-scattering regime.

References and Additional Reading

Angular speckle correlations and the memory effect were initially discussed in the following papers:

S. Feng, C. Kane, P.A. Lee and A.D. Stone, Phys. Rev. Lett. **61**, 834 (1988).
R. Berkovits, M. Kaveh and S. Feng, Phys. Rev. B **40**, 737 (1989).
R. Berkovits and S. Feng, Phys. Rep. **238**, 135 (1994).

This paper reported on the first measurement of the memory effect in volume scattering:

I. Freund, M. Rosenbluh and S. Feng, Phys. Rev. Lett. **61**, 2328 (1988).

In this paper, a connection is made between speckle correlations and mesoscopic fluctuations of electrons in conductors:

S. Feng and P.A. Lee, Science **251**, 633 (1991).

A treatment of angular correlations in speckles produced by volume scattering can be found in:

M.C.W. van Rossum and Th.M. Nieuwenhuizen, Rev. Mod. Phys. **71**, 313 (1999).
E. Akkermans and G. Montambaux, *Mesoscopic Physics of Electrons and Photons* (Cambridge University Press, Cambridge, 2007), chap. 12.

34

Coherent Backscattering

The measurement of the angular dependence of the light intensity reflected from a thick scattering medium reveals the existence of a peak in the backscattering direction. This effect, known as coherent backscattering, is a signature of the underlying coherence of the multiple scattering process and a consequence of the reciprocity theorem.

34.1 Reflected Far-Field

We consider a semi-infinite scattering medium that is illuminated by a plane wave with wavevector $\mathbf{k}_a$, as shown in Fig. 34.1. The reflected intensity is observed in the far-field in the direction defined by the wavevector $\mathbf{k}_b$. We take the z-axis to be perpendicular to the interface and use the following notation: $\mathbf{r}_a = (\boldsymbol{\rho}_a, 0)$, $\mathbf{k}_a = [\mathbf{q}_a, k_z(q_a)]$, etc.

The scattered field in the plane $z = 0$ is linearly related to the incident field $U_0 \exp(i\mathbf{k}_a \cdot \mathbf{r})$. Using the amplitude propagator $h(\mathbf{r}_b, \mathbf{r}_a)$ introduced in Chapter 32,

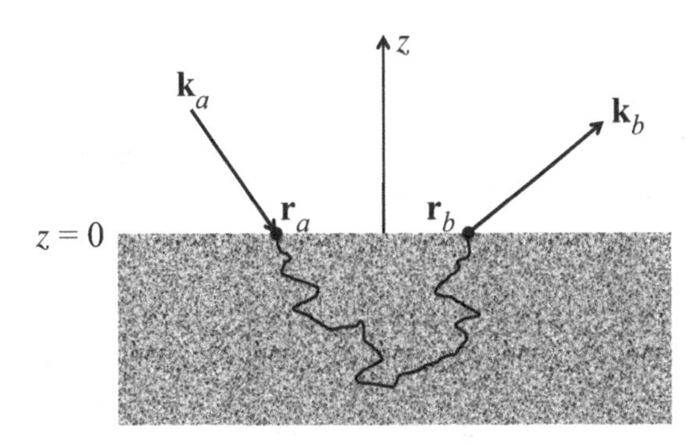

Fig. 34.1 Reflection by a scattering medium of an incident plane wave with wavevector $\mathbf{k}_a$ The far-field intensity is along the direction of $\mathbf{k}_b$.

we have

$$U(\mathbf{r}_b) = \int_{z=0} h(\mathbf{r}_b, \mathbf{r}_a)\, U_0 \, \exp(i\mathbf{q}_a \cdot \boldsymbol{\rho}_a)\, d^2\rho_a \,, \tag{34.1}$$

where $\mathbf{r}_a = (\boldsymbol{\rho}_a, 0)$ and $\mathbf{r}_b = (\boldsymbol{\rho}_b, 0)$ are the input and output points. The reflected far-field has a complex amplitude proportional to the Fourier transform of the field in the plane $z = 0$ [see Chapter 33, Eq. (33.5)]. Denoting by $U(\mathbf{k}_a, \mathbf{k}_b)$ the far-field amplitude reflected along the direction of $\mathbf{k}_b$, we have

$$\begin{aligned} U(\mathbf{k}_a, \mathbf{k}_b) &= k_z(q_b) \int_{z=0} U(\boldsymbol{\rho}_b, z = 0)\, \exp(-i\mathbf{q}_b \cdot \boldsymbol{\rho}_b)\, d^2\rho_b \\ &= k_z(q_b)\, U_0 \int_{z=0} h(\mathbf{r}_b, \mathbf{r}_a)\, \exp(i\mathbf{q}_a \cdot \boldsymbol{\rho}_a - i\mathbf{q}_b \cdot \boldsymbol{\rho}_b)\, d^2\rho_a\, d^2\rho_b \,, \end{aligned} \tag{34.2}$$

where $k_z(q) = (k_0^2 - \mathbf{q}^2)^{1/2}$.

34.2 Reflected Intensity

The reflected intensity $I(\mathbf{k}_a, \mathbf{k}_b) = \langle |U(\mathbf{k}_a, \mathbf{k}_b)|^2 \rangle$ is given by

$$\begin{aligned} I(\mathbf{k}_a, \mathbf{k}_b) = k_z^2(q_b)\, |U_0|^2 \int \langle h(\mathbf{r}_b, \mathbf{r}_a) h^*(\mathbf{r}_{b'}, \mathbf{r}_{a'}) \rangle \, \exp[i\mathbf{q}_a \cdot (\boldsymbol{\rho}_a - \boldsymbol{\rho}_{a'})] \\ \times \exp[-i\mathbf{q}_b \cdot (\boldsymbol{\rho}_b - \boldsymbol{\rho}_{b'})]\, d^2\rho_a d^2\rho_b d^2\rho_{a'} d^2\rho_{b'} \,, \end{aligned} \tag{34.3}$$

where we have used (34.2). In the limit $k_0\ell_s \gg 1$, the ladder approximation gives the leading contribution to the intensity, corresponding to the diagrammatic representation in Fig. 34.2. In the ladder approximation, the correlation function of the amplitude propagator is given by (see Chapter 32):

$$\langle h(\mathbf{r}_b, \mathbf{r}_a) h^*(\mathbf{r}_{b'}, \mathbf{r}_{a'}) \rangle \simeq P(\mathbf{r}_b, \mathbf{r}_a)\, \delta(\mathbf{r}_a - \mathbf{r}_{a'}) \delta(\mathbf{r}_b - \mathbf{r}_{b'}) \,, \tag{34.4}$$

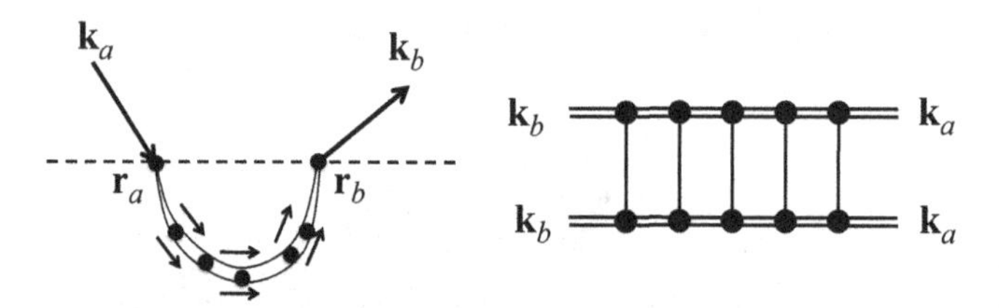

Fig. 34.2 Graphical (left) and diagrammatic (right) representations of the diffuse reflection process in the ladder approximation. The ladder diagram on the right includes five scattering events and is constructed following the rules introduced in Chapter 16, with a double line indicating an average Green's function.

where $P(\mathbf{r}_b, \mathbf{r}_a)$ is the transport probability from $\mathbf{r}_a$ to $\mathbf{r}_b$. In the reflection geometry considered here, $P(\mathbf{r}_b, \mathbf{r}_a)$ depends only on $\boldsymbol{\rho}_b - \boldsymbol{\rho}_a$. We thus find that the intensity is of the form

$$I(\mathbf{k}_a, \mathbf{k}_b) = k_z^2(q_b)\,|U_0|^2 \int P(\boldsymbol{\rho}_b - \boldsymbol{\rho}_a)\, d^2\rho_a d^2\rho_b\,. \tag{34.5}$$

34.3 Reciprocity of the Amplitude Propagator

The amplitude propagator is connected to the reflection coefficient $r(\mathbf{q}, \mathbf{q}')$ which arises in the scattering matrix, as defined in (14.12). Indeed, by definition of the scattering matrix, the Fourier transform of the reflected field in the plane $z = 0$, defined as

$$U_s(\mathbf{q}) = \int U(\rho) \exp(-i\mathbf{q}\cdot\rho)\, d^2\rho \tag{34.6}$$

is connected to the Fourier transform of the incident field U_i by

$$U(\mathbf{q}) = \int r(\mathbf{q}, \mathbf{q}')\, U_i(\mathbf{q}')\, d^2q'. \tag{34.7}$$

Using (34.1) and (34.7), it is easy to see that

$$h(\mathbf{r}, \mathbf{r}') = \int r(\mathbf{q}, \mathbf{q}')\, \exp(i\mathbf{q}\cdot\rho - i\mathbf{q}'\cdot\rho')\, \frac{d^2q\, d^2q'}{(2\pi)^2}\,, \tag{34.8}$$

where the points $\mathbf{r} = (\rho, 0)$ and $\mathbf{r}' = (\rho', 0)$ lie on the plane $z = 0$. The amplitude propagator is therefore the Fourier transform of the reflection coefficient. Inverting (34.8) yields

$$r(\mathbf{q}, \mathbf{q}') = \int_{z=0}\int_{z'=0} h(\mathbf{r}, \mathbf{r}')\, \exp(-i\mathbf{q}\cdot\rho + i\mathbf{q}'\cdot\rho')\, d^2\rho\, d^2\rho'\,. \tag{34.9}$$

In a linear medium with a symmetric permittivity tensor, the reciprocity theorem can be seen to hold (see Chapter 14). In the reflection geometry, we have [see (14.17)]

$$k_z(q)\, r(\mathbf{q}, \mathbf{q}') = k_z(q')\, r(-\mathbf{q}', -\mathbf{q})\,. \tag{34.10}$$

Using (34.9) and (34.10), we obtain the following reciprocity relation for the amplitude propagator:

$$\begin{aligned} k_z(q) \int h(\mathbf{r}, \mathbf{r}')\, \exp(-i\mathbf{q}\cdot\rho + i\mathbf{q}'\cdot\rho')\, d^2\rho\, d^2\rho' &= \\ k_z(q') \int h(\mathbf{r}', \mathbf{r})\, \exp(-i\mathbf{q}\cdot\rho + i\mathbf{q}'\cdot\rho')\, d^2\rho\, d^2\rho'. \end{aligned} \tag{34.11}$$

34.4 Coherent Backscattering Enhancement

The reciprocity relation (34.11) leads to contributions to the specific intensity that are not accounted for within the ladder approximation. Moreover, these contributions cannot be neglected even when $k_0\ell_s \gg 1$. Using (34.11) to transform $h^*(\mathbf{r}_{b'}, \mathbf{r}_{a'})$ into $h^*(\mathbf{r}_{a'}, \mathbf{r}_{b'})$ in (34.3), we find that the intensity is given by

$$I_c(\mathbf{k}_a, \mathbf{k}_b) = k_z(q_b)k_z(q_a)\,|U_0|^2 \int \langle h(\mathbf{r}_b, \mathbf{r}_a)h^*(\mathbf{r}_{a'}, \mathbf{r}_{b'})\rangle \exp[i\mathbf{q}_a \cdot (\boldsymbol{\rho}_a - \boldsymbol{\rho}_{a'})] \times \exp[-i\mathbf{q}_b \cdot (\boldsymbol{\rho}_b - \boldsymbol{\rho}_{b'})]\, d^2\rho_a d^2\rho_b d^2\rho_{a'} d^2\rho_{b'} \,. \quad (34.12)$$

This expression describes the interference between the field produced by a scattering sequence and the field produced by its corresponding reciprocal sequence (the sequence in reverse order), as illustrated in Fig. 34.3. A simple change of variables in (34.12) allows it to be written in the form

$$I_c(\mathbf{k}_a, \mathbf{k}_b) = k_z(q_b)k_z(q_a)\,|U_0|^2 \int \langle h(\mathbf{r}_b, \mathbf{r}_a)h^*(\mathbf{r}_{b'}, \mathbf{r}_{a'})\rangle \exp[i\mathbf{q}_a \cdot (\boldsymbol{\rho}_a - \boldsymbol{\rho}_{b'})] \times \exp[-i\mathbf{q}_b \cdot (\boldsymbol{\rho}_b - \boldsymbol{\rho}_{a'})]\, d^2\rho_a d^2\rho_b d^2\rho_{a'} d^2\rho_{b'}. \quad (34.13)$$

Once again making use of (34.4), we see that

$$I_c(\mathbf{k}_a, \mathbf{k}_b) = k_z(q_b)k_z(q_a)\,|U_0|^2 \int P(\boldsymbol{\rho}_b - \boldsymbol{\rho}_a) \exp[i(\mathbf{q}_a + \mathbf{q}_b)\cdot(\boldsymbol{\rho}_a - \boldsymbol{\rho}_b)] d^2\rho_a d^2\rho_b \,. \quad (34.14)$$

This expression describes a contribution to the reflected intensity that cannot be neglected in the vicinity of the backscattering direction. Indeed, in the backscattering direction defined by $\mathbf{k}_b = -\mathbf{k}_a$, (34.14) is identical to (34.5):

$$I_c(\mathbf{k}_a, -\mathbf{k}_a) = I(\mathbf{k}_a, -\mathbf{k}_a). \quad (34.15)$$

Thus we find that the backscattered intensity is twice the intensity predicted in the ladder approximation. We emphasize that this fundamental result, which is known as coherent backscattering, is a consequence of wave reciprocity. It can be understood as arising from the constructive interference between fields scattered along

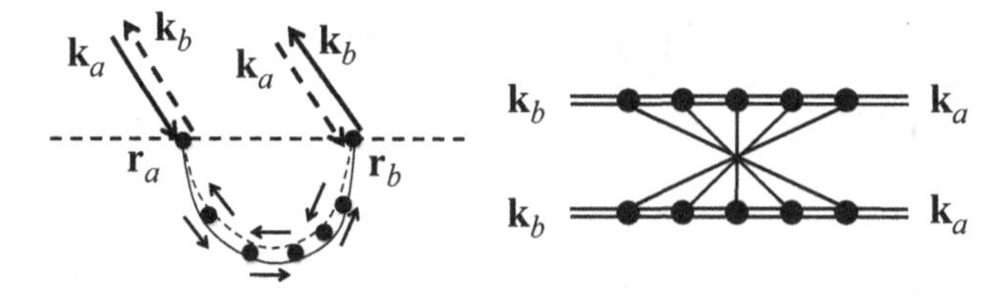

Fig. 34.3 Graphical (left) and diagrammatic (right) representations of the reciprocal scattering sequences contributing to coherent backscattering. The diagram on the right is said to be maximally crossed.

reciprocal scattering sequences. Reciprocity ensures that in the backscattering direction, the fields have identical phases, thus producing constructive interference.

34.5 Coherent Backscattering Cone and Angular Width

We now calculate explicitly the intensity distribution around the backscattering direction $\mathbf{k}_b = -\mathbf{k}_a$. Let us introduce the change of variables $\mathbf{X} = \boldsymbol{\rho}_a - \boldsymbol{\rho}_b$ and $\boldsymbol{\rho} = (\boldsymbol{\rho}_a + \boldsymbol{\rho}_b)/2$ (with unit Jacobian) into (34.14). We thereby obtain

$$I_c(\mathbf{k}_a, \mathbf{k}_b) = k_z(q_b)k_z(q_a)\,|U_0|^2 \left[\int d^2\rho\right] \widetilde{P}(\mathbf{q}_a + \mathbf{q}_b)\,, \qquad (34.16)$$

where $\widetilde{P}(\mathbf{q})$ is the Fourier transform of the transport probability. The remaining integral is not infinite and corresponds to the size S of the illuminated region on the interface. We thus find that

$$I_c(\mathbf{k}_a, \mathbf{k}_b) = k_z(q_b)k_z(q_a)\,S\,|U_0|^2\,\widetilde{P}(\mathbf{q}_a + \mathbf{q}_b)\,, \qquad (34.17)$$

which shows that the coherent backscattering intensity I_c is proportional to the Fourier transform of the probability $P(\mathbf{r} - \mathbf{r}')$ connecting two points $\mathbf{r}$ and $\mathbf{r}'$ on the surface.

In Chapter 32, we have seen that the intensity propagator is given by

$$P(\mathbf{r}_b, \mathbf{r}_a) = \frac{64\,\ell_s^4}{k_0^2}\,L(\mathbf{r}_b, \mathbf{r}_a)\,, \qquad (34.18)$$

where the ladder propagator L obeys the diffusion equation

$$\nabla^2_{\mathbf{r}_a} L(\mathbf{r}_b, \mathbf{r}_a) = -\frac{12\pi}{\ell_s^3}\,\delta(\mathbf{r}_b - \mathbf{r}_a)\,, \qquad (34.19)$$

together with appropriate boundary conditions. To get the propagator in a semi-infinite geometry, we make use of (25.34), with $Q = q$ for a non-absorbing medium, and for $z = 0$ and $z' = 0$, leading to

$$\widetilde{P}(\mathbf{q}) = \frac{64 \times 12\pi\ell_s}{k_0^2}\,\frac{1}{2q}\left(1 - \frac{1 - q\ell}{1 + q\ell}\right) \qquad (34.20)$$

with $\ell = (2/3)\ell_t$. Since $q\ell \ll 1$ in the diffusive limit, this expression can be simplified as

$$\widetilde{P}(\mathbf{q}) = A(1 - q\ell)\,, \qquad (34.21)$$

with $A = 64 \times 12\pi\ell_s\ell/k_0^2$. Inserting this result into (34.17) leads to

$$I_c(\mathbf{k}_a, \mathbf{k}_b) = k_z(q_b)k_z(q_a)\,S\,|U_0|^2\,A(1 - \delta q\,\ell)\,, \qquad (34.22)$$

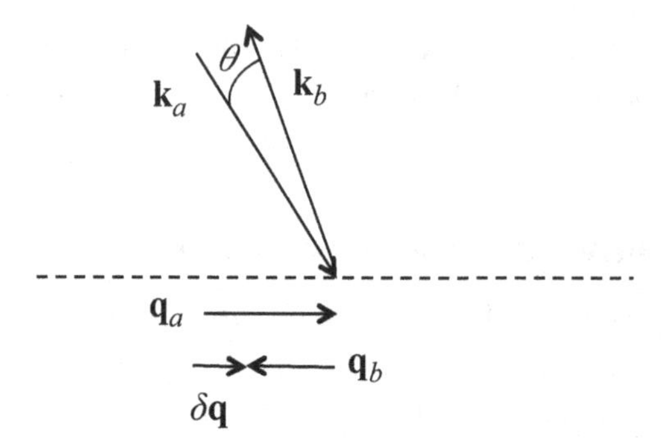

Fig. 34.4 Wavevectors involved in the computation of the backscattered intensity.

where we have defined $\mathbf{q}_b = -\mathbf{q}_a + \delta\mathbf{q}$ and put $\delta q = |\delta\mathbf{q}|$ (see Fig. 34.4). This can also be rewritten as

$$I_c(\mathbf{k}_a, \mathbf{k}_b) \simeq I_c(\delta q = 0)(1 - \delta q\, \ell) \, . \tag{34.23}$$

Equation (34.23) describes the angular dependence of the reflected intensity around the backscattering direction $\delta q = 0$. Writing $\delta q \simeq k_0\theta$, where θ is the angular deviation from the backscattering direction, we see that the backscattering peak has an angular width $\Delta\theta \simeq \lambda/\ell_t$. Equation (34.23) also shows that the backscattering peak exhibits a triangular singularity. This singularity is a signature of the coherent backscattering effect. In the presence of absorption, both the amplitude and the sharpness of the backscattering peak decrease, as has been observed

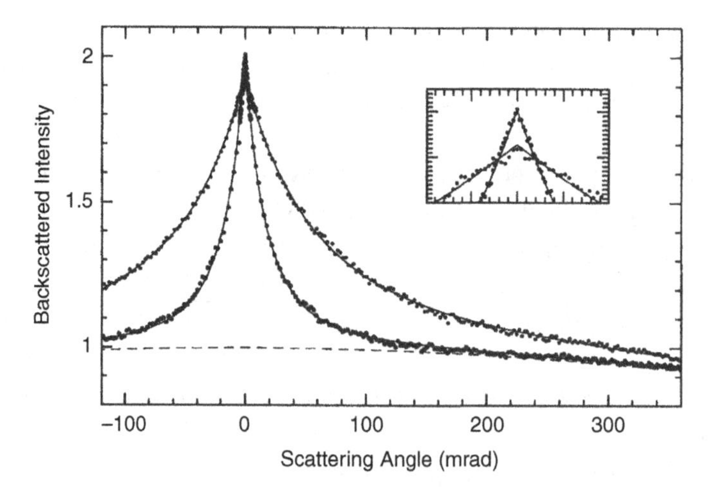

Fig. 34.5 Experimental evidence of coherent backscattering of light from a strongly scattering medium. Narrow cone: $k\ell_t \simeq 23$. Wide cone: $k\ell_t \simeq 6$ (with k the wavenumber in the medium). From D. Wiersma et al., Phys. Rev. Lett. **74**, 4193 (1995).

experimentally. An example of a measured backscattering peak in a non-absorbing medium is shown in Fig. 34.5. The factor of two in the peak height, the triangular shape of the backscattering cone and the dependence of the angular width on ℓ_t are clearly visible.

References and Additional Reading

The first measurements of coherent backscattering peaks produced by volume scattering in optics were reported in:

Y. Kuga and A. Ishimaru, J. Opt. Soc. Am. A **8**, 831 (1984)
M.P. van Albada and A. Lagendijk, Phys. Rev. Lett. **55**, 2692 (1985).
P.E. Wolf and G. Maret, Phys. Rev. Lett. **55**, 2696 (1985).

These papers provided the first theoretical analysis of the coherent backscattering peak lineshape:

E. Akkermans, P.E. Wolf and R. Maynard, Phys. Rev. Lett. **56**, 1471 (1986).
E. Akkermans, P.E. Wolf, R. Maynard and G. Maret, J. Phys. (France) **49**, 77 (1988).

Coherent backscattering was also observed and discussed in the context of scattering from randomly rough surfaces:

K.A. O'Donnell and E.R. Méndez, J. Opt. Soc. Am. A **4**, 1194 (1987).
A.A. Maradudin, T. Michel, A.R. McGurn and E.R. Méndez, Ann. Phys. (NY) **203**, 225 (1990).

35

Dynamic Light Scattering

In this chapter, we study light scattering by particles in motion. In such a situation, the resulting speckle pattern fluctuates in time. We derive the explicit form of the time correlation function of the intensity in the single-scattering and multiple scattering regimes. As an example, we consider scattering by particles undergoing Brownian motion.

35.1 Single Scattering Regime

A typical geometry in a dynamic light scattering experiment is sketched in Fig. 35.1. The figure represents a configuration using transmitted light, but the analysis developed in this chapter is valid for both reflection and transmission geometries.

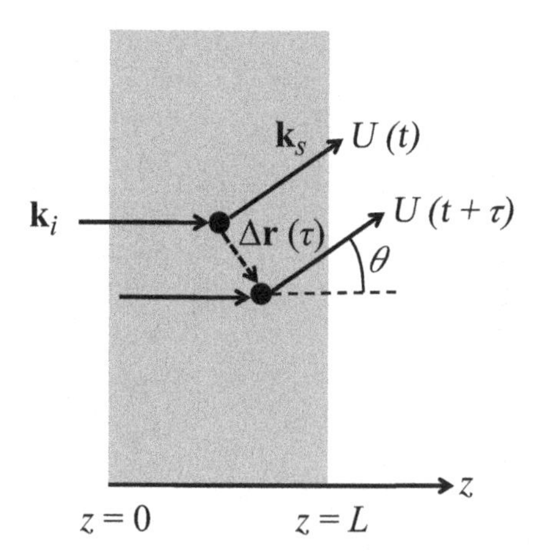

Fig. 35.1 Dynamic light scattering in the single scattering regime. $\Delta\mathbf{r}(\tau)$ is the displacement of a scatterer between times t and $t+\tau$. This displacement induces a phase shift between the scattered fields $U(t)$ and $U(t+\tau)$.

In the single scattering regime, we assume that the sample size is $L \sim \ell_s$, with ℓ_s the scattering mean free path. The medium is illuminated by a monochromatic plane wave with frequency ω, complex amplitude U_0 and wavevector $\mathbf{k}_i$. The field scattered in a direction defined by wavevector $\mathbf{k}_s$ is measured. This field results from the superposition of the waves scattered by all the scatterers. Due to the motion of the scatterers, the phase shifts between the scattered waves change in time, and the amplitude of the scattered field also fluctuates in time. We assume that the fluctuations occur on a time scale much larger than $2\pi/\omega$, so that the field remains quasi-monochromatic, and can be written in the form $U(t)\exp(-i\omega t)$, where $U(t)$ is a slowly varying complex amplitude. Note that since the motion of the scatterers induces Doppler shifts that remain small compared to ω, dynamic light scattering in the single scattering regime is also referred to as quasi-elastic light scattering (QELS).

In order to characterize the field fluctuations in time, we introduce the field correlation function $G_1(\tau) = \langle U(t)\, U^*(t+\tau)\rangle$, where $\langle \cdots \rangle$ denotes an average over the motion of the particles. In the single-scattering regime, the far-zone scattered field in the direction defined by wavevector $\mathbf{k}_s$ is taken to be of the form (see Section 10.3)

$$U(t) = f(\mathbf{q})\, \frac{\exp(ik_0 r)}{r}\, U_0 \sum_j \exp\left[-i\mathbf{q}\cdot\mathbf{r}_j(t)\right] , \tag{35.1}$$

where $\mathbf{q} = \mathbf{k}_s - \mathbf{k}_i$, $k_0 = \omega/c$, $\mathbf{r}_j(t)$ is the position of particle number j at time t, and $f(\mathbf{q})$ is the scattering amplitude of a single scatterer. We thus obtain the following expression for the time correlation function

$$G_1(\tau) = |f(\mathbf{q})|^2 \frac{|U_0|^2}{r^2} \sum_j \left\langle \exp\left[i\mathbf{q}\cdot\Delta\mathbf{r}_j(\tau)\right]\right\rangle , \tag{35.2}$$

where $\Delta\mathbf{r}_j(\tau) = \mathbf{r}_j(t+\tau) - \mathbf{r}_j(t)$ is the displacement of particle number j between time t and time $t+\tau$, and we have assumed that the displacements of different particles are uncorrelated. For a collection of N identical particles, the average value in the summation is the same for all particles, and we find

$$G_1(\tau) = N\, |f(\mathbf{q})|^2 \frac{|U_0|^2}{r^2} \left\langle \exp\left[i\mathbf{q}\cdot\Delta\mathbf{r}(\tau)\right]\right\rangle . \tag{35.3}$$

The average in this equation has to be taken over the random variable $\Delta\mathbf{r}(\tau)$. In the case of three-dimensional Brownian motion, the statistical distribution of the displacements is isotropic and Gaussian, with a probability density

$$P[\Delta r(\tau)] = \frac{1}{(4\pi D_B \tau)^{3/2}} \exp\left[\frac{-\Delta r(\tau)^2}{4 D_B \tau}\right] , \tag{35.4}$$

where D_B is the diffusion constant of the particles and $\Delta r(\tau) = |\Delta \mathbf{r}(\tau)|$. The average in (35.3) is easily performed. Introducing the normalized correlation function $g_1(\tau) = G_1(\tau)/G_1(0)$, we have

$$\begin{aligned} g_1(\tau) &= \left\langle \exp\left[i\mathbf{q}\cdot\Delta\mathbf{r}(\tau)\right]\right\rangle \\ &= \left\langle \exp\left[iq_x\Delta x(\tau)\right]\right\rangle\left\langle \exp\left[iq_y\Delta y(\tau)\right]\right\rangle\left\langle \exp\left[iq_z\Delta z(\tau)\right]\right\rangle , \end{aligned} \tag{35.5}$$

where the averages can be calculated using the result

$$\int_{-\infty}^{+\infty} \exp(ipx)\exp(-ax^2/2)\,dx = (2\pi/a)^{1/2}\exp[-p^2/(2a)] . \tag{35.6}$$

Putting everything together, we find that

$$g_1(\tau) = \exp\left(-D_B\,q^2\,\tau\right). \tag{35.7}$$

The modulus of the scattered wavevector q is $q = 2\,k_0\,\sin(\theta/2)$, with θ being the scattering angle shown in Fig. 35.1. Equation (35.7) shows that a measurement of $g_1(\tau)$ can be used to deduce the diffusion constant D_B, for instance in a colloidal suspension of particles. Indeed, D_B is connected to the radius R of the particles and the viscosity η of the fluid through the Einstein relation $D_B = k_BT/(6\pi\eta R)$, with T being the temperature.

35.2 Measured Signal and Siegert Relation

In practice, one often measures the intensity correlation function $G_2(\tau) = \langle I(t)\,I(t+\tau)\rangle$, with $I(t) = |U(t)|^2$, instead of the field correlation function $G_1(\tau)$. When the fields scattered by different particles can be considered as uncorrelated, a simple relation exists between $G_2(\tau)$ and $G_1(\tau)$.

Using (35.1), the intensity correlation function can be written as

$$\begin{aligned} G_2(\tau) = |f(\mathbf{q})|^4\frac{|U_0|^4}{r^4}\sum_{j,k,l,m}\langle\exp\left[-i\mathbf{q}\cdot\mathbf{r}_j(t)\right]\exp\left[i\mathbf{q}\cdot\mathbf{r}_k(t)\right] \\ \times\exp\left[-i\mathbf{q}\cdot\mathbf{r}_l(\tau)\right]\exp\left[i\mathbf{q}\cdot\mathbf{r}_m(\tau)\right]\rangle . \end{aligned} \tag{35.8}$$

Assuming that the motions of different particles are uncorrelated, the only non-vanishing terms are those corresponding to $j = k = l = m$, to $j = k$ and $l = m$ with $k \neq l$, and $j = m$ and $k = l$ with $j \neq k$, so that

$$G_2(\tau) = |f(\mathbf{q})|^4\frac{|U_0|^4}{r^4}\left[N^2 + \sum_{j\neq k}\left\langle\exp\left[-i\mathbf{q}\cdot\Delta\mathbf{r}_j(\tau)\right]\right\rangle\left\langle\exp\left[i\mathbf{q}\cdot\Delta\mathbf{r}_k(\tau)\right]\right\rangle\right]. \tag{35.9}$$

Since the average is the same for all particles, we obtain

$$G_2(\tau) = |f(\mathbf{q})|^4 \frac{|U_0|^4}{r^4} \left[N^2 + N(N-1)\left\langle \exp\left[-i\mathbf{q}\cdot\Delta\mathbf{r}(\tau)\right]\right\rangle^2\right]. \quad (35.10)$$

Using (35.3), and assuming $N \gg 1$, the above expression can be put into the following form:

$$G_2(\tau) = |G_1(0)|^2 + |G_1(\tau)|^2. \quad (35.11)$$

This result, known as the Siegert relation, shows that the intensity correlation function can be obtained from the field correlation function. In terms of normalized correlation functions $g_1(\tau) = \langle U(t)\, U^*(t+\tau)\rangle / \langle |U(t)|^2\rangle$ and $g_2(\tau) = \langle I(t)\, I(t+\tau)\rangle / \langle I(t)\rangle^2$, the Siegert relation becomes

$$g_2(\tau) = 1 + |g_1(\tau)|^2. \quad (35.12)$$

This relationship is frequently used in the analysis of dynamic light scattering experiments. It has been derived here in the single-scattering regime. In the multiple-scattering regime, this relation holds provided that the field can be considered as Gaussian (see Chapter 29). In this case, the intensity correlation function factorizes into products of second-order correlation functions, leading to the Siegert relation. Note that when $\tau = 0$, the Siegert relation leads to $\langle I^2\rangle = 2\langle I\rangle^2$, which is a feature of Rayleigh statistics derived in Chapter 28.

35.3 Multiple-Scattering Regime and Diffusing-Wave Spectroscopy

When the system size L becomes larger than the scattering mean free path ℓ_s, the single-scattering approximation is no longer valid. As illustrated in Fig. 35.2, we shall now study the time fluctuations of the field resulting from the superposition of multiply-scattered waves.

It is convenient to use the representation of the field as a sum over scattering sequences, as described in Chapter 32. In a heuristic manner, we extend this concept to a time-dependent problem in which the positions of the scattering centers are changing in time, slowly enough so that the field to be considered as quasi-monochromatic.[1] The slowly varying amplitude of the scattered field $U(t)$ can be written as

$$U(t) = U_0 \sum_{n=1}^{\infty} \sum_{\mathcal{S}_n} A_{\mathcal{S}_n}(t)\, \exp[i\phi_{\mathcal{S}_n}(t)]\,, \quad (35.13)$$

[1] The approach we follow was introduced in D.J. Pine, D.A. Weitz, P.M. Chaikin and E. Herbolzheimer, Phys. Rev. Lett. **60**, 1134 (1988), and D.J. Pine, D.A. Weitz, J. X. Zhu and E. Herbolzheimer, J. Phys. **51**, 2101 (1990).

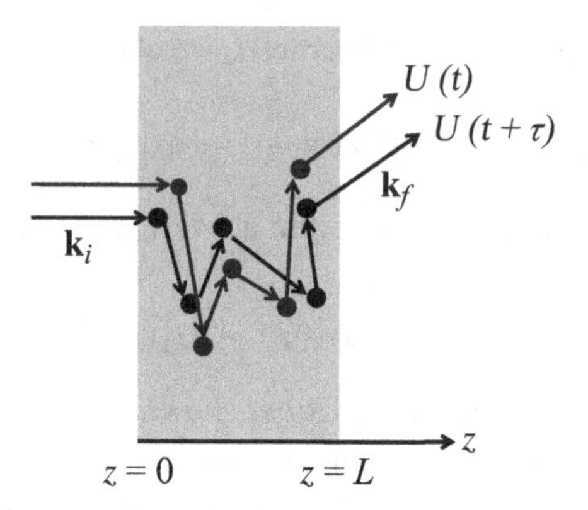

Fig. 35.2 Schematic representation of the field resulting from a scattering sequence with multiple scattering events, at two different times t and $t+\tau$. As a consequence of the motion of scatterers, the accumulated phase shifts along the sequences at t and $t+\tau$ are different from each other.

where U_0 is the amplitude of the incident plane wave. In this representation, a scattering sequence with n scattering events is denoted $\mathcal{S}_n = \{\mathbf{r}_1(t), \mathbf{r}_2(t)...\mathbf{r}_n(t)\}$, where $\mathbf{r}_j(t)$ is the position of scatterer number j at time t. The change in amplitude and phase created by the sequence $\mathcal{S}_n$ are $A_{\mathcal{S}_n}(t)$ and $\phi_{\mathcal{S}_n}(t)$, respectively. The time-correlation function of the field directly follows:

$$G_1(\tau) = |U_0|^2 \sum_n \sum_{\mathcal{S}_n} \sum_{n'} \sum_{\mathcal{S}'_{n'}} \langle A_{\mathcal{S}_n}(t) A_{\mathcal{S}'_{n'}}(t+\tau) \\ \times \exp[i\phi_{\mathcal{S}_n}(t)] \exp[-i\phi_{\mathcal{S}'_{n'}}(t+\tau)] \rangle , \tag{35.14}$$

where $\langle \cdots \rangle$ denotes averaging over the motion of the particles. Under the assumptions of fully developed speckle (see Chapter 28), the complex amplitudes resulting from different sequences are uncorrelated, and the output amplitudes and phase shifts for a given sequence are also uncorrelated. The preceding expression can thus be simplified:

$$G_1(\tau) = |U_0|^2 \sum_n \sum_{\mathcal{S}_n} \langle A_{\mathcal{S}_n}(t) A_{\mathcal{S}_n}(t+\tau) \rangle \langle \exp[i\phi_{\mathcal{S}_n}(t) - i\phi_{\mathcal{S}_n}(t+\tau)] \rangle . \tag{35.15}$$

Due to the random motion of the particles, the phase shift along a scattering sequence will decorrelate much faster than the amplitude when τ increases. We can then assume $\langle A_{\mathcal{S}_n}(t) A_{\mathcal{S}_n}(t+\tau) \rangle \simeq \langle A^2_{\mathcal{S}_n} \rangle$ and write

$$G_1(\tau) = |U_0|^2 \sum_n \sum_{\mathcal{S}_n} \langle A^2_{\mathcal{S}_n} \rangle \langle \exp[i\phi_{\mathcal{S}_n}(t) - i\phi_{\mathcal{S}_n}(t+\tau)] \rangle . \tag{35.16}$$

We are left with the evaluation of the average of the phase term. The phase difference due to the motion of the particles, along a sequence with n scattering events, can be written using the scattered wavevector $\mathbf{q}_j$ of an individual scattering event:

$$G_1(\tau) = |U_0|^2 \sum_n \sum_{\mathcal{S}_n} \langle A_{\mathcal{S}_n}^2 \rangle \left\langle \exp\left[i \sum_{j=1}^{n} \mathbf{q}_j \cdot \Delta\mathbf{r}_j(\tau) \right] \right\rangle . \tag{35.17}$$

In the averaging process, both $\mathbf{q}_j$ and $\Delta\mathbf{r}_j(\tau)$ are random variables (this is a major difference with the single scattering regime). An exact calculation would require handling the correlation between the scattered wavevectors $\mathbf{q}_j$ and $\mathbf{q}_{j+1}$ between successive scattering events, as well as the constraint $\sum_j \mathbf{q}_j = \mathbf{k}_f - \mathbf{k}_i$, with $\mathbf{k}_i$ and $\mathbf{k}_f$ the incident and observation wavevectors. Although this can be done in numerical simulations, a closed-form expression can be obtained by making a series of approximations. We first assume that the average over $\mathbf{q}_j$ is independent on the motion of the particles, and depends only on the phase function of an individual scatterer and on the number of scattering events. Under these conditions, the averaging processes over $\mathbf{q}_j$ and $\Delta\mathbf{r}_j(\tau)$ are considered to be independent. For Brownian motion, the average over $\Delta\mathbf{r}_j(\tau)$ is performed using (35.7). We thereby obtain

$$G_1(\tau) = |U_0|^2 \sum_n \sum_{\mathcal{S}_n} \langle A_{\mathcal{S}_n}^2 \rangle \left\langle \exp\left(-D_B\, \tau \sum_{j=1}^{n} q_j^2 \right) \right\rangle , \tag{35.18}$$

where $q_j = |\mathbf{q}_j|$. The average over q_j remains to be performed. As the final result must be independent on the scattering sequence, it is useful to introduce the quantity

$$P(n) = \sum_{\mathcal{S}_n} \langle A_{\mathcal{S}_n}^2 \rangle , \tag{35.19}$$

which is the fraction of the incident power that has undergone n scattering events. We then rewrite (35.18) in the form

$$G_1(\tau) = |U_0|^2 \sum_n P(n) \left\langle \exp\left(-D_B\, \tau \sum_{j=1}^{n} q_j^2 \right) \right\rangle . \tag{35.20}$$

We evaluate the average over q_j^2 as follows:

$$\begin{aligned} \left\langle \exp\left(-D_B\, \tau \sum_{j=1}^{n} q_j^2 \right) \right\rangle &\simeq \left\langle 1 - D_B\, \tau \sum_{j=1}^{n} q_j^2 \right\rangle \\ &= 1 - D_B\, \tau \sum_{j=1}^{n} \langle q_j^2 \rangle \end{aligned}$$

$$= 1 - n\, D_B\, \tau \langle q_j^2 \rangle$$

$$\simeq \exp(-n\, D_B\, \tau \langle q_j^2 \rangle)\,. \qquad (35.21)$$

If θ denotes the scattering angle of an elementary scattering process, we have

$$\langle q_j^2 \rangle = 2k_0^2 \langle 1 - \cos\theta \rangle = 2k_0^2 (1 - g) = 2k_0^2 \frac{\ell_s}{\ell_t}, \qquad (35.22)$$

where g is the anisotropy factor and $\ell_t = \ell_s/(1-g)$ is the transport mean free path. With these simplifications, the field correlation function becomes

$$G_1(\tau) = |U_0|^2 \sum_n P(n) \exp\left(-2k_0^2 \frac{\ell_s}{\ell_t} n\, D_B\, \tau\right). \qquad (35.23)$$

For practical calculations, it is often easier to manipulate an integral instead of a series. To proceed, we introduce the length s of a scattering sequence with n scattering events, through $s = n\ell_s$. Using the probability density $P(s)$ of a sequence with length s, we obtain the expression for the normalized field-correlation function:

$$g_1(\tau) = \int_0^\infty P(s) \exp\left(-2\frac{\tau}{\tau_0}\frac{s}{\ell_t}\right) ds\,, \qquad (35.24)$$

with $\tau_0 = (k_0^2 D_B)^{-1}$. This expression is widely used in the interpretation of dynamic light scattering experiments in the multiple-scattering regime. The associated technique is often referred to as diffusing-wave spectroscopy (DWS).

For practical calculations, the probability density $P(s)$, also called the path-length distribution, can be obtained as the solution to the radiative transport equation or the diffusion equation. Indeed, from the time-dependent output flux $\phi(t)$ resulting from an incident pulse $I_{inc}\,\delta(t)$, we deduce $P(s) = \phi(t = s/v_E)/I_{inc}$, where v_E is the energy velocity in the medium (assumed to be uniform). Since (35.24) is mathematically a Laplace transform, we can even find $g_1(\tau)$ directly from the solution to the RTE or diffusion equation in the Laplace domain.

References and Additional Reading

This textbook discusses extensively dynamic light scattering in the single scattering regime, and its applications:

P.J. Berne and R. Pecora, *Dynamic Light Scattering* (Wiley, New York, 1976).

These are the papers in which the diffusing wave spectroscopy (DWS) technique and its theory were introduced:

G. Maret and P.E. Wolf, Z. Phys. B **65**, 409 (1987).

D.J. Pine, D.A. Weitz, P.M. Chaikin and E. Herbolzheimer, Phys. Rev. Lett. **60**, 1134 (1988).

This paper contains analytical expressions of the DWS field correlation function in simple geometries:

D.J. Pine, D.A. Weitz, J. X. Zhu and E. Herbolzheimer, J. Phys. **51**, 2101 (1990).

Practical expressions are also given in:

E. Akkermans and G. Montambaux, *Mesoscopic Physics of Electrons and Photons* (Cambridge University Press, Cambridge, 2007), chap. 9.

These papers contain advanced studies that extend the DWS theory beyond the diffusion approximation:

F.C. MacKintosh and S. John, Phys. Rev. B **40**, 2383 (1989).
D.J. Durian, Phys. Rev. E **51**, 3350 (1995).
K.K. Bizheva, A.M. Siegel and D.A. Boas, Phys. Rev. E **58**, 7664 (1998).
G. Popescu and A. Dogaiu, Appl. Opt. **40**, 4215 (2001).
R. Carminati, R. Elaloufi and J.-J. Greffet, Phys. Rev. Lett. **92**, 213903 (2004).
R. Pierrat, J. Opt. Soc. Am. A **25**, 2840 (2008). R. Pierrat, N. Ben Braham, L.F. Rojas-Ochoa, R. Carminati and F. Scheffold, Opt. Commun. **281**, 18 (2008).

Exercises

V.1 Write a computer program that generates random numbers $\phi_j \in [0, 2\pi]$ and computes the intensity $I_N = \left|\sum_{j=1}^{N} \exp(i\phi_j)\right|^2$. Plot the statistical distribution of I_N for increasing values of N. Can you recover Rayleigh statistics for scalar waves as given by (28.9)?

V.2 Write a computer program to generate a random distribution of points in a plane. The coordinates x and y of the points should be taken to be independent and identically distributed random variables. Supposing that each point is a source emitting a spherical scalar wave in free space with complex amplitude $\exp(ik_0 r)/(4\pi r)$, compute the resulting intensity in a plane at a distance of a few wavelengths from the source plane. The intensity pattern forms a speckle. Show that, for a sufficiently large image, the spatial distribution of the intensity follows Rayleigh statistics, as given by (28.9).

V.3 Perform the integration by parts in (28.10) to derive the result $\langle I^2 \rangle = 2\langle I \rangle^2$ valid for a fully developed speckle and scalar waves.

V.4 Consider the Rayleigh statistics (28.9) for the intensity in a fully developed speckle. Show that the moments of the intensity satisfy $\langle I^p \rangle = p!\,\langle I \rangle^p$, with p any positive integer.

V.5 Show that by inserting (29.2) into (29.3), Rayleigh statistics as described by (28.9) are recovered.

V.6 Using the modified Rayleigh statistics (28.14) for the intensity in an unpolarized fully developed speckle, show that the speckle contrast is $\sigma_I/\langle I \rangle = 1/\sqrt{2}$.

V.7 Consider a far-field speckle pattern with unpolarized light produced as the superposition of two independent speckle patterns with intensities I_1 and I_2, each obeying the Rayleigh statistics with the same average intensity.

(a) Express the second moment $\langle I^2 \rangle$ of the intensity $I = I_1 + I_2$ in terms of $\langle I_1 \rangle$ and $\langle I_2 \rangle$ (use only properties of second moments).

(b) Deduce the value of the speckle contrast $\sigma_I / \langle I \rangle$. This is another way to derive the result of the previous exercise.

V.8 The scalar Green function in an infinite homogeneous medium with complex dielectric function $\varepsilon(\omega)$ obeys

$$\nabla^2 G(\mathbf{r}, \mathbf{r}') + \varepsilon(\omega) k_0^2 G(\mathbf{r}, \mathbf{r}') = -\delta(\mathbf{r} - \mathbf{r}'), \tag{35.25}$$

together with the outgoing wave condition. Green's second identity states that

$$\int_V (u \nabla^2 v - v \nabla^2 u)\, d^3 r = \int_S \left(u \frac{\partial v}{\partial n} - v \frac{\partial u}{\partial n} \right) d^2 r, \tag{35.26}$$

where V is a volume bounded by a surface S with outward unit normal $\hat{\mathbf{n}}$. Applying this identity with $u(\mathbf{r}) = G(\mathbf{r}, \mathbf{r}_1)$ and $v(\mathbf{r}) = G^*(\mathbf{r}, \mathbf{r}_2)$, derive the following relation:

$$k_0^2 \,\mathrm{Im}\varepsilon(\omega) \int_V G(\mathbf{r}_1, \mathbf{r})\, G^*(\mathbf{r}_2, \mathbf{r})\, d^3 r = \mathrm{Im} G(\mathbf{r}_1, \mathbf{r}_2), \tag{35.27}$$

which is very useful in the calculation of speckle correlations. Hint: Use a far-field approximation in the surface integral and the reciprocity theorem.

V.9 A liquid placed in a transparent container with thickness $L = 5$ mm contains a suspension of non-absorbing spheres with radii $R = 5\ \mu$m) and number density ρ. The particles undergo Brownian motion with a diffusion coefficient $D_B = 5 \times 10^{-14}\ \mathrm{m^2 s^{-1}}$. The medium is illuminated at normal incidence with a laser ($\lambda = 600$ nm), and the transmitted light is observed in the far-field at an angle $\theta = 30°$.

(a) What is the scattering cross section of a single particle?

(b) Which condition on the density ρ must be satisfied in order for the system to be in the single-scattering regime?

(c) The intensity $I(t)$ fluctuates in time. Assuming that the fluctuations result from scattering by a large number of particles, can you estimate $\langle I^2(t) \rangle$ in terms of $\langle I(t) \rangle$ where $\langle \cdots \rangle$ denotes time averaging? What is the dynamic speckle contrast?

(d) The intensity fluctuates on a time scale given by the correlation time of the field $U(t)$. Express this time scale in the single-scattering regime and compute its numerical value.

V.10 Under the same conditions as in the previous exercise, now assume that the density of particles is such that the transport mean free path is $\ell_t = 0.5$ mm.

(a) Is the diffusion approximation valid to describe light transport through the suspension?
(b) Estimate the average path length for diffusion through the container.
(c) Derive an approximate expression for the time-correlation function of the intensity $I(t)$ by assuming that the path-length distribution is dominated by the average path length.
(d) Evaluate the smallest time scale that can be probed using the temporal intensity fluctuations.

Part VI

Electromagnetic Waves and Near-Field Scattering

36

Vector Waves

Maxwell's equations and the electromagnetic wave equation were introduced in Chapter 2. Here we present some essential features of vector waves that will prove to be useful in the chapters that follow.

36.1 Vector Wave Equation

We recall from Chapter 2 that in the presence of external charge and current densities ρ_{ext} and $\mathbf{j}_{\text{ext}}$, the macroscopic Maxwell's equations are

$$\nabla \times \mathbf{E} = i\omega \mathbf{B} \,, \tag{36.1}$$

$$\nabla \cdot \mathbf{D} = \rho_{\text{ext}} \,, \tag{36.2}$$

$$\nabla \times \mathbf{H} = \mathbf{j}_{\text{ext}} + i\omega \mathbf{D} \,, \tag{36.3}$$

$$\nabla \cdot \mathbf{B} = 0 \,. \tag{36.4}$$

Here we have assumed that the fields are monochromatic with frequency ω. We also impose the following constitutive relations, which hold in linear and isotropic material media:

$$\mathbf{D} = \varepsilon_0 \, \varepsilon \, \mathbf{E} \,, \tag{36.5}$$

$$\mathbf{B} = \mu_0 \, \mu \, \mathbf{H} \,, \tag{36.6}$$

where ε is the dielectric function and μ is the relative magnetic permeability. In inhomogeneous and dispersive media, ε and μ depend on both position and frequency. In the following, we will consider non-magnetic materials with $\mu = 1$.

By taking the curl of (36.1), and making use of (36.3), (36.5) and (36.6), we see that the electric field obeys

$$\nabla \times \nabla \times \mathbf{E} - \varepsilon \, k_0^2 \, \mathbf{E} = i \mu_0 \omega \, \mathbf{j}_{\text{ext}} \,, \tag{36.7}$$

where $k_0 = \omega/c = 2\pi/\lambda$ is the free-space wavenumber, with λ being the wavelength in vacuum. This equation is the vector form of Helmholtz's equation, in which the external current density appears as a source term.

36.2 Energy Conservation

We consider the conservation of energy for monochromatic electromagnetic waves and derive the corresponding Poynting theorem. We begin with the macroscopic Maxwell's equations in the absence of external sources, that is with $\rho_{\text{ext}} = 0$ and $\mathbf{j}_{\text{ext}} = 0$. Multiplying (36.3) by $\mathbf{E}^*$ leads to

$$\mathbf{E}^* \cdot \nabla \times \mathbf{H} = -i\omega \mathbf{D} \cdot \mathbf{E}^* . \tag{36.8}$$

The left-hand side can be rewritten using the identity

$$\nabla \cdot (\mathbf{E}^* \times \mathbf{H}) = \mathbf{H} \cdot \nabla \times \mathbf{E}^* - \mathbf{E}^* \cdot \nabla \times \mathbf{H} , \tag{36.9}$$

along with (36.1). We thus obtain

$$- i\omega \mathbf{H} \cdot \mathbf{B}^* - \nabla \cdot (\mathbf{E}^* \times \mathbf{H}) = -i\omega \mathbf{D} \cdot \mathbf{E}^* . \tag{36.10}$$

Introducing the time-averaged Poynting vector

$$\mathbf{\Pi} = \frac{1}{2}\text{Re}(\mathbf{E}^* \times \mathbf{H}) , \tag{36.11}$$

and considering a non-magnetic medium with $\mathbf{D} = \varepsilon_0 \varepsilon \mathbf{E}$ and $\mathbf{B} = \mu_0 \mathbf{H}$, we immediately see that

$$\nabla \cdot \mathbf{\Pi} + \frac{\omega \epsilon_0}{2} \text{Im}\varepsilon |\mathbf{E}|^2 = 0 . \tag{36.12}$$

This result is similar to (2.38), which expresses the conservation of energy for scalar waves. Here the energy current $\mathbf{J}$ is replaced by the Poynting vector, and $\mathcal{P}_a = (\omega\epsilon_0/2)\text{Im}\varepsilon|\mathbf{E}|^2$ is the absorbed power per unit volume.

Integrating (36.12) over a volume V bounded by a closed surface S with unit outward normal $\hat{\mathbf{n}}$, and applying the divergence theorem, we obtain

$$\int_S \mathbf{\Pi} \cdot \hat{\mathbf{n}}\, d^2 r + P_a = 0 , \tag{36.13}$$

where $P_a = \int_V \mathcal{P}_a \, d^3 r$ is the power absorbed in volume V.

In the presence of an external source with current density $\mathbf{j}_{\text{ext}}$, it can be easily verified that the local form of the energy conservation Eq. (36.12) becomes

$$\nabla \cdot \mathbf{\Pi} + \mathcal{P}_a = \mathcal{P}_s , \tag{36.14}$$

where $\mathcal{P}_s = -(1/2)\text{Re}(\mathbf{j}_{\text{ext}} \cdot \mathbf{E}^*)$ is the power by unit volume transferred from the source to the field.

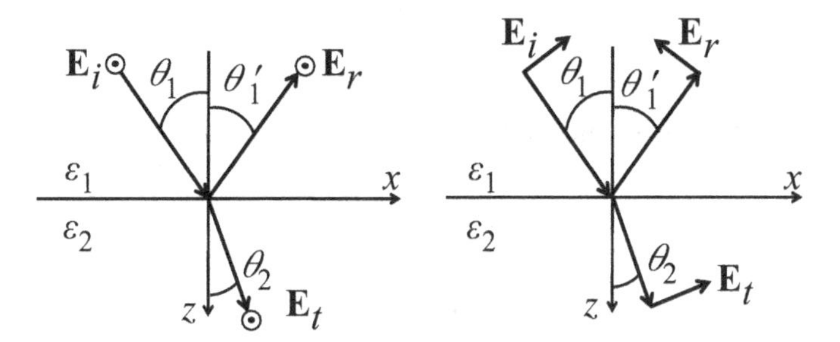

Fig. 36.1 Illustrating the canonical polarizations for vector electromagnetic waves at a planar interface separating two homogeneous and isotropic media with dielectric constants ε_1 and ε_2. Left: s polarization (electric field perpendicular to the plane of incidence). Right: p polarization (electric field in the plane of incidence).

36.3 Reflection and Transmission of Electromagnetic Waves

The reflection and transmission of scalar waves at a flat interface were studied in Section 4.2. Here we revisit this problem for vector waves and calculate the Fresnel coefficients that will be used in Chapter 45. We begin by defining the two canonical linear s and p polarizations, as shown in Fig. 36.1.

s Polarization

For s-polarized waves, the electric field is perpendicular to the $x - z$ plane, defined as the plane of incidence. By working with the y-component of the electric field, we can reduce the problem to one involving a scalar wave field. Assuming that the incident plane wave has a unit amplitude in medium 1, the total field in medium 1 is written as

$$E_{1y}(\mathbf{r}) = \exp(i\mathbf{k}_1 \cdot \mathbf{r}) + R_s \exp(i\mathbf{k}_1' \cdot \mathbf{r}) , \tag{36.15}$$

where R_s is the amplitude reflection coefficient. Likewise, the field in medium 2 is

$$E_{2y}(\mathbf{r}) = T_s \exp(i\mathbf{k}_2 \cdot \mathbf{r}) , \tag{36.16}$$

where T_s is the amplitude transmission coefficient. Making use of the boundary conditions (2.28) and (2.31), and Maxwell's equation $\nabla \times \mathbf{E} = i\omega\mu_0\mathbf{H}$, we can easily show that the boundary conditions at the interface $z = 0$ are

$$E_{1y}(\boldsymbol{\rho}, 0) = E_{2y}(\boldsymbol{\rho}, 0) , \tag{36.17}$$

$$\frac{\partial E_{1y}}{\partial z}(\boldsymbol{\rho}, 0) = \frac{\partial E_{2y}}{\partial z}(\boldsymbol{\rho}, 0) . \tag{36.18}$$

It is then easy to obtain the Fresnel coefficients:

$$R_s = \frac{k_{1z} - k_{2z}}{k_{1z} + k_{2z}} , \tag{36.19}$$

$$T_s = \frac{2k_{1z}}{k_{1z} + k_{2z}} . \tag{36.20}$$

We note that the reflection and transmission coefficients for s-polarized waves are identical to those derived for scalar waves in Section 4.2.

p Polarization

For p-polarized waves, the electric field is in the plane of incidence. Since the magnetic field has only one component, H_y, perpendicular to this plane, it can be treated as a scalar field. The field in medium 1 is written as

$$H_{1y}(\mathbf{r}) = \exp(i\mathbf{k}_1 \cdot \mathbf{r}) + R_p \exp(i\mathbf{k}_1' \cdot \mathbf{r}) , \tag{36.21}$$

while the field in medium 2 is

$$H_{2y}(\mathbf{r}) = T_p \exp(i\mathbf{k}_2 \cdot \mathbf{r}) . \tag{36.22}$$

These expressions define the Fresnel coefficients R_p and T_p for p-polarized waves.

From the boundary conditions (2.28) and (2.31), and by making use of the Maxwell equation $\nabla \times \mathbf{H} = -i\omega\mathbf{D}$, it can be seen that the boundary conditions at the interface $z = 0$ are of the form

$$H_{1y}(\boldsymbol{\rho}, 0) = H_{2y}(\boldsymbol{\rho}, 0) , \tag{36.23}$$

$$\frac{1}{\varepsilon_1}\frac{\partial H_{1y}}{\partial z}(\boldsymbol{\rho}, 0) = \frac{1}{\varepsilon_2}\frac{\partial H_{2y}}{\partial z}(\boldsymbol{\rho}, 0) . \tag{36.24}$$

Applying the boundary conditions, we obtain the following expressions for the Fresnel coefficients:

$$R_p = \frac{\varepsilon_2 k_{1z} - \varepsilon_1 k_{2z}}{\varepsilon_2 k_{1z} + \varepsilon_1 k_{2z}} , \tag{36.25}$$

$$T_p = \frac{2\varepsilon_2 k_{1z}}{\varepsilon_2 k_{1z} + \varepsilon_1 k_{2z}} . \tag{36.26}$$

An important property of p-polarized waves is that $R_p = 0$ for the incidence angle θ_B, known as the Brewster angle. We easily find that $\tan\theta_B = n_2/n_1$.

Note that we could choose to define Fresnel coefficients in p polarization from the amplitude of the electric field. Working with the amplitude of $\mathbf{E}$ (instead of H_y) changes the boundary conditions and therefore the Fresnel coefficients. The expression for R_p is unchanged, but the expression of T_p is modified and becomes

$$T_p' = \sqrt{\frac{\varepsilon_1}{\varepsilon_2}}\, T_p . \tag{36.27}$$

References and Additional Reading

For an introduction to the basic equations governing electromagnetic fields, see for example:

J.D. Jackson, *Classical Electrodynamics* (Wiley, New York, 1962), chap. 6.

L.D. Landau, E.M. Lifshitz and L.P. Pitaevskii, *Electrodynamics of Continuous Media* (Pergamon Press, Oxford, 1984), chap. 9.

L.D. Landau and E.M. Lifshitz, *The Classical Theory of Fields* (Butterworth Heinemann, Amsterdam, 1994), chap. 2 and 6.

A. Zangwill, *Modern Electrodynamics* (Cambridge University Press, Cambridge, 2013), chap. 2 and 15.

The reflection and transmission of electromagnetic waves at flat interfaces is treated in many textbooks. See for example:

J.D. Jackson, *Classical Electrodynamics* (Wiley, New York, 1962), chap. 7.

E. Hecht, *Theory and Problems in Optics* (McGraw Hill, New York, 1975), chap. 3.

L.D. Landau, E.M. Lifshitz and L.P. Pitaevskii, *Electrodynamics of Continuous Media* (Pergamon Press, Oxford, 1984), chap. 10.

M. Born and E. Wolf, *Principles of Optics* (Cambridge University Press, Cambridge, 1999), chap. 1.

37

Electromagnetic Green's Functions

Green's functions are essential mathematical tools in scattering theory. The Green's function for the scalar wave equation was introduced in Chapter 5. Here we consider the Green's function for the vector wave equation.

37.1 Tensor Green's Function

In free space, the complex amplitude of a monochromatic electric field with frequency ω satisfies the vector Helmholtz equation

$$\nabla \times \nabla \times \mathbf{E}(\mathbf{r}) - k_0^2\,\mathbf{E}(\mathbf{r}) = i\mu_0\omega\,\mathbf{j}(\mathbf{r})\,, \tag{37.1}$$

where $k_0 = \omega/c = 2\pi/\lambda$ and $\mathbf{j}$ is the current density of the source. The Green's function $\mathbf{G}_0$ for (37.1) is a second-rank tensor, which is defined as the solution to

$$\nabla \times \nabla \times \mathbf{G}_0(\mathbf{r}, \mathbf{r}') - k_0^2\,\mathbf{G}_0(\mathbf{r}, \mathbf{r}') = \delta(\mathbf{r} - \mathbf{r}')\,\mathbf{I}\,, \tag{37.2}$$

where $\mathbf{I}$ is the unit second-rank tensor. Here the curl of a second-rank tensor is defined in the following way. Given $\mathbf{T} = T_{ij}\hat{\mathbf{e}}_i \otimes \hat{\mathbf{e}}_j$, $\nabla \times \mathbf{T} = \partial_j T_{kl}\,\epsilon_{jli}\,\hat{\mathbf{e}}_k \otimes \hat{\mathbf{e}}_i$, where ϵ_{jli} is the Levi-Civita symbol, and summation over repeated indices is assumed. The retarded Green's function also obeys the Sommerfeld radiation condition

$$\lim_{R\to\infty} R\left[\nabla \times \mathbf{G}_0(\mathbf{r}, \mathbf{r}') - ik_0\hat{\mathbf{R}} \times \mathbf{G}_0(\mathbf{r}, \mathbf{r}')\right] = 0\,, \tag{37.3}$$

where $\mathbf{R} = \mathbf{r} - \mathbf{r}'$ and $R = |\mathbf{R}|$. This condition forces the Green's function to behave as an outgoing spherical wave for $R \to \infty$. The solution to (37.1) can be written in terms of the Green's function as

$$\mathbf{E}(\mathbf{r}) = i\mu_0\omega \int \mathbf{G}_0(\mathbf{r}, \mathbf{r}')\,\mathbf{j}(\mathbf{r}')\,d^3r'\,, \tag{37.4}$$

which is immediately verified from (37.1) and (37.2). This relation leads to a simple interpretation of the Green's function. Consider a point electric-dipole located at position $\mathbf{r}_s$. The current density is $\mathbf{j}(\mathbf{r}) = -i\omega\mathbf{p}\,\delta(\mathbf{r}-\mathbf{r}_s)$, with $\mathbf{p}$ the dipole moment. Using (37.4), we immediately see that

$$\mathbf{E}(\mathbf{r}) = \mu_0\omega^2\,\mathbf{G}_0(\mathbf{r}, \mathbf{r}_s)\mathbf{p}\,. \tag{37.5}$$

That is, the Green's function is proportional to the electric field radiated by a point electric-dipole.

The scalar Green's function in free space is the solution to

$$\nabla^2 G_0(\mathbf{r}, \mathbf{r}') + k_0^2\, G_0(\mathbf{r}, \mathbf{r}') = -\delta(\mathbf{r} - \mathbf{r}')\,, \tag{37.6}$$

which additionally satisfies the radiation condition (5.8). As shown in Section 5.2, G_0 is given by

$$G_0(\mathbf{r}, \mathbf{r}') = \frac{\exp(ik_0|\mathbf{r} - \mathbf{r}'|)}{4\pi\,|\mathbf{r} - \mathbf{r}'|}\,, \tag{37.7}$$

which corresponds to an outgoing spherical wave. There is a relation between the tensor Green's function and the scalar Green's function. To see this, we first rewrite (37.1) as

$$\nabla^2\mathbf{E}(\mathbf{r}) + k_0^2\,\mathbf{E}(\mathbf{r}) = -i\mu_0\omega\,\mathbf{j}(\mathbf{r}) + \nabla(\nabla\cdot\mathbf{E}(\mathbf{r}))\,. \tag{37.8}$$

From the Maxwell equation

$$\nabla\times\mathbf{B} = \mu_0\mathbf{j} - i\omega\varepsilon_0\mu_0\mathbf{E}\,, \tag{37.9}$$

we find that $\nabla\cdot\mathbf{j} = i\omega\varepsilon_0\nabla\cdot\mathbf{E}$, which allows us to rewrite (37.8) in the form

$$\nabla^2\mathbf{E}(\mathbf{r}) + k_0^2\,\mathbf{E}(\mathbf{r}) = -i\mu_0\omega\left[\mathbf{j}(\mathbf{r}) + \frac{1}{k_0^2}\nabla(\nabla\cdot\mathbf{j}(\mathbf{r}))\right]. \tag{37.10}$$

Making use of (37.6), the solution to the above equation can be written as

$$\mathbf{E}(\mathbf{r}) = i\mu_0\omega\int G_0(\mathbf{r}, \mathbf{r}')\,\mathbf{j}(\mathbf{r}')\,d^3r' + \frac{i\mu_0\omega}{k_0^2}\int G_0(\mathbf{r}, \mathbf{r}')\nabla(\nabla\cdot\mathbf{j}(\mathbf{r}'))\,d^3r'. \tag{37.11}$$

After an integration by parts of the second term, we obtain

$$\mathbf{E}(\mathbf{r}) = i\mu_0\omega\int\left(\mathbf{I} + \frac{1}{k_0^2}\nabla\otimes\nabla\right)G_0(\mathbf{r}, \mathbf{r}')\,\mathbf{j}(\mathbf{r}')\,d^3r'. \tag{37.12}$$

It is important to note that the second term in the integrand is hypersingular and thus (37.12) must be understood in the distributional sense. Using (37.4) and (37.12), we immediately find that the tensor Green's function is of the form

$$\mathbf{G}_0(\mathbf{r}, \mathbf{r}') = \left(\mathbf{I} + \frac{1}{k_0^2}\nabla\otimes\nabla\right)G_0(\mathbf{r}, \mathbf{r}')\,. \tag{37.13}$$

In order to compute $\mathbf{G}_0$ explicitly, we first note that the second term on the right-hand side must be of the form

$$\partial_i \partial_j G_0(\mathbf{r}, \mathbf{r}') = A(\mathbf{R})\,\delta_{ij} + B(\mathbf{R})\hat{R}_i \hat{R}_j \,, \tag{37.14}$$

which transforms correctly under rotations. Taking $i = j$ and summing over i, we find that

$$3A(\mathbf{R}) + B(\mathbf{R}) = -k_0^2 G_0(\mathbf{r}, \mathbf{r}') - \delta(\mathbf{R}) \,, \tag{37.15}$$

where we have used (37.6). For $i \neq j$, we have

$$\partial_i \partial_j G_0(\mathbf{r}, \mathbf{r}') = B(\mathbf{R})\hat{R}_i \hat{R}_j \,. \tag{37.16}$$

Calculating the left-hand side of the above and solving for $A(\mathbf{R})$ and $B(\mathbf{R})$, we obtain

$$\mathbf{G}_0(\mathbf{r}, \mathbf{r}') = \frac{\exp(ik_0 R)}{4\pi R}\left[\mathbf{I} - \hat{\mathbf{R}} \otimes \hat{\mathbf{R}} - \left(\frac{1}{ik_0 R} + \frac{1}{k_0^2 R^2}\right)\left(\mathbf{I} - 3\hat{\mathbf{R}} \otimes \hat{\mathbf{R}}\right)\right] - \frac{\mathbf{I}}{3k_0^2}\delta(\mathbf{R}) \,. \tag{37.17}$$

The above result can be rewritten as a principal value and a singular part:

$$\mathbf{G}_0(\mathbf{r}, \mathbf{r}') = \mathrm{P}\left(\mathbf{I} + \frac{1}{k_0^2}\nabla \otimes \nabla\right)\frac{\exp(ik_0 R)}{4\pi R} - \frac{\mathbf{I}}{3k_0^2}\delta(\mathbf{R}) \,, \tag{37.18}$$

which is consistent with (37.12).

37.2 Far-Field and Near-Field Asymptotics

Considering $\mathbf{r}'$ as the source point and $\mathbf{r}$ as the observation point, the far-field regime corresponds to $r \gg r'$. In this case, we can expand (37.17) to first order in r'/r, noting that

$$R = |\mathbf{r} - \mathbf{r}'| \simeq r - \hat{\mathbf{r}} \cdot \mathbf{r}' \,. \tag{37.19}$$

We find that

$$\mathbf{G}_0(\mathbf{r}, \mathbf{r}') \sim \frac{\exp(ik_0 r)}{4\pi r}\exp(-ik_0 \hat{\mathbf{r}} \cdot \mathbf{r}')\left(\mathbf{I} - \hat{\mathbf{r}} \otimes \hat{\mathbf{r}}\right) \,, \tag{37.20}$$

which is the far-field expression of the tensor Green's function. As already shown in Section 5.3 for scalar waves, the first-order expansion in the phase term requires the condition $r \gg r'^2/\lambda$ to be satisfied. The near-field corresponds to $R \to 0$, and the expression for the Green's function in this regime becomes

$$\mathbf{G}_0(\mathbf{r}, \mathbf{r}') \sim \mathrm{P}\left(\frac{\mathbf{I} - 3\hat{\mathbf{R}} \otimes \hat{\mathbf{R}}}{4\pi k_0^2 R^3}\right) - \frac{\mathbf{I}}{2k_0^2}\delta(\mathbf{R}) \,. \tag{37.21}$$

37.3 Far-Field Radiated Power

Consider a monochromatic source with frequency ω and current density $\mathbf{j}$ that occupies a volume V in free space. The radiated power is obtained by integrating the flux of the Poynting vector over a large sphere enclosing the source. We begin by calculating the electric field in the far-zone, using (37.4) and (37.20). We thus obtain

$$\mathbf{E}(\mathbf{r}) = i\mu_0\omega \frac{\exp(ik_0 r)}{4\pi r} (\mathbf{I} - \hat{\mathbf{r}} \otimes \hat{\mathbf{r}}) \int \mathbf{j}(\mathbf{r}') \exp(-ik_0 \hat{\mathbf{r}} \cdot \mathbf{r}') \, d^3r' \, . \qquad (37.22)$$

We note that the electric field is transverse to the direction of observation $\hat{\mathbf{r}}$. The magnetic field can be deduced from the Maxwell equation $\nabla \times \mathbf{E} = i\mu_0\omega\mathbf{H}$. Using (37.4) and (37.17) and taking the far-field limit, we find that the magnetic field is given by

$$\mathbf{H}(\mathbf{r}) = \frac{k_0}{\mu_0\omega} \hat{\mathbf{r}} \times \mathbf{E}(\mathbf{r}) \, , \qquad (37.23)$$

which is also transverse.

The power radiated in the direction $\hat{\mathbf{r}}$ per unit solid angle is

$$\frac{dP}{d\Omega} = \mathbf{\Pi}(\mathbf{r}) \cdot \hat{\mathbf{r}} \, r^2 \, , \qquad (37.24)$$

where $\mathbf{\Pi} = (1/2)\mathrm{Re}(\mathbf{E} \times \mathbf{H}^*)$ is the time-averaged Poynting vector. Using (37.23), we find that

$$\frac{dP}{d\Omega} = \frac{\epsilon_0 c}{2} |\mathbf{E}(\mathbf{r})|^2 \, r^2 \, , \qquad (37.25)$$

with $\mathbf{E}$ given by (37.22). This expression, in conjunction with (37.22), can be used to calculate the angular distribution of the power in terms of the current density of a radiating source. It will also be used to compute the power scattered by a particle in later chapters.

37.4 Plane-Wave Expansion

By taking advantage of translational invariance, we can write $\mathbf{G}_0(\mathbf{r}, \mathbf{r}')$ as the Fourier integral

$$\mathbf{G}_0(\mathbf{r}, \mathbf{r}') = \int \mathbf{G}_0(\mathbf{k}) \, \exp[i\mathbf{k} \cdot (\mathbf{r} - \mathbf{r}')] \, \frac{d^3k}{(2\pi)^3} \, , \qquad (37.26)$$

which defines the Green's function $\mathbf{G}_0(\mathbf{k})$ in Fourier space. Introducing (37.26) into (37.2) leads to

$$i\mathbf{k} \times [i\mathbf{k} \times \mathbf{G}_0(\mathbf{k})] - k_0^2 \, \mathbf{G}_0(\mathbf{k}) = \mathbf{I} \, , \qquad (37.27)$$

or equivalently

$$(k^2 - k_0^2)\,\mathbf{G}_0(\mathbf{k}) - (\mathbf{k}\otimes\mathbf{k})\,\mathbf{G}_0(\mathbf{k}) = \mathbf{I}\,. \tag{37.28}$$

Solving for $\mathbf{G}_0$ gives

$$\mathbf{G}_0(\mathbf{k}) = \left[(k^2 - k_0^2 - i\epsilon)\mathbf{I} - \mathbf{k}\otimes\mathbf{k}\right]^{-1}\,, \tag{37.29}$$

where the limit $\epsilon \to 0^+$ is implied (see Section 5.2). The tensor on the right-hand side above can be inverted by noting that $(\mathbf{k}\otimes\mathbf{k})^n = k^{2n-2}\,\mathbf{k}\otimes\mathbf{k}$ for $n \geq 1$. We thus obtain

$$\mathbf{G}_0(\mathbf{k}) = \frac{1}{k^2 - k_0^2 - i\epsilon}\left(\mathbf{I} - \frac{\mathbf{k}\otimes\mathbf{k}}{k_0^2}\right)\,, \tag{37.30}$$

which after insertion into (37.26) leads to

$$\mathbf{G}_0(\mathbf{r},\mathbf{r}') = \int \frac{1}{k^2 - k_0^2 - i\epsilon}\left(\mathbf{I} - \frac{\mathbf{k}\otimes\mathbf{k}}{k_0^2}\right)\exp[i\mathbf{k}\cdot(\mathbf{r}-\mathbf{r}')]\,\frac{d^3k}{(2\pi)^3}. \tag{37.31}$$

In order to obtain a plane-wave expansion, we define the quantity

$$\mathbf{g}_0(\mathbf{q}, z, z') = \int \frac{1}{k_z^2 - (k_0^2 - q^2)}\left(\mathbf{I} - \frac{\mathbf{k}\otimes\mathbf{k}}{k_0^2}\right)\exp[ik_z(z-z')]\,\frac{dk_z}{2\pi}. \tag{37.32}$$

The Green's function thus becomes

$$\mathbf{G}_0(\mathbf{r},\mathbf{r}') = \int \mathbf{g}_0(\mathbf{q}, z, z')\,\exp[i\mathbf{q}\cdot(\boldsymbol{\rho}-\boldsymbol{\rho}')]\,\frac{d^2q}{(2\pi)^2}\,, \tag{37.33}$$

where $\boldsymbol{\rho} = (x, y)$, $\boldsymbol{\rho}' = (x', y')$ and $\mathbf{q} = (k_x, k_y)$. The integral in (37.32) can be performed by contour integration. We note that the case $z = z'$ is handled by adding a small imaginary part to k_0. A convenient way to write the result is to use the basis corresponding to the s and p polarizations, defined by the vectors

$$\hat{\mathbf{s}} = \hat{\mathbf{q}}\times\hat{\mathbf{z}}\,, \tag{37.34}$$

$$\hat{\mathbf{p}} = \frac{1}{k_0}\left[q\hat{\mathbf{z}} - \mathrm{sgn}(z-z')k_z(q)\,\hat{\mathbf{q}}\right]\,. \tag{37.35}$$

Here we have defined

$$k_z(q) = (k_0^2 - q^2)^{1/2} \tag{37.36}$$

with the requirement $\mathrm{Re}[k_z(q)] > 0$ and $\mathrm{Im}[k_z(q)] > 0$. We find that

$$\mathbf{g}_0(\mathbf{q}, z, z') = \frac{i}{2k_z(q)}(\hat{\mathbf{s}}\otimes\hat{\mathbf{s}}+\hat{\mathbf{p}}\otimes\hat{\mathbf{p}})\,\exp[ik_z(q)|z-z'|]-\frac{1}{k_0^2}\,\delta(z-z')\,\hat{\mathbf{z}}\otimes\hat{\mathbf{z}}\,. \tag{37.37}$$

Equations (37.33) and (37.37) can be seen as the analog of the Weyl formula (6.13) for vector waves.

37.5 Transverse and Longitudinal Green's Function

Introducing the transverse and longitudinal projection operators $\mathbf{P}_\perp(\mathbf{k}) = \mathbf{I} - \hat{\mathbf{k}} \otimes \hat{\mathbf{k}}$ and $\mathbf{P}_\parallel(\mathbf{k}) = \hat{\mathbf{k}} \otimes \hat{\mathbf{k}}$, we can rewrite (37.30) as

$$\mathbf{G}_0(\mathbf{k}) = \mathbf{G}_0^\perp(\mathbf{k}) + \mathbf{G}_0^\parallel(\mathbf{k}) \tag{37.38}$$

$$= \frac{\mathbf{P}_\perp(\mathbf{k})}{k^2 - k_0^2 - i\epsilon} - \frac{\mathbf{P}_\parallel(\mathbf{k})}{k_0^2}, \tag{37.39}$$

which defines the transverse and longitudinal Green's function $\mathbf{G}_0^\perp(\mathbf{k})$ and $\mathbf{G}_0^\parallel(\mathbf{k})$.

The real-space form of the longitudinal Green's function can be obtained from the Fourier integral

$$\mathbf{G}_0^\parallel(\mathbf{r}, \mathbf{r}') = \int -\frac{\hat{\mathbf{k}} \otimes \hat{\mathbf{k}}}{k_0^2} \exp(i\mathbf{k} \cdot \mathbf{R}) \frac{d^3k}{(2\pi)^3}. \tag{37.40}$$

In order to perform the above integration, we first consider the off-diagonal terms, for example,

$$G_{0,xy}^\parallel(\mathbf{r}, \mathbf{r}') = \frac{-1}{k_0^2} \int \frac{k_x k_y}{k^2} \exp(i\mathbf{k} \cdot \mathbf{R}) \frac{d^3k}{(2\pi)^3}$$

$$= \frac{1}{k_0^2} \partial_x \partial_y \int \frac{1}{k^2} \exp(i\mathbf{k} \cdot \mathbf{R}) \frac{d^3k}{(2\pi)^3}. \tag{37.41}$$

The integral is easily calculated using cylindrical coordinates, leading to

$$G_{0,xy}^\parallel(\mathbf{r}, \mathbf{r}') = \frac{1}{4\pi k_0^2} \partial_x \partial_y \frac{1}{R} = \frac{1}{4\pi k_0^2} \frac{3R_x R_y}{R^5}. \tag{37.42}$$

For the diagonal terms, we need to compute terms of the form

$$G_{0,zz}^\parallel(\mathbf{r}, \mathbf{r}') + \frac{1}{3k_0^2}\delta(\mathbf{R}) = \frac{-1}{k_0^2} \int \left(\frac{k_z^2}{k^2} - \frac{1}{3}\right) \exp(i\mathbf{k} \cdot \mathbf{R}) \frac{d^3k}{(2\pi)^3}$$

$$= \frac{-1}{k_0^2} \int \frac{2k_z^2 - k_x^2 - k_y^2}{3k^2} \exp(i\mathbf{k} \cdot \mathbf{R}) \frac{d^3k}{(2\pi)^3}, \tag{37.43}$$

where the singular term, derived previously, has been added on the left-hand side in order to regularize the resulting integral. Proceeding in a similar manner as for the off-diagonal terms, we obtain

$$G_{0,zz}^\parallel(\mathbf{r}, \mathbf{r}') + \frac{1}{3k_0^2}\delta(\mathbf{R}) = \frac{1}{4\pi k_0^2} \frac{2R_z^2 - R_x^2 - R_y^2}{R^5}. \tag{37.44}$$

We see that the final expression for the longitudinal Green's function in real space takes the form

$$\mathbf{G}_0^{\|}(\mathbf{r}, \mathbf{r}') = \mathrm{P}\left(\frac{3\hat{\mathbf{R}} \otimes \hat{\mathbf{R}} - \mathbf{I}}{4\pi R^3}\right) - \frac{\mathbf{I}}{3k_0^2}\,\delta(\mathbf{R})\,. \tag{37.45}$$

We find that the longitudinal Green's function coincides with the near-field Green's function (37.21). The transverse Green's function is obtained by subtracting the longitudinal part of the full Green's function in (37.17), and contains terms decaying as R^{-3}, R^{-2} and R^{-1}.

37.6 Half-Space Green's Function

Here we derive the plane-wave expansion of the Green's function in the half-space geometry. We define the plane $z = 0$ as the interface between two homogeneous, isotropic and non-magnetic media, with ε_1 and ε_2 being the dielectric constants of the half-spaces $z < 0$ and $z > 0$, respectively. The Green's function $\mathbf{G}(\mathbf{r}, \mathbf{r}')$ obeys

$$\nabla \times \nabla \times \mathbf{G}(\mathbf{r}, \mathbf{r}') - \varepsilon(\mathbf{r})\, k_0^2\, \mathbf{G}(\mathbf{r}, \mathbf{r}') = \delta(\mathbf{r} - \mathbf{r}')\,\mathbf{I}\,, \tag{37.46}$$

where $\varepsilon(\mathbf{r}) = \varepsilon_1$ for $z < 0$ and $\varepsilon(\mathbf{r}) = \varepsilon_2$ for $z > 0$. The Green's function satisfies the outgoing wave condition for $|z - z'| \to \infty$ and boundary conditions on the interface $z = 0$. Making use of translational invariance, we introduce the Fourier integral representation of the Green's function:

$$\mathbf{G}(\mathbf{r}, \mathbf{r}') = \int \mathbf{g}(\mathbf{q}, z, z')\, \exp[i\mathbf{q} \cdot (\boldsymbol{\rho} - \boldsymbol{\rho}')]\, \frac{d^2q}{(2\pi)^2}\,, \tag{37.47}$$

which defines $\mathbf{g}(\mathbf{q}, z, z')$. We consider a source point $\mathbf{r}'$ in medium 1 ($z < 0$) and a point $\mathbf{r}$ either in medium 1 (reflection geometry) or in medium 2 (transmission geometry). It is convenient to employ the basis of s and p polarizations, defined in this geometry by the unit vectors[1]

$$\hat{\mathbf{s}} = \hat{\mathbf{q}} \times \hat{\mathbf{z}} \tag{37.48}$$

$$\hat{\mathbf{p}}_j^{\pm} = \frac{1}{\sqrt{\varepsilon_j}\, k_0}\left[q\hat{\mathbf{z}} \mp k_z^j(q)\,\hat{\mathbf{q}}\right]. \tag{37.49}$$

Here we use the notation $k_z^j(q) = [\varepsilon_j k_0^2 - q^2]^{1/2}$ for the z-component of the wavevector in medium j, with $j = 1, 2$, and we require $\mathrm{Re}[k_z^j(q)] > 0$ and $\mathrm{Im}[k_z^j(q)] > 0$.

To determine $\mathbf{g}(\mathbf{q}, z, z')$, we make use of the plane-wave expansion (37.37) and apply the boundary conditions at the interface $z = 0$. The boundary conditions

[1] We follow the approach used in J. Sipe, J. Opt. Soc. Am. B **4**, 481 (1987).

can be directly written in terms of the reflection and transmission coefficients introduced in the previous section.

Reflection

In the case of the reflection geometry, both $\mathbf{r}$ and $\mathbf{r}'$ lie in medium 1, and we find that

$$\mathbf{g}(\mathbf{q}, z, z') = \mathbf{g}_1(\mathbf{q}, z, z') + \frac{i}{2k_z^1(q)}[R_s^{11}(q)\,\hat{\mathbf{s}} \otimes \hat{\mathbf{s}} + R_p^{11}(q)\,\hat{\mathbf{p}}_1^- \otimes \hat{\mathbf{p}}_1^+] \exp[ik_z^1(q)|z + z'|]. \tag{37.50}$$

The first term on the right-hand side corresponds to propagation in medium 1 from $\mathbf{r}'$ to $\mathbf{r}$, and is given by

$$\mathbf{g}_1(\mathbf{q}, z, z') = \frac{i}{2k_z^1(q)}(\hat{\mathbf{s}} \otimes \hat{\mathbf{s}} + \hat{\mathbf{p}}_1^\pm \otimes \hat{\mathbf{p}}_1^\pm) \exp[ik_z^1(q)|z - z'|] - \frac{1}{\varepsilon_1 k_0^2}\,\delta(z - z')\,\mathbf{I}_{zz}\,. \tag{37.51}$$

In this expression, we employ $\hat{\mathbf{p}}_1^+$ if $z > z'$ and $\hat{\mathbf{p}}_1^-$ if $z < z'$. The second term on the right-hand side corresponds to propagation from $\mathbf{r}'$ to the interface $z = 0$, reflection (with polarization-dependent reflection coefficients) and propagation from the interface to $\mathbf{r}$. The Fresnel reflection coefficients in medium 1 were calculated in the previous section, and are given by

$$R_s^{11}(q) = \frac{k_z^1(q) - k_z^2(q)}{k_z^1(q) + k_z^2(q)}, \quad R_p^{11}(q) = \frac{\varepsilon_2 k_z^1(q) - \varepsilon_1 k_z^2(q)}{\varepsilon_2 k_z^1(q) + \varepsilon_1 k_z^2(q)}. \tag{37.52}$$

Transmission

In the transmission problem, $\mathbf{r}'$ and $\mathbf{r}$ are located in medium 1 and medium 2, respectively. We find that

$$\mathbf{g}(\mathbf{q}, z, z') = \frac{i}{2k_z^1(q)}[T_s^{12}(q)\,\hat{\mathbf{s}} \otimes \hat{\mathbf{s}} + T_p^{12}(q)\,\hat{\mathbf{p}}_2^+ \otimes \hat{\mathbf{p}}_1^+] \exp[ik_z^2(q)z - ik_z^1(q)z']. \tag{37.53}$$

The Fresnel transmission coefficients introduced in the previous section are given by

$$T_s^{12}(q) = \frac{2k_z^1(q)}{k_z^1(q) + k_z^2(q)}, \quad T_p^{12}(q) = \frac{2\sqrt{\varepsilon_1}\sqrt{\varepsilon_2}\,k_z^1(q)}{\varepsilon_2 k_z^1(q) + \varepsilon_1 k_z^2(q)}. \tag{37.54}$$

Note that the change from $\hat{\mathbf{p}}_1^+$ to $\hat{\mathbf{p}}_2^+$ for the basis vector of p polarization results from refraction at the interface.

References and Additional Reading

These textbooks contain advanced presentations of Green's functions and their use:

P.M. Morse and H. Feshbach, *Methods of Theoretical Physics* (McGraw-Hill, New York, 1953).

Chen-To Tai, *Dyadic Green's Functions in Electromagnetic Theory*, 2nd edition (IEEE Press Series on Electromagnetic Waves, Piscataway, 1994).

A treatment of the singular behavior of the electromagnetic Green's function can be found in:

A.D. Yaghjian, Proc. IEEE **68** 248, (1980).

J. van Bladel, *Singular Electromagnetic Fields and Sources* (Clarendon, Oxford, 1991), chap. 3.

This paper contains an explicit calculation of the transverse and longitudinal Green's functions:

H.F. Arnoldus, J. Mod. Opt. **50**, 755 (2003).

The reflection and transmission of electromagnetic waves at a flat interfaces is treated in many textbooks. See for example:

J.D. Jackson, *Classical Electrodynamics* (Wiley, New York, 1962), chap. 7.

E. Hecht, *Theory and Problems in Optics* (McGraw Hill, New York, 1975), chap. 3.

L.D. Landau, E.M. Lifshitz and L.P. Pitaevskii, *Electrodynamics of Continuous Media* (Pergamon Press, Oxford, 1984), chap. 10.

M. Born and E. Wolf, *Principles of Optics* (Cambridge University Press, Cambridge, 1999), chap. 1.

In the presentation of the electromagnetic Green's function in the semi-infinite geometry, we have followed the approach used in this paper:

J. Sipe, J. Opt. Soc. Am. B **4**, 481 (1987).

This paper develops a short-distance expansion for the Green's function in the half-space geometry:

G. Panasyuk, J.C. Schotland and V.M. Markel, J. Phys. A **42**, 275203 (2009).

38
Electric Dipole Radiation

Electric dipole radiation plays an important role in light scattering, including scattering by small particles and the problem of light emission by small sources. In this chapter, we describe its essential features.

38.1 Far-Field, Near-Field and Quasi-static Limit

As we have seen in Chapter 37, the electric field radiated at the point $\mathbf{r}$ by an electric point dipole $\mathbf{p}$, oscillating at frequency ω and located at the point $\mathbf{r}'$, is given by $\mathbf{E}(\mathbf{r}) = \mu_0\omega^2\,\mathbf{G}_0(\mathbf{r},\mathbf{r}')\mathbf{p}$. In this expression, $\mathbf{G}_0$ is the free-space Green's function. Using (37.17), we immediately find that

$$\mathbf{E}(\mathbf{r}) = k_0^2\,\frac{\exp(ik_0R)}{4\pi\varepsilon_0\,R}\left\{\mathbf{p}-(\mathbf{p}\cdot\hat{\mathbf{R}})\hat{\mathbf{R}}-\left(\frac{1}{ik_0R}+\frac{1}{k_0^2R^2}\right)\left[\mathbf{p}-3(\mathbf{p}\cdot\hat{\mathbf{R}})\hat{\mathbf{R}}\right]\right\}, \tag{38.1}$$

where $k_0=\omega/c=2\pi/\lambda$, $\mathbf{R}=\mathbf{r}-\mathbf{r}'$ and $\hat{\mathbf{R}}=\mathbf{R}/|\mathbf{R}|$.

Far-Field Amplitude

In the far-field, which corresponds to $k_0R\gg 1$, the electric field assumes the form

$$\mathbf{E}(\mathbf{r}) \simeq k_0^2\,\frac{\exp(ik_0R)}{4\pi\varepsilon_0\,R}\left[\mathbf{p}-(\mathbf{p}\cdot\hat{\mathbf{R}})\hat{\mathbf{R}}\right], \tag{38.2}$$

which decays as $1/R$. The vector $\mathbf{p}-(\mathbf{p}\cdot\hat{\mathbf{R}})\hat{\mathbf{R}}$ is the projection of $\mathbf{p}$ onto the plane perpendicular to $\hat{\mathbf{R}}$. Hence the electric field is transverse with respect to the direction of observation and, in particular, vanishes along the vector $\mathbf{p}$.

Near-Field Amplitude

In the near-field, which corresponds to an observation distance much smaller than the wavelength ($k_0 R \ll 1$), the electric field is given by

$$\mathbf{E}(\mathbf{r}) \simeq \frac{\mathbf{p} - 3(\mathbf{p}\cdot\hat{\mathbf{R}})\hat{\mathbf{R}}}{4\pi\varepsilon_0 R^3} . \tag{38.3}$$

We note that in this situation, the electric field decays as $1/R^3$, as in electrostatics. In fact, (38.3) has the same form as the field of an electrostatic dipole and is said to be quasi-static. Near-fields are often computed in this so-called quasi-static limit.

38.2 Radiated Power

The time-averaged power transferred from a source with current density $\mathbf{j}(\mathbf{r})$ to the electromagnetic field is

$$P = -\frac{1}{2}\,\mathrm{Re}\int \mathbf{j}^*(\mathbf{r})\cdot\mathbf{E}(\mathbf{r})\,d^3r. \tag{38.4}$$

For a point electric dipole located at $\mathbf{r}_s$, the current density is $\mathbf{j}(\mathbf{r}) = -i\omega\,\mathbf{p}\,\delta(\mathbf{r}-\mathbf{r}_s)$ and

$$P = \frac{\omega}{2}\,\mathrm{Im}\left[\mathbf{p}^*\cdot\mathbf{E}(\mathbf{r}_s)\right] . \tag{38.5}$$

In a medium with Green's function $\mathbf{G}(\mathbf{r},\mathbf{r}_s)$, P can be rewritten as

$$P = \frac{\mu_0\,\omega^3}{2}\,|\mathbf{p}|^2\,\mathrm{Im}\left[\hat{\mathbf{p}}\cdot\mathbf{G}(\mathbf{r}_s,\mathbf{r}_s)\hat{\mathbf{p}}\right] , \tag{38.6}$$

where we have used the relation $\mathbf{E}(\mathbf{r}) = \mu_0\omega^2\,\mathbf{G}(\mathbf{r},\mathbf{r}_s)\mathbf{p}$ and $\hat{\mathbf{p}} = \mathbf{p}/|\mathbf{p}|$. In this result, the imaginary part of the Green's function $\mathrm{Im}\mathbf{G}(\mathbf{r}_s,\mathbf{r}_s)$ is not singular, provided that the source is located in a nonabsorbing medium. The power calculated this way accounts both for far-field radiation and absorption by the medium. For the case of a dipole in free space, the power is only dissipated by radiation and can be obtained from the free-space Green's function $\mathbf{G}_0(\mathbf{r},\mathbf{r}_s)$. The imaginary part can be calculated from (37.18) and gives

$$\mathrm{Im}\left[\mathbf{G}_0(\mathbf{r}_s,\mathbf{r}_s)\right] = \frac{k_0}{6\pi}\,\mathbf{I}, \tag{38.7}$$

with $\mathbf{I}$ being the unit tensor. Inserting (38.7) into (38.6), we find that

$$P_0 = \frac{\mu_0\,\omega^4}{12\pi\,c}\,|\mathbf{p}|^2 = \frac{\omega^4}{12\pi\,\varepsilon_0\,c^3}\,|\mathbf{p}|^2 , \tag{38.8}$$

which is the well-known expression for the power emitted by an electric dipole in free space.

38.3 Local Density of States

In this section, we introduce the concept of the local density of states in the idealized situation of a non-absorbing and non-dispersive medium described by a real dielectric function $\varepsilon(\mathbf{r})$ and enclosed in a cavity. In this context, we consider the eigenmodes $\mathbf{e}_n$ with eigenfrequency ω_n which obey the vector wave equation

$$\nabla \times \nabla \times \mathbf{e}_n(\mathbf{r}) - \varepsilon(\mathbf{r})\frac{\omega_n^2}{c^2}\,\mathbf{e}_n(\mathbf{r}) = 0\,. \tag{38.9}$$

The above equation can be rewritten as

$$\left[\frac{1}{\sqrt{\varepsilon(\mathbf{r})}}\nabla \times \nabla \times \frac{1}{\sqrt{\varepsilon(\mathbf{r})}}\right]\mathbf{u}_n(\mathbf{r}) - \frac{\omega_n^2}{c^2}\,\mathbf{u}_n(\mathbf{r}) = 0\,, \tag{38.10}$$

where $\mathbf{u}_n(\mathbf{r}) = \sqrt{\varepsilon(\mathbf{r})}\,\mathbf{e}_n(\mathbf{r})$. Equation (38.10) defines an eigenvalue problem for a Hermitian operator. Such problems have real eigenfrequencies ω_n and eigenfunctions that satisfy the orthonormality condition

$$\int \mathbf{u}_m(\mathbf{r}) \cdot \mathbf{u}_n^*(\mathbf{r})\,d^3r = \delta_{mn}\,. \tag{38.11}$$

In terms of the eigenmodes of (38.9), the orthonormality condition is of the form

$$\int \varepsilon(\mathbf{r})\,\mathbf{e}_m(\mathbf{r}) \cdot \mathbf{e}_n^*(\mathbf{r})\,d^3r = \delta_{mn}\,. \tag{38.12}$$

The Green's function satisfies

$$\nabla \times \nabla \times \mathbf{G}(\mathbf{r},\mathbf{r}') - \varepsilon(\mathbf{r})\frac{\omega^2}{c^2}\,\mathbf{G}(\mathbf{r},\mathbf{r}') = \delta(\mathbf{r}-\mathbf{r}')\mathbf{I}\,. \tag{38.13}$$

We now expand the Green's function in the basis of eigenmodes $\mathbf{e}_n(\mathbf{r})$:

$$\mathbf{G}(\mathbf{r},\mathbf{r}') = \sum_n \mathbf{e}_n(\mathbf{r}) \otimes \mathbf{g}_n(\mathbf{r}')\,, \tag{38.14}$$

where $\mathbf{g}_n$ are suitable coefficients. Inserting (38.14) into (38.13) yields

$$\sum_n \left[\nabla \times \nabla \times \mathbf{e}_n(\mathbf{r}) - \varepsilon(\mathbf{r})\frac{\omega^2}{c^2}\,\mathbf{e}_n(\mathbf{r})\right] \otimes \mathbf{g}_n(\mathbf{r}') = \delta(\mathbf{r}-\mathbf{r}')\mathbf{I}\,, \tag{38.15}$$

which, using (38.9), can be transformed into

$$\sum_n \left(\frac{\omega_n^2}{c^2} - \frac{\omega^2}{c^2}\right)\varepsilon(\mathbf{r})\,\mathbf{e}_n(\mathbf{r}) \otimes \mathbf{g}_n(\mathbf{r}') = \delta(\mathbf{r}-\mathbf{r}')\mathbf{I}\,. \tag{38.16}$$

Multiplying both sides of the above equation by $\mathbf{e}_m^*(\mathbf{r})$, integrating over $\mathbf{r}$ and using the orthogonality condition (38.12) leads to

$$(\omega_n^2 - \omega^2)\,\mathbf{g}_n(\mathbf{r}') = c^2\,\mathbf{e}_n^*(\mathbf{r}'). \tag{38.17}$$

The general solution of this equation can be written as

$$\mathbf{g}_n(\mathbf{r}') = c^2\,\mathbf{e}_n^*(\mathbf{r}')\left[\mathrm{P}\left(\frac{1}{\omega_n^2 - \omega^2}\right) + A\,\delta(\omega - \omega_n) + B\,\delta(\omega + \omega_n)\right], \tag{38.18}$$

where P denotes the principal value and A and B are complex-valued constants. The constants can be determined by making use of the identities

$$\frac{1}{\omega_n^2 - \omega^2} = \frac{1}{2\omega_n}\left(\frac{1}{\omega_n - \omega} + \frac{1}{\omega_n + \omega}\right), \tag{38.19}$$

$$\lim_{\eta \to 0} \frac{1}{x - x_0 - i\eta} = \mathrm{P}\frac{1}{x - x_0} + i\pi\,\delta(x - x_0)\,. \tag{38.20}$$

We thereby obtain $A = i\pi/(2\omega_n)$ and $B = -i\pi/(2\omega_n)$. Inserting the final expression for $\mathbf{g}_n$ into (38.14), we obtain the mode expansion of the Green's function:

$$\mathbf{G}(\mathbf{r}, \mathbf{r}') = \sum_n c^2 \mathbf{e}_n(\mathbf{r}) \otimes \mathbf{e}_n^*(\mathbf{r}')\left[\mathrm{P}\left(\frac{1}{\omega_n^2 - \omega^2}\right) + \frac{i\pi}{2\omega_n}\delta(\omega - \omega_n)\right]. \tag{38.21}$$

Here we have dropped the term containing the delta function $\delta(\omega + \omega_n)$ since it does not contribute to positive frequencies. Note that fields and Green's functions are real quantities in the time domain, and have Hermitian symmetry in the frequency domain, so that their spectra can be restricted to positive frequencies. The expression above is sometimes written in the simpler form

$$\mathbf{G}(\mathbf{r}, \mathbf{r}') = \sum_n c^2 \frac{\mathbf{e}_n(\mathbf{r}) \otimes \mathbf{e}_n^*(\mathbf{r}')}{\omega_n^2 - \omega^2}, \tag{38.22}$$

in which the expansion into a principal value and a delta function is implicit. Using (38.21), we readily deduce that

$$\mathrm{Im}\left[\mathrm{Tr}\,\mathbf{G}(\mathbf{r}, \mathbf{r})\right] = \frac{\pi c^2}{2\omega}\sum_n |\mathbf{e}_n(\mathbf{r})|^2\,\delta(\omega - \omega_n)\,, \tag{38.23}$$

where ω_n has been replaced by ω in the prefactor due to the presence of the delta function. The local density of states (LDOS) $\rho(\mathbf{r}, \omega)$ is defined as

$$\rho(\mathbf{r}, \omega) = \sum_n |\mathbf{e}_n(\mathbf{r})|^2\,\delta(\omega - \omega_n)\,. \tag{38.24}$$

The LDOS counts the number of eigenmodes per unit volume and per unit frequency, weighted by their contribution at $\mathbf{r}$. By comparison with (38.23), we obtain the relation between the LDOS and the imaginary part of the Green's function:

$$\rho(\mathbf{r},\omega) = \frac{2\omega}{\pi c^2}\,\mathrm{Im}\,[\mathrm{Tr}\,\mathbf{G}(\mathbf{r},\mathbf{r})]\,. \tag{38.25}$$

In free space, the imaginary part of the Green's function is given by (38.7) and the LDOS is simply

$$\rho_0(\omega) = \frac{\omega^2}{\pi^2 c^3}. \tag{38.26}$$

38.4 Local Density of States and Dipole Radiation

For a dipole oriented along a fixed direction $\hat{\mathbf{p}}$, the power transferred from the source is given by (38.6). Defining the projected LDOS by

$$\rho_{\hat{\mathbf{p}}}(\mathbf{r},\omega) = \frac{2\omega}{\pi c^2}\,\mathrm{Im}\left[\hat{\mathbf{p}}\cdot\mathbf{G}(\mathbf{r},\mathbf{r})\hat{\mathbf{p}}\right], \tag{38.27}$$

we can rewrite the power in the form

$$P = \frac{\pi\,\omega^2}{4\varepsilon_0}|\mathbf{p}|^2\,\rho_{\hat{\mathbf{p}}}(\mathbf{r}_s,\omega). \tag{38.28}$$

For free space, the projected LDOS can be deduced from (38.7), leading to

$$\rho_{0,\hat{\mathbf{p}}}(\omega) = \frac{\omega^2}{3\pi^2 c^3}. \tag{38.29}$$

The projected LDOS accounts for the radiation produced by an electric dipole with a given orientation. The full LDOS in (38.25) corresponds to a sum over three orthogonal dipole orientations:

$$\rho(\mathbf{r},\omega) = \sum_{\hat{\mathbf{p}}}\rho_{\hat{\mathbf{p}}}(\mathbf{r},\omega) = \frac{2\omega}{\pi c^2}\,\mathrm{Im}\,[\mathrm{Tr}\,\mathbf{G}(\mathbf{r},\mathbf{r})]\,. \tag{38.30}$$

It is important to note that the eigenmode expansion of the Green's function which is used to calculate the LDOS is restricted to non-dissipative and non-dispersive media. Nevertheless, the expression (38.28) for the power emitted by an electric dipole can be used to provide a more general notion of the LDOS based on (38.27) and (38.30). Indeed, we can consider these two equations as the definitions of the partial and full LDOS.

38.5 A Simple Classical to Quantum Correspondence

The imaginary part of the Green's function, or equivalently the LDOS, determines the spontaneous decay rate Γ of a quantum electric-dipole emitter. The full quantum electrodynamics calculation of the decay rate in a structured medium is beyond the scope of this book. Here we use simple correspondence principle arguments to deduce the expression for the decay rate from the expression of the power emitted by a classical electric dipole.

Beginning from (38.6), we can transform the emitted power (averaged emitted energy per unit time) into a decay rate (averaged number of transitions per unit time) by dividing the power by the quantum of energy $\hbar\omega$. The classical dipole moment $\mathbf{p}$ must then be replaced by the transition dipole $\mathbf{p}_{eg} = \langle g|\mathbf{D}|e\rangle$, where $|e\rangle$ and $|g\rangle$ are the excited and ground states, and $\mathbf{D}$ is the electric dipole operator, with the frequency ω becoming the Bohr transition frequency of the quantum dipole. More precisely, we make the replacement $\mathbf{p} \to 2\mathbf{p}_{eg}$, since the quantum mechanical transition dipole is defined with positive frequencies only, while the classical dipole $\mathbf{p}$ involves both positive and negative frequencies. This simple approach leads to the following expression for the decay rate:

$$\Gamma = \frac{2\mu_0\,\omega^2}{\hbar}\,|\mathbf{p}_{eg}|^2\,\mathrm{Im}\left[\hat{\mathbf{p}}\cdot\mathbf{G}(\mathbf{r}_s,\mathbf{r}_s)\hat{\mathbf{p}}\right], \tag{38.31}$$

where $\mathbf{r}_s$ is the position of the emitter. This result coincides with that obtained from quantum electrodynamics. In terms of the projected LDOS, it can be rewritten as

$$\Gamma = \frac{\pi\,\omega}{\hbar\,\varepsilon_0}|\mathbf{p}_{eg}|^2\,\rho_{\hat{\mathbf{p}}}(\mathbf{r},\omega), \tag{38.32}$$

which takes the form of Fermi's golden rule. For the case of free space, the decay rate is given by

$$\Gamma_0 = \frac{\omega^3}{3\hbar\,\varepsilon_0\,c^3}|\mathbf{p}_{eg}|^2. \tag{38.33}$$

This is the well-known expression for the spontaneous decay rate of a two-level atom in vacuum.

38.6 Purcell Factor

The change in the spontaneous decay rate Γ/Γ_0 due to a structured environment is known as the Purcell effect, since it was described by E.M. Purcell in 1946 in the case of an emitter in a single mode cavity. We will show that the formalism introduced above yields Purcell's expression for Γ/Γ_0, known as the Purcell factor. To proceed, we begin from the expression for the Green's function (38.22) for a perfect lossless cavity. In the presence of weak dissipation, mode attenuation can be accounted for introducing a mode damping rate γ_n and modifying (38.22) according to

$$\mathbf{G}(\mathbf{r}, \mathbf{r}') = \sum_n c^2 \frac{\mathbf{e}_n(\mathbf{r}) \otimes \mathbf{e}_n^*(\mathbf{r}')}{\omega_n^2 - \omega^2 - i\omega\gamma_n} \,. \tag{38.34}$$

This qualitative approach is restricted to eigenmodes with high-quality factors $Q_n = \omega/\gamma_n$. Following Purcell, we assume the emitter is on resonance with one of the eigenmodes ($\omega = \omega_n$), so that the Green's function is dominated by the contribution of this specific eigenmode. From (38.31) and (38.34), we obtain

$$\Gamma = \frac{2}{\varepsilon_0 \hbar} |\mathbf{p}_{eg}|^2 \, Q \, |\mathbf{e}_n(\mathbf{r}_s) \cdot \hat{\mathbf{p}}|^2 \,. \tag{38.35}$$

We can also define the mode volume by the relation

$$\frac{1}{V} = |\mathbf{e}_n(\mathbf{r}_s) \cdot \hat{\mathbf{p}}|^2 \,, \tag{38.36}$$

where we recall that $|\mathbf{e}_n(\mathbf{r}_s)|^2$ has the unit of inverse volume due to the orthogonality condition (38.12). Using the expression (38.33) for the spontaneous decay rate in vacuum, the normalized decay rate finally becomes

$$\frac{\Gamma}{\Gamma_0} = \frac{3}{4\pi^2} \lambda^3 \frac{Q}{V} \,. \tag{38.37}$$

The right-hand side is the Purcell factor. Note that in the original derivation by Purcell, the emitter is located at the point $\mathbf{r}_{\max}$, coinciding with the maximum amplitude of the mode, so that $V^{-1} = |\mathbf{e}_n(\mathbf{r}_{\max}, \omega) \cdot \hat{\mathbf{p}}|^2$. This maximizes the Purcell factor.

Although the definition of the mode volume is clear in the case of a medium for which a discrete basis of eigenmodes can be defined (as a closed and non-absorbing cavity), its definition in the general case of an open and absorbing medium is problematic. Moreover, the splitting of Γ/Γ_0 into a contribution from a quality factor Q and a mode volume V is arbitrary, since these two parameters are not independent. In fact, the relevant parameter for the change in the decay rate is the LDOS (including radiative and non-radiative contributions), which in the case of a single mode cavity is translational to the ratio Q/V.

A general expression for the normalized decay rate Γ/Γ_0 can be obtained from (38.31) and (38.33):

$$\frac{\Gamma}{\Gamma_0} = \frac{6\pi c}{\omega} \, \mathrm{Im}\left[\hat{\mathbf{p}} \cdot \mathbf{G}(\mathbf{r}_s, \mathbf{r}_s)\hat{\mathbf{p}}\right] \,. \tag{38.38}$$

The right-hand side can be understood as a generalized Purcell factor. This expression applies to any system, including open and/or absorbing media. It is also interesting to note that using the power emitted by a classical dipole, the same result is obtained. It follows from (38.6) and (38.8) that the normalized emitted power P/P_0 is readily obtained:

$$\frac{P}{P_0} = \frac{6\pi c}{\omega} \, \mathrm{Im}\left[\hat{\mathbf{p}} \cdot \mathbf{G}(\mathbf{r}_s, \mathbf{r}_s)\hat{\mathbf{p}}\right] . \tag{38.39}$$

This expression shows that the generalized Purcell factor also appears in the expression for the normalized power emitted by a classical dipole. In the classical antenna formalism, this factor can be understood as a change in the impedance of the medium.

38.7 Cross Density of States

In this section, we consider the radiation by an extended source in an arbitrary medium and introduce the concept of the cross density of states (CDOS). In a structured medium described by a dielectric function $\varepsilon(\mathbf{r})$ occupying a finite volume V, the electric field obeys the vector Helmholtz equation

$$\nabla \times \nabla \times \mathbf{E}(\mathbf{r}) - \varepsilon(\mathbf{r})\frac{\omega^2}{c^2}\mathbf{E}(\mathbf{r}) = i\mu_0\omega\mathbf{j}(\mathbf{r}) , \tag{38.40}$$

where $\mathbf{j}(\mathbf{r})$ is the source current density. Following the same steps as in Section 37.1, it is easy to show that

$$\mathbf{E}(\mathbf{r}) = i\mu_0\omega \int \mathbf{G}(\mathbf{r}, \mathbf{r}')\, \mathbf{j}(\mathbf{r}')\, d^3r' . \tag{38.41}$$

Using (38.4), we can write the power transferred from the source to the medium as

$$P = \frac{\mu_0\omega}{2} \int \mathbf{j}^*(\mathbf{r}) \cdot \mathrm{Im}[\mathbf{G}(\mathbf{r}, \mathbf{r}')]\, \mathbf{j}(\mathbf{r}')\, d^3r d^3r' . \tag{38.42}$$

Consider now the particular case of an extended source compose of two coherent electric point dipoles $\mathbf{p}_1$ and $\mathbf{p}_2$ located at positions $\mathbf{r}_1$ and $\mathbf{r}_2$. The current density is $\mathbf{j}(\mathbf{r}) = -i\omega\mathbf{p}_1\delta(\mathbf{r} - \mathbf{r}_1) - i\omega\mathbf{p}_2\delta(\mathbf{r} - \mathbf{r}_2)$. Inserting this expression into (38.42) leads to

$$P = \frac{\mu_0\omega^3}{2} \left[|\mathbf{p}_1|^2 \mathrm{Im}G_{11} + |\mathbf{p}_2|^2 \mathrm{Im}G_{22} + 2\mathrm{Re}(p_1 p_2^*)\mathrm{Im}G_{12}\right] , \tag{38.43}$$

where we have used the simplified notations $\mathrm{Im}G_{ij} = \hat{\mathbf{p}}_i \cdot \mathrm{Im}[\mathbf{G}(\mathbf{r}_i, \mathbf{r}_j)]\hat{\mathbf{p}}_j$, and $\mathbf{p}_i = p_i\hat{\mathbf{p}}_i$. We see that the interference term in the power emitted by the two dipoles is controlled by the imaginary part of the two point Green's function $\mathrm{Im}G_{12}$.

Proceeding along the same lines as for the LDOS, we define the CDOS as

$$\rho(\mathbf{r}, \mathbf{r}', \omega) = \frac{2\omega}{\pi c^2} \operatorname{Im}\left[\operatorname{Tr} \mathbf{G}(\mathbf{r}, \mathbf{r}')\right]. \tag{38.44}$$

The CDOS defined this way averages the polarization degrees of freedom, as in the full LDOS in (38.30). In order to account for polarization, we can also define a partial CDOS

$$\rho_{ij}(\mathbf{r}, \mathbf{r}', \omega) = \frac{2\omega}{\pi c^2} \operatorname{Im} G_{ij}(\mathbf{r}, \mathbf{r}'), \tag{38.45}$$

such that $\rho(\mathbf{r}, \mathbf{r}', \omega) = \sum_i \rho_{ii}(\mathbf{r}, \mathbf{r}', \omega)$. The partial CDOS governs, for example, the power emitted by two linearly polarized dipole sources, as in (38.43).

The physical meaning of the CDOS can be understood by considering the situation of a non-absorbing medium in a closed cavity. Using the expression (38.21) for the Green's function, we find that

$$\begin{aligned} \rho(\mathbf{r}, \mathbf{r}', \omega) = &\sum_n \operatorname{Re}[\mathbf{e}_n(\mathbf{r}) \otimes \mathbf{e}_n^*(\mathbf{r}')]\delta(\omega - \omega_n) \\ &+ \frac{2\omega}{\pi} \sum_n \mathrm{P}\left(\frac{1}{\omega_n^2 - \omega^2}\right) \operatorname{Im}[\mathbf{e}_n(\mathbf{r}) \otimes \mathbf{e}_n^*(\mathbf{r}')]. \end{aligned} \tag{38.46}$$

This result can be simplified by using the reciprocity theorem (40.10), which in terms of the eigenmode expansion (38.21) is of the form

$$\sum_n \left[\mathrm{PV}\left(\frac{1}{\omega_n^2 - \omega^2}\right) + \frac{i\pi}{2\omega_n}\delta(\omega - \omega_n)\right] [\mathbf{e}_n(\mathbf{r}) \otimes \mathbf{e}_n^*(\mathbf{r}') - \mathbf{e}_n^*(\mathbf{r}) \otimes \mathbf{e}_n(\mathbf{r}')] = 0. \tag{38.47}$$

Noting that $\mathbf{e}_n(\mathbf{r}) \otimes \mathbf{e}_n^*(\mathbf{r}') - \mathbf{e}_n^*(\mathbf{r}) \otimes \mathbf{e}_n(\mathbf{r}') = 2i \operatorname{Im}[\mathbf{e}_n(\mathbf{r}) \otimes \mathbf{e}_n^*(\mathbf{r}')]$, and taking the imaginary part of (38.47), we immediately see that the second term on the right-hand side in Eq. (38.46) vanishes. We finally obtain a simple expression of the CDOS for the particular case of a non-dissipative medium:

$$\rho(\mathbf{r}, \mathbf{r}', \omega) = \sum_n \operatorname{Re}[\mathbf{e}_n(\mathbf{r}) \otimes \mathbf{e}_n^*(\mathbf{r}')]\delta(\omega - \omega_n). \tag{38.48}$$

This expression shows that the CDOS counts the number of eigenmodes at a given frequency connecting two different points in the medium. For this reason, the CDOS enters expressions for the power emitted by two point sources. It is also frequently encountered in the calculation of the spatial correlation function between fields at two different points (an example will be seen in Chapter 44). In this case, the connection between the observation points, provided by the electromagnetic eigenmodes, generates spatial coherence, independent of the incident field.

References and Additional Reading

The basic concepts of near-field optics are presented in:

J.-J. Greffet and R. Carminati, Prog. Surf. Sci. **56**, 133 (1997).

L. Novotny and B. Hecht, *Principles of Nano-Optics* (Cambridge University Press, Cambridge, 2006).

The canonical procedure for the introduction of eigenmodes of Maxwell's equations in non-dissipative and non-dispersive environments is presented in:

R.J. Glauber and M. Lewenstein, Phys. Rev. A **43**, 467 (1991).

A description of local and cross density of states in terms of quasi-normal modes (QNMs) can be found in:

C. Sauvan, J. P. Hugonin, I. S. Maksymov and P. Lalanne, Phys. Rev. Lett. **110**, 237401 (2013).

C. Sauvan, J. P. Hugonin, R. Carminati and P. Lalanne, Phys. Rev. A **89**, 89, 043825 (2014).

The introduction of QNMs extends the expansion (38.21) to open and/or dissipative systems.

A treatment of non-Hermitian quantum mechanics, based on the concept of resonances instead of usual eigenmodes, is presented in:

N. Moiseyev, *Non-Hermitian Quantum Mechanics* (Cambridge University Press, 2011).

The correspondence between classical radiation and quantum spontaneous decay rate is discussed in:

M. Born, *Atomic Physics* (Dover, New York, 1936), chap. 5.

A method to calculate spontaneous decay rates from QED in arbitrary environments, using the fluctuation-dissipation theorem, is given in:

G.S. Agarwal, Phys. Rev. A **11**, 253 (1975).

J.M. Wylie and J.E. Sipe, Phys. Rev. A **30**, 1185 (1984).

The following paper discusses the singularity of the Green's function in arbitrary environments. In particular, this paper shows that $\mathrm{Im}\mathbf{G}(\mathbf{r}_s, \mathbf{r}_s)$ is not singular, provided that the observation point $\mathbf{r}_s$ lies in a *locally* non-absorbing dielectric (whatever its proximity to absorbing materials):

C.A. Guérin, B. Gralak and A. Tip, Phys. Rev. E **75**, 056601 (2007).

The definition of the LDOS including electric and magnetic contributions is discussed in:

K. Joulain, R. Carminati, J.P. Mulet and J.-J. Greffet, Phys. Rev. B **68**, 245405 (2003).

K. Joulain, J.P. Mulet, F. Marquier, R. Carminati and J.-J. Greffet, Surf. Sci. Rep. **57**, 59 (2005).

This review paper presents the concepts of LDOS and CDOS in near-field optics, and the measurement techniques:

R. Carminati, A. Cazé, D. Cao, F. Peragut, V. Krachmalnicoff, R. Pierrat and Y. De Wilde, Surf. Sci. Rep. **70**, 1 (2015).

This is the original paper on the Purcell factor:
E.M. Purcell, Phys. Rev. **69**, 681 (1946).

The cross density of states and its interpretation in terms of eigenmodes have been first discussed in the following paper:
A. Cazé, R. Pierrat and R. Carminati, Phys. Rev. Lett. **110**, 063903 (2013).

39

Scattering of Electromagnetic Waves

In this chapter, we introduce the basic equations of electromagnetic scattering theory. The approach is similar to that used in Chapter 10 for scalar waves. We apply the formalism to the study of Rayleigh–Gans scattering.

39.1 Integral Equations

We consider the scattering problem shown in Fig. 39.1, in which a scattering medium with dielectric function $\varepsilon(\mathbf{r})$ occupies a volume V and is illuminated by a monochromatic source with current density $\mathbf{j}_{\text{ext}}$. The total electric field $\mathbf{E}$ obeys the vector Helmholtz equation

$$\nabla \times \nabla \times \mathbf{E} - \varepsilon\, k_0^2\, \mathbf{E} = i\mu_0\omega\, \mathbf{j}_{\text{ext}}\,, \tag{39.1}$$

where $k_0 = \omega/c = 2\pi/\lambda$ is the free-space wavenumber. The incident field $\mathbf{E}_i$ obeys the Helmholtz equation in free space

$$\nabla \times \nabla \times \mathbf{E}_i - k_0^2\, \mathbf{E}_i = i\mu_0\omega\, \mathbf{j}_{\text{ext}}\,. \tag{39.2}$$

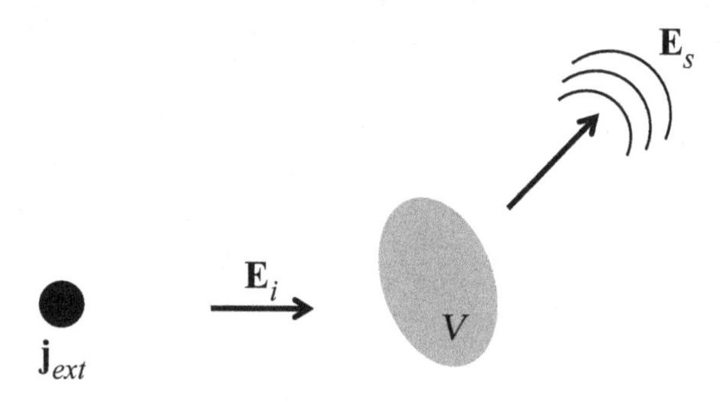

Fig. 39.1 Geometry of the scattering problem.

Subtracting the above equations, we find that the scattered field $\mathbf{E}_s = \mathbf{E} - \mathbf{E}_i$ satisfies

$$\nabla \times \nabla \times \mathbf{E}_s - k_0^2 \mathbf{E}_s = k_0^2 \delta\varepsilon \mathbf{E}, \tag{39.3}$$

where $\delta\varepsilon = \varepsilon - 1$. The solution to (39.3) can be written in terms of the free-space Green's function introduced in Section 37.1. Using (37.1) and (37.4), we immediately find that

$$\mathbf{E}_s(\mathbf{r}) = k_0^2 \int \mathbf{G}_0(\mathbf{r}, \mathbf{r}')\, \delta\varepsilon(\mathbf{r}')\, \mathbf{E}(\mathbf{r}')\, d^3r' . \tag{39.4}$$

This expression shows that the scattered field is the field radiated by the induced polarization $\mathbf{P} = \varepsilon_0 \delta\varepsilon \mathbf{E}$ inside the scattering medium. Finally, we see that the total field satisfies the integral equation

$$\mathbf{E}(\mathbf{r}) = \mathbf{E}_i(\mathbf{r}) + k_0^2 \int \mathbf{G}_0(\mathbf{r}, \mathbf{r}')\, \delta\varepsilon(\mathbf{r}')\, \mathbf{E}(\mathbf{r}')\, d^3r' , \tag{39.5}$$

which is known as the Lippmann–Schwinger equation.

39.2 Scattering Amplitude and Cross Sections

When the observation point $\mathbf{r}$ is in the far-field, we can use the asymptotic expression for the Green's function given by (37.20), and rewrite the scattered field in the form

$$\mathbf{E}_s(\mathbf{r}) \sim k_0^2 \frac{\exp(ik_0 r)}{4\pi r} (\mathbf{I} - \hat{\mathbf{r}} \otimes \hat{\mathbf{r}}) \int \delta\varepsilon(\mathbf{r}')\, \mathbf{E}(\mathbf{r}') \exp(-ik_0 \hat{\mathbf{r}} \cdot \mathbf{r}')\, d^3r' . \tag{39.6}$$

The scattered field behaves as a transverse spherical wave in the far-field. By writing

$$\mathbf{E}(\mathbf{r}) \sim \mathbf{A}(\hat{\mathbf{r}}) \frac{\exp(ik_0 r)}{r} , \tag{39.7}$$

which defines the vector scattering amplitude $\mathbf{A}$, we obtain

$$\mathbf{A}(\hat{\mathbf{r}}) = \frac{k_0^2}{4\pi} (\mathbf{I} - \hat{\mathbf{r}} \otimes \hat{\mathbf{r}}) \int \delta\varepsilon(\mathbf{r}')\, \mathbf{E}(\mathbf{r}') \exp(-ik_0 \hat{\mathbf{r}} \cdot \mathbf{r}')\, d^3r' . \tag{39.8}$$

From (37.25) and (39.7), we find that the power scattered in direction $\hat{\mathbf{r}}$, per unit solid angle, is given by

$$\frac{dP_s}{d\Omega} = \frac{\epsilon_0 c}{2} |\mathbf{A}(\hat{\mathbf{r}})|^2 . \tag{39.9}$$

The differential scattering cross section is defined as

$$\frac{d\sigma_s}{d\Omega} = \frac{1}{I_i} \frac{dP_s}{d\Omega} , \tag{39.10}$$

where I_i is the incident flux per unit area carried by the incident wave. The scattering cross section is

$$\sigma_s = \int \frac{d\sigma_s}{d\Omega}\, d\hat{\mathbf{r}}\,. \tag{39.11}$$

It is easy to see that $\sigma_s I_i = P_s$, with P_s the total scattered power, meaning that the scattering cross section can be understood as the effective cross-sectional area of the scatterer. Likewise, we define the absorption cross section σ_a by $\sigma_a I_i = P_a$, where P_a the power absorbed by the scatterer, as defined in Section 36.2.

When the incident field is a plane wave with complex amplitude

$$\mathbf{E}_i(\mathbf{r}) = \mathbf{E}_0 \exp(i\mathbf{k}\cdot\mathbf{r})\,, \tag{39.12}$$

the scattering amplitude explicitly depends on the incident wavevector $\mathbf{k}$ and on the wavevector $\mathbf{k}'$ and is denoted by $\mathbf{A}(\mathbf{k},\mathbf{k}')$. The incident flux per unit area across a plane perpendicular to $\mathbf{k}$ can be deduced from the time-averaged Poynting vector $\mathbf{\Pi}_i = (\varepsilon_0 c/2)|\mathbf{E}_0|^2\hat{\mathbf{k}}$, and we find that $I_i = (\varepsilon_0 c/2)|\mathbf{E}_0|^2$. Hence, the differential scattering cross section is given by

$$\frac{d\sigma_s}{d\Omega} = \frac{|\mathbf{A}(\mathbf{k},\mathbf{k}')|^2}{|\mathbf{E}_0|^2}\,. \tag{39.13}$$

It will prove to be useful to define the second-order tensor $\mathbf{S}(\mathbf{k},\mathbf{k}')$ by $\mathbf{A}(\mathbf{k},\mathbf{k}') = \mathbf{S}(\mathbf{k},\mathbf{k}')\mathbf{E}_0$, known as the scattering matrix, which is independent of the amplitude of the incident field. For a linearly polarized incident plane wave, with polarization along the unit vector $\hat{\mathbf{e}}_0$, we have $\mathbf{E}_0 = E_0\hat{\mathbf{e}}_0$ and

$$\frac{d\sigma_s}{d\Omega} = |\mathbf{S}(\mathbf{k},\mathbf{k}')\hat{\mathbf{e}}_0|^2\,. \tag{39.14}$$

39.3 Born Approximation and Rayleigh–Gans Scattering

For a weakly scattering object, the scattered field $\mathbf{E}_s$ may be calculated in the Born approximation, by writing

$$\mathbf{E}_s(\mathbf{r}) = k_0^2 \int \mathbf{G}_0(\mathbf{r},\mathbf{r}')\,\delta\varepsilon(\mathbf{r}')\,\mathbf{E}_i(\mathbf{r}')\,d^3r'\,. \tag{39.15}$$

In the special case of scattering from a sphere with radius a made of a homogeneous material with dielectric constant ϵ, the Born approximation is expected to be accurate, provided that the condition $k_0 a|\delta\varepsilon| \ll 1$ is satisfied. This regime is known as Rayleigh–Gans scattering.

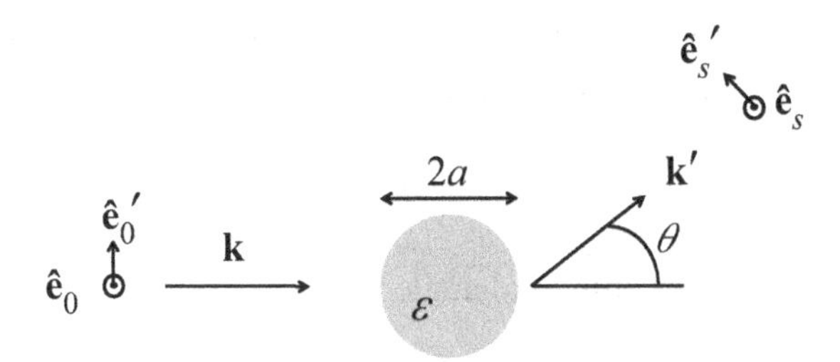

Fig. 39.2 Scattering from a homogeneous sphere with radius a and dielectric function ε.

As shown in Fig. 39.2, we consider the spherical scatterer to be illuminated by a plane wave of the form (39.12). The scattering amplitude is

$$\mathbf{A}(\mathbf{k},\mathbf{k}') = \frac{k_0^2}{4\pi}\delta\varepsilon\,(\mathbf{I}-\hat{\mathbf{r}}\hat{\mathbf{r}})qa\mathbf{E}_0\int_V \exp(-i\mathbf{q}\cdot\mathbf{r}')\,d^3r', \tag{39.16}$$

where $\mathbf{q} = \mathbf{k}' - \mathbf{k}$ and the integral is over the volume of the sphere. The integral can be calculated in spherical coordinates:

$$\int_V \exp(-i\mathbf{q}\cdot\mathbf{r}')\,d^3r' = 3V\left[\frac{\sin(qa) - qa\cos(qa)}{(qa)^3}\right]. \tag{39.17}$$

Hence the scattering amplitude is given by

$$\mathbf{A}(\mathbf{k},\mathbf{k}') = \frac{3V}{4\pi}k_0^2\,\delta\varepsilon\,(\mathbf{I}-\hat{\mathbf{r}}\hat{\mathbf{r}})\mathbf{E}_0\left[\frac{\sin(qa) - qa\cos(qa)}{(qa)^3}\right]. \tag{39.18}$$

It is interesting to characterize the scattering cross section for unpolarized light. To proceed, we write $\mathbf{E}_0 = E_0\hat{\mathbf{e}}_0$, with $\hat{\mathbf{e}}_0$ a unit vector defining the polarization of the incident field. We also assume that the scattering amplitude is detected by a polarizer oriented along the direction $\hat{\mathbf{e}}_s$. Next, we consider two orthogonal directions $\mathbf{e}_0$ and $\mathbf{e}_0'$ for the incident field, and two orthogonal directions $\mathbf{e}_s$ and $\mathbf{e}_s'$ for the scattering amplitude, both being orthogonal to the observation direction $\hat{\mathbf{r}}$, as shown in Fig. 39.2. In terms of the scattering matrix $\mathbf{S}(\mathbf{k},\mathbf{k}')$ defined in the previous section, we find that

$$\mathbf{e}_s\cdot\mathbf{S}(\mathbf{k},\mathbf{k}')\mathbf{e}_0 = \frac{3V}{4\pi}k_0^2\,\delta\varepsilon\left[\frac{\sin(qa) - qa\cos(qa)}{(qa)^3}\right]\cos\theta, \tag{39.19}$$

$$\mathbf{e}_s'\cdot\mathbf{S}(\mathbf{k},\mathbf{k}')\mathbf{e}_0' = \frac{3V}{4\pi}k_0^2\,\delta\varepsilon\left[\frac{\sin(qa) - qa\cos(qa)}{(qa)^3}\right], \tag{39.20}$$

$$\mathbf{e}_s'\cdot\mathbf{S}(\mathbf{k},\mathbf{k}')\mathbf{e}_0 = 0, \tag{39.21}$$

$$\mathbf{e}_s\cdot\mathbf{S}(\mathbf{k},\mathbf{k}')\mathbf{e}_0' = 0, \tag{39.22}$$

where θ is the scattering angle defined in Fig. 39.2 such that $q = 2k_0 \sin\theta$. For unpolarized light, the incident flux per unit area is

$$I_i = \frac{\epsilon_0 c}{2}\left[|E_0\mathbf{e}_0|^2 + |E_0\mathbf{e}_0'|^2\right] = \epsilon_0 c|E_0|^2 \tag{39.23}$$

and the scattered power per unit solid angle is

$$\begin{aligned}\frac{dP_s}{d\Omega} &= \frac{\epsilon_0 c}{2}|E_0|^2\left[|\mathbf{e}_s \cdot \mathbf{S}(\mathbf{k},\mathbf{k}')\mathbf{e}_0|^2 + |\mathbf{e}_s' \cdot \mathbf{S}(\mathbf{k},\mathbf{k}')\mathbf{e}_0'|^2\right] \\ &= \frac{\epsilon_0 c}{2}\frac{V^2}{(4\pi)^2}k_0^4|\delta\varepsilon|^2 P(q)(1+\cos^2\theta)|E_0|^2 .\end{aligned} \tag{39.24}$$

Here

$$P(q) = 9\left[\frac{\sin(qa) - qa\cos(qa)}{(qa)^3}\right]^2 , \tag{39.25}$$

and is defined so that $P(0) = 1$. From (39.23) and (39.24), we find that the differential scattering cross section is

$$\frac{d\sigma_s}{d\Omega} = \frac{V^2}{32\pi^2}k_0^4|\delta\varepsilon|^2 P(q)(1+\cos^2\theta) . \tag{39.26}$$

We then find that the scattering cross section is

$$\sigma_s = \int \frac{d\sigma_s}{d\Omega}d\hat{\mathbf{q}} = \frac{V^2}{16\pi}k_0^4|\delta\varepsilon|^2 \int_0^\pi P(2k_0\sin\theta)(1+\cos^2\theta)\sin\theta d\theta . \tag{39.27}$$

In general, the above integral must be calculated numerically. In the Rayleigh scattering regime, where $k_0 a \ll 1$, we have that

$$\sigma_s = \frac{V^2}{6\pi}k_0^4|\delta\varepsilon|^2 , \tag{39.28}$$

which has a characteristic ω^4 frequency dependence. Rayleigh scattering is discussed in more detail in Chapter 41.

References and Additional Reading

Light scattering by particles is extensively covered in the following textbooks:

H.C. van de Hulst, *Light Scattering by Small Particles* (Dover, New York, 1981).

C.F. Bohren and D.R. Huffman, *Absorption and Scattering of Light by Small Particles* (Wiley, New York, 1998).

M.I. Mishchenko, L.D. Travis and A.A. Lacis, *Scattering, Absorption, and Emission of Light by Small Particles* (Cambridge University Press, Cambridge, 2002).

Scattering of scalar waves from homogeneous spheres of arbitrary size is studied in Chapter 12. The exact solution of scattering of an electromagnetic plane wave from a homogeneous sphere was given in:

G. Mie, Ann. Phys. **25**, 377 (1908).

The so-called regime of Mie scattering is well documented, and open-access solvers can easily be found online.

The following review articles address the problem of single and multiple scattering from small particles:

A. Lagendijk and B.A. van Tiggelen, Phys. Rep. **270**, 143 (1996).
P. de Vries, D.V. van Coevorden and A. Lagendijk, Rev. Mod. Phys. **70**, 447 (1998).
M.C.W. van Rossum and Th.M. Nieuwenhuizen, Rev. Mod. Phys. **71**, 313 (1999).

40

Electromagnetic Reciprocity and the Optical Theorem

In this chapter, we derive the reciprocity theorem and the optical theorem for electromagnetic waves. These theorems were established for scalar waves in Chapters 11 and 14. We also derive two integral theorems for the Green's function, which turns out to be useful in various scattering problems.

40.1 Lorentz Reciprocity Relation

We consider a medium described by a dielectric function $\varepsilon(\mathbf{r})$ and a relative magnetic permeability $\mu(\mathbf{r})$, which are assumed to be complex-valued. Let V_1 be a volume containing a source with current density $\mathbf{j}_1(\mathbf{r})$ at frequency ω. The corresponding electric and magnetic fields are denoted by $\mathbf{E}_1(\mathbf{r})$ and $\mathbf{H}_1(\mathbf{r})$. Likewise, let V_2 be a volume with current density $\mathbf{j}_2(\mathbf{r})$, and $\mathbf{E}_2(\mathbf{r})$ and $\mathbf{H}_2(\mathbf{r})$ be the corresponding electric and magnetic fields. The situation is illustrated in Fig. 40.1. The

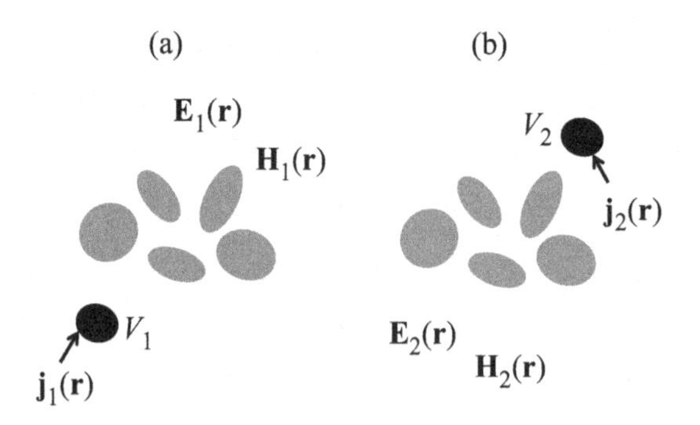

Fig. 40.1 Illustrating the geometry of the reciprocity theorem.

fields satisfy the Maxwell equations

$$\nabla \times \mathbf{E}_j = i\omega \mathbf{B}_j \qquad \nabla \times \mathbf{H}_j = \mathbf{j}_j - i\omega \mathbf{D}_j \tag{40.1}$$

with $j = 1, 2$, together with the constitutive relations $\mathbf{D}_j = \varepsilon_0 \varepsilon\, \mathbf{E}_j$ and $\mathbf{B}_j = \mu_0 \mu\, \mathbf{H}_j$. The following identity is easily deduced from the Maxwell equations:

$$\begin{aligned} &\mathbf{H}_2 \cdot \nabla \times \mathbf{E}_1 - \mathbf{E}_1 \cdot \nabla \times \mathbf{H}_2 + \mathbf{E}_2 \cdot \nabla \times \mathbf{H}_1 - \mathbf{H}_1 \cdot \nabla \times \mathbf{E}_2 = \\ &i\omega(\mathbf{B}_1 \cdot \mathbf{H}_2 - \mathbf{H}_1 \cdot \mathbf{B}_2) - i\omega(\mathbf{D}_1 \cdot \mathbf{E}_2 - \mathbf{E}_1 \cdot \mathbf{D}_2) + \mathbf{j}_1 \cdot \mathbf{E}_2 - \mathbf{j}_2 \cdot \mathbf{E}_1 \,. \end{aligned} \tag{40.2}$$

The left-hand side can be seen to be equal to $\nabla \cdot (\mathbf{E}_1 \times \mathbf{H}_2 - \mathbf{E}_2 \times \mathbf{H}_1)$. The first two terms on the right-hand side vanish due to the constitutive relations. Equation (40.2) thus becomes

$$\nabla \cdot (\mathbf{E}_1 \times \mathbf{H}_2 - \mathbf{E}_2 \times \mathbf{H}_1) = \mathbf{j}_1 \cdot \mathbf{E}_2 - \mathbf{j}_2 \cdot \mathbf{E}_1 \,. \tag{40.3}$$

Integrating this relation over a volume V enclosed by a surface S containing the scattering medium and the sources, and making use of the divergence theorem, we obtain

$$\begin{aligned} &\int_S [\mathbf{E}_1(\mathbf{r}) \times \mathbf{H}_2(\mathbf{r}) - \mathbf{E}_2(\mathbf{r}) \times \mathbf{H}_1(\mathbf{r})] \cdot \hat{\mathbf{n}}\, d^2 r = \\ &\qquad \int_V \left[\mathbf{j}_1(\mathbf{r}) \cdot \mathbf{E}_2(\mathbf{r}) - \mathbf{j}_2(\mathbf{r}) \cdot \mathbf{E}_1(\mathbf{r})\right] d^3 r \,, \end{aligned} \tag{40.4}$$

where $\hat{\mathbf{n}}$ is the outward unit normal on S. This result is known as the Lorentz reciprocity theorem for vector fields.

40.2 Consequences of the Reciprocity Theorem

For a medium of finite size in free space, we can choose the volume V to be a sphere of radius $R \to \infty$. In the far-field, using (37.23), we find that

$$\omega \mu_0 \mathbf{H}_j(\mathbf{r}) = k_0 \hat{\mathbf{n}} \times \mathbf{E}_j(\mathbf{r}) \,, \tag{40.5}$$

so the term $\mathbf{E}_1 \times \mathbf{H}_2 - \mathbf{E}_2 \times \mathbf{H}_1$ vanishes identically on the surface S. Equation (40.4) thus becomes

$$\int \mathbf{j}_1(\mathbf{r}) \cdot \mathbf{E}_2(\mathbf{r})\, d^3 r = \int \mathbf{j}_2(\mathbf{r}) \cdot \mathbf{E}_1(\mathbf{r})\, d^3 r \,, \tag{40.6}$$

where the integrals are over all of space. For the special case of illumination by point electric-dipoles located at $\mathbf{r}_1$ and $\mathbf{r}_2$, the current densities are $\mathbf{j}_1(\mathbf{r}) = -i\omega \mathbf{p}_1\, \delta(\mathbf{r} - \mathbf{r}_1)$ and $\mathbf{j}_2(\mathbf{r}) = -i\omega \mathbf{p}_2\, \delta(\mathbf{r} - \mathbf{r}_2)$. (40.6) then yields

$$\mathbf{p}_1 \cdot \mathbf{E}_2(\mathbf{r}_1) = \mathbf{p}_2 \cdot \mathbf{E}_1(\mathbf{r}_2) \,, \tag{40.7}$$

which is the reciprocity theorem for point sources.

Next, we derive the reciprocity relation obeyed by the Green's function. Making use of (37.5), we have $\mathbf{E}_1(\mathbf{r}_2) = \mu_0\omega^2\,\mathbf{G}(\mathbf{r}_2,\mathbf{r}_1)\mathbf{p}_1$ and $\mathbf{E}_2(\mathbf{r}_1) = \mu_0\omega^2\,\mathbf{G}(\mathbf{r}_1,\mathbf{r}_2)\mathbf{p}_2$. Inserting these expression into (40.7) leads to

$$\mathbf{p}_1\cdot\mathbf{G}(\mathbf{r}_1,\mathbf{r}_2)\mathbf{p}_2 = \mathbf{p}_2\cdot\mathbf{G}(\mathbf{r}_2,\mathbf{r}_1)\mathbf{p}_1\,, \tag{40.8}$$

or equivalently

$$\mathbf{p}_1\cdot\mathbf{G}(\mathbf{r}_1,\mathbf{r}_2)\mathbf{p}_2 = \mathbf{p}_1\cdot\mathbf{G}^T(\mathbf{r}_2,\mathbf{r}_1)\mathbf{p}_2\,, \tag{40.9}$$

where the superscript T denotes the matrix transpose. Since this equality must be satisfied for any $\mathbf{p}_1$ and $\mathbf{p}_2$, the Green's function obeys the relation

$$\mathbf{G}(\mathbf{r}_1,\mathbf{r}_2) = \mathbf{G}^T(\mathbf{r}_2,\mathbf{r}_1)\,. \tag{40.10}$$

This result is known as the reciprocity theorem for the Green's function.

40.3 Conservation of Energy in a Scattering Problem

In this section, we review the conservation of energy for a scattering problem. We consider the scattering geometry in Fig. 39.1, in which a scattering medium with dielectric function $\varepsilon(\mathbf{r})$, confined in a volume V, is illuminated by an incident field generated by an external source with current density $\mathbf{j}_{\text{ext}}$. The incident magnetic field obeys the Maxwell equation

$$\nabla\times\mathbf{H}_i = \mathbf{j}_{ext} - i\omega\varepsilon_0\mathbf{E}_i\,, \tag{40.11}$$

while the total magnetic field satisfies

$$\nabla\times\mathbf{H} = \mathbf{j}_{ext} - i\omega\mathbf{P} - i\omega\varepsilon_0\mathbf{E}\,. \tag{40.12}$$

Here $\mathbf{P} = \varepsilon_0\delta\varepsilon\mathbf{E}$ is the induced polarization density in the scattering medium, with $\delta\varepsilon = \varepsilon - 1$. Subtracting the above equations, we obtain

$$\nabla\times\mathbf{H}_s = -i\omega\mathbf{P} - i\omega\varepsilon_0\mathbf{E}_s\,, \tag{40.13}$$

where $\mathbf{E}_s = \mathbf{E} - \mathbf{E}_i$ and $\mathbf{H}_s = \mathbf{H} - \mathbf{H}_i$ are the scattered electric and magnetic fields, respectively.

We now establish a form of Poynting's theorem that is well suited to scattering problems. Multiplying (40.13) by $\mathbf{E}_s^*$ yields

$$\mathbf{E}_s^*\cdot\nabla\times\mathbf{H}_s = -i\omega\mathbf{P}\cdot\mathbf{E}_s^* - i\omega\varepsilon_0|\mathbf{E}_s|^2\,. \tag{40.14}$$

The left-hand side can be rewritten using the vector identity $\nabla\cdot(\mathbf{A}\times\mathbf{B}) = \mathbf{B}\cdot\nabla\times\mathbf{A} - \mathbf{A}\cdot\nabla\times\mathbf{B}$, leading to

$$\mathbf{H}_s\cdot\nabla\times\mathbf{E}_s^* - \nabla\cdot(\mathbf{E}_s^*\times\mathbf{H}_s) = -i\omega\mathbf{P}\cdot\mathbf{E}_s^* - i\omega\varepsilon_0|\mathbf{E}_s|^2\,. \tag{40.15}$$

Making use of the Maxwell equation $\nabla \times \mathbf{E}_s = i\omega\mu_0 \mathbf{H}_s$, we find that

$$-i\omega\mu_0|\mathbf{H}_s|^2 - \nabla \cdot (\mathbf{E}_s^* \times \mathbf{H}_s) = -i\omega\mathbf{P} \cdot \mathbf{E}_s^* - i\omega\varepsilon_0|\mathbf{E}_s|^2 . \tag{40.16}$$

Upon taking the real part of both sides of the above, we obtain

$$\nabla \cdot \mathbf{\Pi}_s = -\frac{\omega}{2} \operatorname{Im}(\mathbf{P} \cdot \mathbf{E}_s^*) , \tag{40.17}$$

where $\mathbf{\Pi}_s = (1/2)\mathrm{Re}(\mathbf{E}_s^* \times \mathbf{H}_s)$ is the time-averaged Poynting vector of the scattered field. The right-hand side can be rewritten using $\mathbf{E} = \mathbf{E}_i + \mathbf{E}_s$. We then obtain the conservation law

$$\frac{\omega}{2} \operatorname{Im}(\mathbf{P} \cdot \mathbf{E}_i^*) = \frac{\omega}{2} \operatorname{Im}(\mathbf{P} \cdot \mathbf{E}^*) + \nabla \cdot \mathbf{\Pi}_s . \tag{40.18}$$

Integrating this equation over a surface S enclosing the scatterer, we obtain

$$P_e = P_a + P_s , \tag{40.19}$$

where

$$P_e = \frac{\omega}{2} \int_V \operatorname{Im}(\mathbf{P} \cdot \mathbf{E}_i^*) \, d^3r , \tag{40.20}$$

$$P_a = \frac{\omega}{2} \int_V \operatorname{Im}(\mathbf{P} \cdot \mathbf{E}^*) \, d^3r , \tag{40.21}$$

$$P_s = \int_S \mathbf{\Pi}_s \cdot \hat{\mathbf{n}} \, d^2r . \tag{40.22}$$

The extinguished power P_e is the power transferred from the incident field to the scatterer. The quantity P_a is the power absorbed by the scatterer, and P_s is the power carried out by the scattered field. Evidently, the extinguished power is either radiated to the far-field or absorbed in the scatterer.

Finally, we note that following the definition of the scattering and absorption cross-sections in Section 39.2, we can introduce the extinction cross section σ_e such that $\sigma_e I_i = P_e$, with I_i being the power per unit surface carried by the incident wave. From (40.19), we immediately see that $\sigma_e = \sigma_s + \sigma_a$.

40.4 Optical Theorem for Electromagnetic Waves

In order to derive the optical theorem, we start by writing the extinguished power in terms of the dielectric function of the scattering medium:

$$P_e = \frac{\omega\varepsilon_0}{2} \operatorname{Im} \int \delta\varepsilon(\mathbf{r})\mathbf{E}(\mathbf{r}) \cdot \mathbf{E}_i^*(\mathbf{r}) \, d^3r . \tag{40.23}$$

This expression is similar to (11.7) derived for scalar waves. Next, we consider the incident field to be a plane wave, with unit amplitude, of the form $\mathbf{E}_i(\mathbf{r}) =$

$\hat{\mathbf{e}}_0 \exp(i\mathbf{k}\cdot\mathbf{r})$. Here the unit vector $\hat{\mathbf{e}}_0$ defines the polarization of the incident field. From (40.23), we find that

$$P_e = \frac{\omega\varepsilon_0}{2}\mathrm{Im}\int \delta\varepsilon(\mathbf{r})\,\hat{\mathbf{e}}_0\cdot\mathbf{E}(\mathbf{r})\,\exp(-i\mathbf{k}\cdot\mathbf{r})\,d^3r\,. \tag{40.24}$$

In Section 39.2, we have shown that the scattering amplitude $\mathbf{A}(\mathbf{k},\mathbf{k}')$ observed in direction $\mathbf{k}' = k_0\hat{\mathbf{r}}$ is

$$\mathbf{A}(\mathbf{k},\mathbf{k}') = \frac{k_0^2}{4\pi}\left(\mathbf{I} - \hat{\mathbf{r}}\otimes\hat{\mathbf{r}}\right)\int \delta\varepsilon(\mathbf{r}')\,\mathbf{E}(\mathbf{r}')\exp(-ik_0\hat{\mathbf{r}}\cdot\mathbf{r}')\,d^3r'\,. \tag{40.25}$$

Taking $\mathbf{k}' = \mathbf{k}$, and projecting the scattering amplitude along $\hat{\mathbf{e}}_0$, we obtain

$$\hat{\mathbf{e}}_0\cdot\mathbf{A}(\mathbf{k},\mathbf{k}) = \frac{k_0^2}{4\pi}\int \delta\varepsilon(\mathbf{r}')\,\hat{\mathbf{e}}_0\cdot\mathbf{E}(\mathbf{r}')\exp(-ik_0\hat{\mathbf{r}}\cdot\mathbf{r}')\,d^3r'\,. \tag{40.26}$$

Note that we have removed the transverse projection operator since $\hat{\mathbf{e}}_0$ is necessarily perpendicular to $\mathbf{k}$. Upon comparing (40.24) and (40.26), we see that the extinguished power is

$$P_e = \frac{2\pi}{\mu_0\,\omega}\,\mathrm{Im}[\hat{\mathbf{e}}_0\cdot\mathbf{A}(\mathbf{k},\mathbf{k})]\,. \tag{40.27}$$

This is the optical theorem for vector waves. It shows that the extinguished power is given by the scattering amplitude in the forward direction, projected onto the polarization of the incident wave. The result can also be written in terms of the scattering matrix $\mathbf{S}$ introduced in Section 39.2, in the form

$$P_e = \frac{2\pi}{\mu_0\,\omega}\,\mathrm{Im}[\hat{\mathbf{e}}_0\cdot\mathbf{S}(\mathbf{k},\mathbf{k})\hat{\mathbf{e}}_0]\,, \tag{40.28}$$

which is independent of the amplitude of the incident field.

Noting that $I_i = (\varepsilon_0\,c/2)\,|E_0|^2$ for an incident plane wave, we also find that

$$\sigma_e = \frac{4\pi}{k_0}\,\mathrm{Im}[\hat{\mathbf{e}}_0\cdot\mathbf{S}(\mathbf{k},\mathbf{k})\hat{\mathbf{e}}_0]\,, \tag{40.29}$$

which is the optical theorem.

40.5 Integral Theorems

In this section, we derive two integral theorems that prove to be useful in scattering problems. We consider a medium with dielectric function $\varepsilon(\mathbf{r})$ that is generally frequency-dependent and complex-valued. The Green's function $\mathbf{G}$ obeys

$$\nabla\times\nabla\times\mathbf{G}(\mathbf{r},\mathbf{r}') - k_0^2\,\varepsilon(\mathbf{r})\,\mathbf{G}(\mathbf{r},\mathbf{r}') = \delta(\mathbf{r}-\mathbf{r}')\,\mathbf{I} \tag{40.30}$$

and satisfies the radiation condition at infinity. We note that if the volume in which $\varepsilon(\mathbf{r}) \neq 1$ is not finite, the outgoing Green's function is understood to be the retarded solution, meaning that $\mathbf{G}$ is the Fourier transform of the retarded Green's function in the time domain. To make further progress, we require the vector Green's identity

$$\int_V [\mathbf{A}(\mathbf{r}) \cdot \nabla \times \nabla \times \mathbf{B}(\mathbf{r}) - \mathbf{B}(\mathbf{r}) \cdot \nabla \times \nabla \times \mathbf{A}(\mathbf{r})]\, d^3r$$
$$= \int_S [\mathbf{B}(\mathbf{r}) \times \nabla \times \mathbf{A}(\mathbf{r}) - \mathbf{A}(\mathbf{r}) \times \nabla \times \mathbf{B}(\mathbf{r})] \cdot \hat{\mathbf{n}}\, d^2r\,, \tag{40.31}$$

where V is a volume bounded by a surface S and $\hat{\mathbf{n}}$ is the outward normal to S. Applying this result with $\mathbf{A}(\mathbf{r}) = \mathbf{G}(\mathbf{r}, \mathbf{r}_1)\mathbf{C}_1$ and $\mathbf{B}(\mathbf{r}) = \mathbf{G}^*(\mathbf{r}, \mathbf{r}_2)\mathbf{C}_2$, where $\mathbf{C}_1$ and $\mathbf{C}_2$ are constant vectors, and making use of (40.30), we obtain

$$k_0^2 \int_V \left[\varepsilon^*(\mathbf{r})\mathbf{G}(\mathbf{r}, \mathbf{r}_1)\mathbf{C}_1 \cdot \mathbf{G}^*(\mathbf{r}, \mathbf{r}_2)\mathbf{C}_2 - \varepsilon(\mathbf{r})\mathbf{G}^*(\mathbf{r}, \mathbf{r}_2)\mathbf{C}_2 \cdot \mathbf{G}(\mathbf{r}, \mathbf{r}_1)\mathbf{C}_1\right] d^3r$$
$$+ \mathbf{G}(\mathbf{r}_2, \mathbf{r}_1)\mathbf{C}_1 \cdot \mathbf{C}_2 - \mathbf{G}^*(\mathbf{r}_1, \mathbf{r}_2)\mathbf{C}_2 \cdot \mathbf{C}_1$$
$$= \int_S \left[\mathbf{G}^*(\mathbf{r}, \mathbf{r}_2)\mathbf{C}_2 \times \nabla \times \mathbf{G}(\mathbf{r}, \mathbf{r}_1)\mathbf{C}_1 - \mathbf{G}(\mathbf{r}, \mathbf{r}_1)\mathbf{C}_1 \times \nabla \times \mathbf{G}^*(\mathbf{r}, \mathbf{r}_2)\mathbf{C}_2\right] \cdot \hat{\mathbf{n}}\, d^2r\,. \tag{40.32}$$

For the case of a medium with uniform absorption, the Green's function vanishes at infinity and the surface integral does not contribute. Equation (40.32) thus becomes

$$\mathbf{C}_1 \cdot \mathrm{Im}\mathbf{G}(\mathbf{r}_1, \mathbf{r}_2)\mathbf{C}_2 = k_0^2\, \mathrm{Im}(\varepsilon) \int_V \mathbf{G}(\mathbf{r}, \mathbf{r}_1)\mathbf{C}_1 \cdot \mathbf{G}^*(\mathbf{r}, \mathbf{r}_2)\mathbf{C}_2\, d^3r\,, \tag{40.33}$$

which in tensor form can be rewritten as

$$\mathrm{Im}G_{ij}(\mathbf{r}_1, \mathbf{r}_2) = k_0^2\, \mathrm{Im}(\varepsilon) \int_V G_{ki}(\mathbf{r}, \mathbf{r}_1) G^*_{kj}(\mathbf{r}, \mathbf{r}_2)\, d^3r\,. \tag{40.34}$$

This identity will be employed for the calculation of field correlations of speckle patterns.

Another relation can be derived when the medium is non-absorbing, and contained in a finite volume. Given that the dielectric function is real, the first term in (40.32) vanishes. Using the reciprocity theorem (40.10), the second term on the left-hand side can be rewritten as $2i\, \mathbf{C}_1 \cdot \mathrm{Im}\mathbf{G}(\mathbf{r}_1, \mathbf{r}_2)\mathbf{C}_2$. The surface integral can also be simplified, by taking the surface S to be a sphere whose radius tends to infinity. Since the electric field is transverse in the far-field, we have $\mathbf{G}(\mathbf{r}, \mathbf{r}')\mathbf{C} \cdot \hat{\mathbf{n}} = 0$, for any vector $\mathbf{C}$ with $\mathbf{r} \in S$. Equation (40.32) thus becomes

$$\mathbf{C}_1 \cdot \mathrm{Im}\mathbf{G}(\mathbf{r}_1, \mathbf{r}_2)\mathbf{C}_2 = k_0 \int_S \mathbf{G}(\mathbf{r}, \mathbf{r}_1)\mathbf{C}_1 \cdot \mathbf{G}^*(\mathbf{r}, \mathbf{r}_2)\mathbf{C}_2\, d^2r\,. \tag{40.35}$$

Since the above relation must hold for any vectors $\mathbf{C}_1$ and $\mathbf{C}_2$, it can be rewritten as

$$\mathrm{Im}G_{ij}(\mathbf{r}_1, \mathbf{r}_2) = k_0 \int_S G_{ki}(\mathbf{r}, \mathbf{r}_1) G^*_{kj}(\mathbf{r}, \mathbf{r}_2)\, d^2r\,. \tag{40.36}$$

References and Additional Reading

These are the original papers on Lorentz's reciprocity relation:

H.A. Lorentz, Versl. Gewone Vergad. Afd. Natuurkd. K. Ned. Akad. Wet. **4**, 176188 (1896).

H.A. Lorentz, *Collected Papers* (Nijhoff, Den Haag, The Netherlands, 1936), Vol. 3.

A general discussion of reciprocity for electromagnetic waves can be found in:

L.D. Landau, E.M. Lifshitz, and L.P. Pitaevskii, *Electrodynamics of Continuous Media* (Pergamon, Oxford, 1984), section 89.

Reciprocity relations for electromagnetic waves in terms of the scattering matrix are derived in:

R. Carminati, M. Nieto-Vesperinas and J.-J. Greffet, J. Opt. Soc. Am. A **15**, 706 (1998).

Our presentation of the optical theorem was modeled after the derivation in:

D. Lytle, P.S. Carney, J. Schotland, and E. Wolf Phys. Rev. E **71**, 056610 (2005).

41

Electromagnetic Scattering by Subwavelength Particles

In this chapter, we address the problem of scattering of an electromagnetic wave by a particle much smaller than the incident wavelength. We treat the particle in the electric-dipole limit and discuss the regimes of Rayleigh and resonant scattering. We also address the regime of near-field scattering, in which the source or the observation point is located subwavelength distance from the scatterer.

41.1 Polarizability

A scatterer that is much smaller in size compared to the wavelength can be described in the electric dipole approximation. In this approach, the response of the scatterer is written in terms of the electric polarizability. For didactic reasons, we assume the scatterer to be a homogeneous sphere, with radius a and dielectric constant ε that is centered at the position $\mathbf{r}_0$. The incident field is monochromatic with complex amplitude $\mathbf{E}_0$ and frequency ω. The total field obeys the Lippmann–Schwinger equation (see 39.5):

$$\mathbf{E}(\mathbf{r}) = \mathbf{E}_0(\mathbf{r}) + k_0^2 \delta\varepsilon \int_V \mathbf{G}_0(\mathbf{r}, \mathbf{r}')\, \mathbf{E}(\mathbf{r}')\, d^3r' \,, \tag{41.1}$$

with V the volume of the scatterer. Assuming $a \ll \lambda$, we can consider the field inside the particle to be uniform. To determine this field, we apply (41.1) at $\mathbf{r} = \mathbf{r}_0$, in the limit $a \to 0$. Here we must account for the singular behavior of the Green's function when $\mathbf{r}' = \mathbf{r}_0$. The real part of the Green's function has a non-integrable singularity (see Chapter 37), and we can write

$$\begin{aligned} \mathbf{E}(\mathbf{r}_0) = \mathbf{E}_0(\mathbf{r}_0) &+ k_0^2\, \delta\varepsilon \int_V \mathrm{Re}[\mathbf{G}_0(\mathbf{r}_0, \mathbf{r}')]\, \mathbf{E}(\mathbf{r}')\, d^3r' \\ &+ i k_0^2\, \delta\varepsilon\, \mathrm{Im}[\mathbf{G}_0(\mathbf{r}_0, \mathbf{r}_0)]\, \mathbf{E}(\mathbf{r}_0)\, V \,. \end{aligned} \tag{41.2}$$

Using (37.18), we find that

$$\int_V \mathrm{Re}[\mathbf{G}_0(\mathbf{r}_0,\mathbf{r}')]\,\mathbf{E}(\mathbf{r}')\,d^3r' \simeq -\frac{\mathbf{E}(\mathbf{r}_0)}{3k_0^2} + \frac{a^2}{3}\mathbf{E}(\mathbf{r}_0)\,, \tag{41.3}$$

when $a \to 0$. The first term results from the delta function in (37.18), while the second term is the lowest-order contribution of the principal value. The imaginary part is non-singular and is simply [see (38.7)]

$$\mathrm{Im}[\mathbf{G}_0(\mathbf{r}_0,\mathbf{r}_0)] = \frac{k_0}{6\pi}\mathbf{I}. \tag{41.4}$$

Inserting these two results into (41.2), we obtain

$$\begin{aligned}\mathbf{E}(\mathbf{r}_0) = {} & \frac{3}{\varepsilon + 2 - (k_0 a)^2[\varepsilon - 1]} \\ & \times \left[1 - i\,\frac{k_0^3}{6\pi}\frac{4\pi a^3[\varepsilon - 1]}{\varepsilon + 2 - (k_0 a)^2[\varepsilon - 1]}\right]^{-1} \mathbf{E}_0(\mathbf{r}_0)\,,\end{aligned} \tag{41.5}$$

which relates the field inside the scatterer to the external field. Note that in the static limit $k_0 \to 0$, we have $\mathbf{E}(\mathbf{r}_0) = 3\mathbf{E}_0(\mathbf{r}_0)/(\varepsilon + 2)$, which is the relationship between the field inside a homogeneous sphere and the applied external field known in electrostatics. Using the above, we can compute the induced electric dipole moment:

$$\begin{aligned}\mathbf{p} &= \int_V \varepsilon_0 \delta\varepsilon \mathbf{E}(\mathbf{r})\,d^3r \\ &\simeq \varepsilon_0\delta\varepsilon\,\mathbf{E}(\mathbf{r}_0)\,\delta V \\ &= \varepsilon_0\,\alpha_0(\omega)\left[1 - i\frac{k_0^3}{6\pi}\alpha_0(\omega)\right]^{-1} \mathbf{E}_0(\mathbf{r}_0).\end{aligned} \tag{41.6}$$

In the last line, we made use of (41.5) and introduced

$$\alpha_0(\omega) = 4\pi a^3 \frac{\varepsilon(\omega) - 1}{\varepsilon(\omega) + 2 - (k_0 a)^2[\varepsilon(\omega) - 1]}\,. \tag{41.7}$$

We also made explicit the dependence on ω for reasons that will become clear below. By definition of the dynamic polarizability $\alpha(\omega)$, we also have $\mathbf{p} = \alpha(\omega)\,\varepsilon_0\,\mathbf{E}_0(\mathbf{r}_0)$. From (41.6), we directly obtain

$$\alpha(\omega) = \frac{\alpha_0(\omega)}{1 - i\dfrac{k_0^3}{6\pi}\alpha_0(\omega)}\,. \tag{41.8}$$

The denominator in (41.8) accounts for radiation by the induced dipole and is often referred to as the radiative correction. This radiative correction ensures that energy conservation is satisfied, as discussed in the next section. By neglecting the term proportional to $(k_0 a)^2$ in (41.7), we get $\alpha_0 = \alpha_{qs}$, with α_{qs} the quasi-static polarizability defined as

$$\alpha_{qs}(\omega) = 4\pi a^3 \frac{\varepsilon(\omega) - 1}{\varepsilon(\omega) + 2} . \tag{41.9}$$

The quasi-static polarizability is similar to the electrostatic polarizability, the essential difference being the frequency dependence of the dielectric constant $\varepsilon(\omega)$ that is here taken at optical frequencies. Using α_{qs} instead of α_0 in (41.8) reduces the accuracy of the expression of the dynamic polarizability but leads to an expression that remains consistent with energy conservation.

41.2 Energy Conservation

For a non-absorbing scatterer, the extinguished power P_e and the scattered power P_s must be equal. Using (40.20) and (40.22), and noting that the polarization density here is $\mathbf{P}(\mathbf{r}) = \mathbf{p}\delta(\mathbf{r} - \mathbf{r}_0) = \alpha\varepsilon_0\mathbf{E}_0\delta(\mathbf{r} - \mathbf{r}_0)$, we find that

$$P_e = \frac{\varepsilon_0 \omega}{2} \operatorname{Im} \alpha(\omega) \, |E_0|^2 \tag{41.10}$$

and

$$P_s = \frac{\varepsilon_0 \omega^4}{12\pi c^3} |\alpha(\omega)|^2 \, |E_0|^2 . \tag{41.11}$$

In the absence of absorption, we see that the dynamic polarizability must satisfy

$$\operatorname{Im} \alpha(\omega) = \frac{k_0^3}{6\pi} |\alpha(\omega)|^2 . \tag{41.12}$$

This relation can be rewritten as

$$\operatorname{Im} \left[\frac{1}{\alpha^*(\omega)} - i \frac{k_0^3}{6\pi} \right] = 0 , \tag{41.13}$$

showing that the term in brackets must be real. Evidently,

$$\frac{1}{\alpha(\omega)} = \frac{1}{\alpha_r(\omega)} - i \frac{k_0^3}{6\pi} , \tag{41.14}$$

where $\alpha_r(\omega)$ is real. Therefore, for a non-absorbing scatterer, the dynamic polarizability must necessarily be of the form

$$\alpha(\omega) = \frac{\alpha_r(\omega)}{1 - i \frac{k_0^3}{6\pi} \alpha_r(\omega)} , \tag{41.15}$$

where $\alpha_r(\omega)$ is a real quantity. Equation (41.8) derived in the previous section satisfies this condition and is consistent with energy conservation.

41.3 Rayleigh and Resonant Scattering

By definition of the scattering cross section σ_s, the scattered power is written as $P_s = \sigma_s I_i$, with $I_i = (\varepsilon_0 c/2)|\mathbf{E}_0|^2$ the flux per unit surface carried by the incident wave. Using (41.11), we immediately find that

$$\sigma_s(\omega) = \frac{k_0^4}{6\pi}|\alpha(\omega)|^2. \tag{41.16}$$

There are two features of the scattering cross section of a small particle that deserve to be mentioned. In a frequency range far from internal resonances of the particle, $\sigma_s(\omega) \sim \omega^4$. This regime is known as Rayleigh scattering. It is also interesting to note that since $\alpha_0(\omega) \sim a^3$, the scattering cross section $\sigma_s(\omega) \sim a^6$ when $a \to 0$.

In a similar way, the extinction cross section σ_e is defined by the relation $P_e = \sigma_e I_i$. Using (41.10), we readily obtain

$$\sigma_e(\omega) = k_0 \, \mathrm{Im}[\alpha(\omega)]. \tag{41.17}$$

The extinction cross section is proportional to the imaginary part of the polarizability. Here the importance of the radiative correction becomes apparent. For a non-absorbing particle, the dielectric constant $\varepsilon(\omega)$ is real, so that $\alpha_0(\omega)$ defined in (41.7) is also real. The imaginary part of the polarizability is determined only by the radiative correction in this case. By making the approximation $\alpha(\omega) \simeq \alpha_0(\omega)$, or $\alpha(\omega) \simeq \alpha_{qs}(\omega)$, we neglect extinction by scattering.

The absorption cross-section σ_a is directly obtained by subtraction, since $\sigma_e = \sigma_s + \sigma_a$. This leads to:

$$\sigma_a(\omega) = k_0 \left[\mathrm{Im}[\alpha(\omega)] - \frac{k_0^3}{6\pi}|\alpha(\omega)|^2 \right]. \tag{41.18}$$

It is interesting to note that to leading order when $a \to 0$, the absorption cross-section scales as $\sigma_a(\omega) \sim a^3$.

Finally, we examine the behavior of the polarizability on resonance. A resonance occurs when $\mathrm{Re}[\alpha(\omega)^{-1}] = 0$, which can be seen as a general resonance condition. For a non-absorbing dipole scatterer, since (41.15) must hold, we deduce that the polarizability on resonance is purely imaginary and is given by

$$\alpha(\omega_0) = \frac{i6\pi c^3}{\omega_0^3}, \tag{41.19}$$

where ω_0 is the resonance frequency. As a result, the on-resonance scattering cross section is

$$\sigma_s(\omega_0) = \frac{6\pi c^2}{\omega_0^2} = \frac{3\lambda_0^2}{2\pi}, \tag{41.20}$$

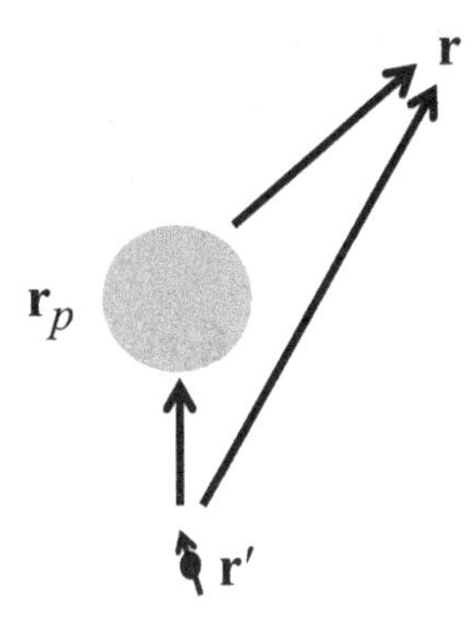

Fig. 41.1 Schematic diagram showing the two contributions involved in the calculation of the Green's function $\mathbf{G}(\mathbf{r}, \mathbf{r}')$ in the presence of a subwavelength scatterer centered at point $\mathbf{r}_p$.

with $\lambda_0 = 2\pi c/\omega_0$ as the resonance wavelength. We see that on resonance, a non-absorbing electric-dipole scatterer has a scattering cross section on the order of λ_0^2. It is interesting to note that the same result is obtained for resonant scattering by a two-level atom in quantum optics.

41.4 Near-Field Scattering

Next, we consider the problem of near-field scattering,[1] in the geometry represented in Fig. 41.1. A subwavelength scatterer, described by its polarizability $\alpha(\omega)$, is located at the position $\mathbf{r}_p$ and is illuminated by a point electric-dipole source at the point $\mathbf{r}'$. The source is assumed to have constant amplitude $\mathbf{p}$. The Green's function must account for the two contributions represented in Fig. 41.1:

$$\mathbf{G}(\mathbf{r}, \mathbf{r}') = \mathbf{G}_0(\mathbf{r}, \mathbf{r}') + \mathbf{G}_0(\mathbf{r}, \mathbf{r}_p)\alpha(\omega)\varepsilon_0\, \mathbf{G}_0(\mathbf{r}_p, \mathbf{r}') \,. \tag{41.21}$$

The first term on the right-hand side accounts for direct radiation from the source to the observation point. The second term describes radiation from the source to the scatterer, followed by radiation from the induced dipole to the observation point.

The power P transferred from the source to the environment can be calculated using the formalism introduced in Chapter 38 [see (38.27) and (38.28)]. It can be written as

$$P = \frac{\pi\,\omega^2}{4\varepsilon_0}|\mathbf{p}|^2\,\rho_{\hat{\mathbf{p}}}(\mathbf{r}', \omega)\,, \tag{41.22}$$

where

$$\rho_{\hat{\mathbf{p}}}(\mathbf{r}', \omega) = \frac{2\omega}{\pi\,c^2}\,\mathrm{Im}\left[\hat{\mathbf{p}} \cdot \mathbf{G}(\mathbf{r}', \mathbf{r}')\hat{\mathbf{p}}\right] \tag{41.23}$$

[1] We follow the approach presented in R. Carminati, J.-J. Greffet, C. Henkel and J.M. Vigoureux, Opt. Commun. **261**, 368 (2006).

is the projected local density of states (LDOS).

The total emitted power P can be split into radiative and non-radiative contributions. The radiative contribution P^R corresponds to the power radiated to the far-field and is given by

$$P^R = \int \frac{\varepsilon_0 c}{2} |\mathbf{E}(\mathbf{r})|^2 r^2 \, d\hat{\mathbf{r}} , \tag{41.24}$$

according to (37.25). The non-radiative contribution P^{NR} corresponds to absorption by the particle and is given by

$$P^{NR} = \int_V \frac{\omega \varepsilon_0}{2} \, \mathrm{Im}(\varepsilon) \, |\mathbf{E}(\mathbf{r})|^2 \, d^3 r . \tag{41.25}$$

Energy conservation requires that $P = P^R + P^{NR}$.

Since the power is proportional to the LDOS, it is also possible to define radiative and non-radiative contributions to the projected LDOS, and write

$$\rho_{\hat{\mathbf{p}}}(\mathbf{r}, \omega) = \rho^R_{\hat{\mathbf{p}}}(\mathbf{r}, \omega) + \rho^{NR}_{\hat{\mathbf{p}}}(\mathbf{r}, \omega) . \tag{41.26}$$

As discussed in Chapter 38, the projected LDOS also controls the spontaneous decay rate of a quantum emitter. In the following, we will compute changes in the radiative and non-radiative LDOS due to the presence of the scatterer in the near-field. The results can be understood equivalently in terms of scattering of the light emitted by a classical electric dipole source, or in terms of changes in the spontaneous decay rate of a quantum emitter.

41.5 Near-Field Local Density of States

From (41.21), (37.17) and (41.23), the projected LDOS $\rho_{\hat{\mathbf{p}}}$ can be calculated in the vicinity of the particle. Knowledge of the Green's function for all points $\mathbf{r}$ and $\mathbf{r}'$ permits a calculation of the electric field $\mathbf{E}(\mathbf{r})$ everywhere, so that the radiative and non-radiative contributions to the projected LDOS can also be calculated using (41.24)–(41.26). For a source dipole oriented along the z-direction pointing toward the particle (along the direction connecting $\mathbf{r}'$ to $\mathbf{r}_p$ in Fig. 41.1), we obtain

$$\frac{\rho_z}{\rho_{0,\hat{\mathbf{p}}}} = 1 + \frac{3k_0}{2\pi} \mathrm{Im}\left[\alpha(\omega) \exp(2ik_0 z) \left(\frac{-1}{k_0^2 z^4} + \frac{2}{i k_0^3 z^5} + \frac{1}{k_0^4 z^6} \right)\right] , \tag{41.27}$$

where $\rho_{0,\hat{\mathbf{p}}}$ is the projected LDOS in free-space given by (38.29). For a source dipole oriented along the x-direction (perpendicular to the direction connecting $\mathbf{r}'$ to $\mathbf{r}_p$ in Fig. 41.1), we obtain

$$\frac{\rho_x}{\rho_{0,\hat{\mathbf{p}}}} = 1 + \frac{3k_0}{8\pi} \mathrm{Im}\left[\alpha(\omega) \exp(2ik_0 z) \left(\frac{1}{z^2} - \frac{2}{i k_0 z^3} - \frac{3}{k_0^2 z^4} + \frac{2}{i k_0^3 z^5} + \frac{1}{k_0^4 z^6} \right)\right] . \tag{41.28}$$

In the limit $k_0 z \ll 1$, the above expressions can be simplified. Expanding the exponential term $\exp(2ik_0 z)$ to third order in $(k_0 z)^{-1}$, we obtain expressions valid to order $(k_0 z)^{-3}$. The normalized projected LDOS along the z and x directions are

$$\frac{\rho_z}{\rho_{0,\hat{\mathbf{p}}}} = 1 + \frac{3k_0^3}{2\pi}\mathrm{Im}\left[\alpha(\omega)\left(\frac{1}{(k_0 z)^6} + \frac{1}{(k_0 z)^4} + \frac{2i}{3(k_0 z)^3}\right)\right], \tag{41.29}$$

$$\frac{\rho_x}{\rho_{0,\hat{\mathbf{p}}}} = 1 + \frac{3k_0^3}{8\pi}\mathrm{Im}\left[\alpha(\omega)\left(\frac{1}{(k_0 z)^6} - \frac{1}{(k_0 z)^4} - \frac{4i}{3(k_0 z)^3}\right)\right]. \tag{41.30}$$

In the same limit, the radiative component of the normalized projected LDOS can also be calculated:

$$\frac{\rho_z^R}{\rho_{0,\hat{\mathbf{p}}}} = 1 + \frac{k_0^6}{4\pi^2}|\alpha(\omega)|^2\left[\frac{1}{(k_0 z)^6} + \frac{1}{(k_0 z)^4}\right] + \frac{k_0^3}{\pi}\mathrm{Re}[\alpha(\omega)]\frac{1}{(k_0 z)^3}, \tag{41.31}$$

$$\frac{\rho_x^R}{\rho_{0,\hat{\mathbf{p}}}} = 1 + \frac{k_0^6}{16\pi^2}|\alpha(\omega)|^2\left[\frac{1}{(k_0 z)^6} - \frac{1}{(k_0 z)^4}\right] - \frac{k_0^3}{2\pi}\mathrm{Re}[\alpha(\omega)]\frac{1}{(k_0 z)^3}. \tag{41.32}$$

The non-radiative component of the normalized projected LDOS can be deduced by subtraction since $\rho_{\hat{\mathbf{p}}}^{NR}(\mathbf{r},\omega) = \rho_{\hat{\mathbf{p}}}(\mathbf{r},\omega) - \rho_{\hat{\mathbf{p}}}^{R}(\mathbf{r},\omega)$, and we obtain

$$\frac{\rho_z^{NR}}{\rho_{0,\hat{\mathbf{p}}}} = \frac{3k_0^3}{2\pi}\left[\mathrm{Im}[\alpha(\omega)] - \frac{k_0^3}{6\pi}|\alpha(\omega)|^2\right]\left[\frac{1}{(k_0 z)^6} + \frac{1}{(k_0 z)^4}\right], \tag{41.33}$$

$$\frac{\rho_x^{NR}}{\rho_{0,\hat{\mathbf{p}}}} = \frac{3k_0^3}{8\pi}\left[\mathrm{Im}[\alpha(\omega)] - \frac{k_0^3}{6\pi}|\alpha(\omega)|^2\right]\left[\frac{1}{(k_0 z)^6} - \frac{1}{(k_0 z)^4} + \frac{1}{(k_0 z)^2}\right]. \tag{41.34}$$

41.6 Discussion

Equations (41.33) and (41.34) show that at short distances ($k_0 z \ll 1$), the leading term in the non-radiative projected LDOS scales as z^{-6}. This is the usual dependence found in non-radiative energy transfer due to dipole-dipole interaction in Förster's theory.

In the expressions for the radiative rates (41.31) and (41.32), a term proportional to z^{-3} survives. A term proportional to z^{-6} still contributes, together with a z^{-4} term which ensures a crossover between the two regimes. The relative weight of the z^{-6} and z^{-3} terms can be evaluated. Their ratio is $|\alpha|^2/(4\pi\mathrm{Re}(\alpha)\,z^3) \simeq (a/z)^3\,(\mathrm{Re}(\beta)^2 + \mathrm{Im}(\beta)^2)/\mathrm{Re}(\beta)$, where $\beta = [\varepsilon(\omega) - 1]/[\varepsilon(\omega) + 2]$ and a is the radius of the particle. This simple expression shows qualitatively that for a very small particle, the ratio should be very small so that the z^{-3} term dominates. Note that the situation can be different at plasmon resonance (i.e., at a frequency such that $|\beta| \gg 1$). In this case, the ratio may become larger than unity so that the z^{-6} contribution may still contribute.

The physical origin of the z^{-3} term in the expression of the radiative projected LDOS can be understood as follows. The power radiated by the source-particle system is proportional to $|\mathbf{E}_1 + \mathbf{E}_2|^2$, where $\mathbf{E}_1$ is the far-field radiated by the source (proportional to the dipole moment $\mathbf{p}$) and $\mathbf{E}_2$ is the far-field radiated by the induced dipole in the nanoparticle (proportional to the polarisability and to z^{-3}). The radiative rate thus contains three contributions: the power directly radiated by

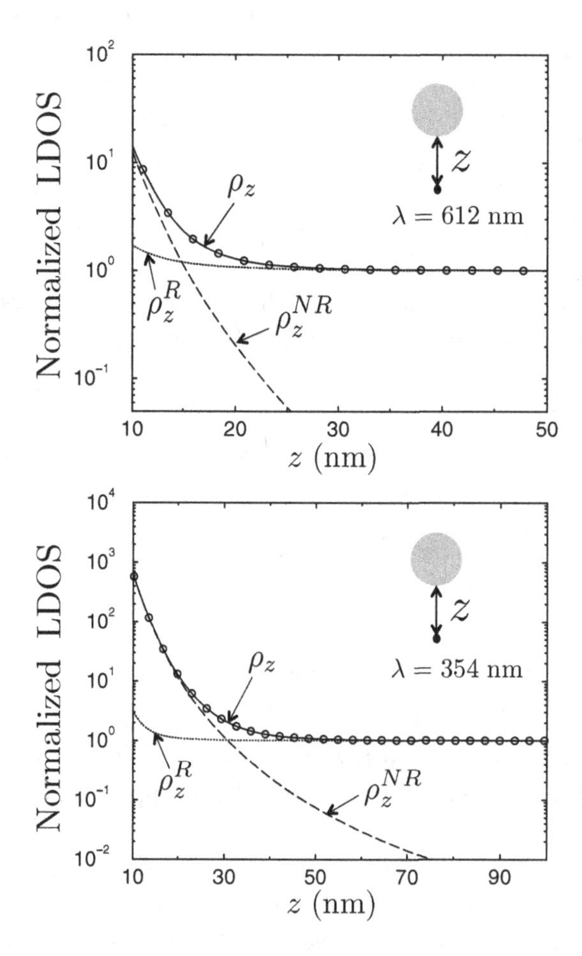

Fig. 41.2 Normalized projected LDOS $\rho_z/\rho_{0,\hat{\mathbf{p}}}$, $\rho_z^R/\rho_{0,\hat{\mathbf{p}}}$ and $\rho_z^{NR}/\rho_{0,\hat{\mathbf{p}}}$ versus the distance z to a silver nanoparticle with radius $a = 5$ nm. The z direction is pointing towards the nanoparticle. (a) Wavelength $\lambda = 612$ nm (off-resonance). Dielectric function of silver $\varepsilon(612nm) = -15.04 + 1.02i$. (b) Wavelength $\lambda = 354$ nm (plasmon resonance). Dielectric function of silver $\varepsilon(354nm) = -2.03 + 0.6i$. Adapted from R. Carminati et al., Opt. Comm. **261**, 368 (2006).

the emitter ($\propto |\mathbf{E}_1|^2$), the power radiated by the induced dipole in the nanoparticle ($\propto |\mathbf{E}_2|^2 \propto z^{-6}$) and an interference term ($\propto \mathbf{E}_1 \cdot \mathbf{E}_2^* \propto z^{-3}$). The existence of the z^{-3} term in the radiative LDOS is a direct consequence of the interference between the two radiative channels shown in Fig. 41.1.

The distance dependence of the projected LDOS, and of its radiative and non-radiative components, in the vicinity of a silver particle is plotted in Fig. 41.2. The particle has a diameter $d = 10$ nm, and the emission wavelength is $\lambda = 612$ nm in Fig. 41.2(a) (off-resonance) and $\lambda = 354$ nm in Fig. 41.2(b) (plasmon resonance).

In Fig. 41.2(a), the projected LDOS $\rho_{\hat{\mathbf{p}}}$ increases at short distance from the particle. From the distance dependence of the radiative ($\rho_{\hat{\mathbf{p}}}^{R}$) and non-radiative ($\rho_{\hat{\mathbf{p}}}^{NR}$) LDOS, it is also clear that the increase at short distance is dominated by non-radiative coupling. The trends remain identical in Fig. 41.2(b) where the plasmon resonance of the particle is excited, with enhanced modifications of the LDOS.

References and Additional Reading

Light scattering by subwavelength particles is addressed in the following textbooks:

H.C. van de Hulst, *Light Scattering by Small Particles* (Dover, New York, 1981), chap. 6.
C.F. Bohren and D.R. Huffman, *Absorption and Scattering of Light by Small Particles* (Wiley, New York, 1998), chap. 5.

These papers introduced electromagnetic scattering by small particles based on the coupled-dipole method:

E.M. Purcell and C.R. Pennypacker, Astrophys. J. **186**, 705 (1973).
B.T. Draine, Astrophys. J. **333**, 848 (1988).
A. Laktakhia, Opt. Commun. **79**, 1 (1990).

The following review articles address the problem of single and multiple scattering from small particles, including resonant scattering:

A. Lagendijk and B.A. van Tiggelen, Phys. Rep. **270**, 143 (1996).
P. de Vries, D.V. van Coevorden and A. Lagendijk, Rev. Mod. Phys. **70**, 447 (1998).
M.C.W. van Rossum and Th. M. Nieuwenhuizen, Rev. Mod. Phys. **71**, 313 (1999).

This paper derives the general form of the dynamic polarizability, including the case of anisotropic particles, based on energy conservation:

S. Albaladejo R. Gómez-Medina, L.S. Froufe-Pérez, H. Marinchio, R. Carminati, J.F. Torrado, G. Armelles, A. García-Martín and J.J. Sáenz, Opt. Express. **18**, 3556 (2010).

The presentation of near-field scattering was adapted from the following paper:

R. Carminati, J.-J. Greffet, C. Henkel and J.M. Vigoureux, Opt. Commun. **261**, 368 (2006).

Near-field scattering from subwavelength objects, and the interaction of an electric-dipole emitter with a small particle, are basic problems in nano-optics. See for example:

V.M. Shalaev, *Non Linear Optics of Random Media* (Springer, Berlin, 2000).
L. Novotny and B. Hecht, *Principles of Nano-Optics* (Cambridge University Press, Cambridge, 2006), chap. 8, 9, 10.
O. Keller, *Quantum Theory of Near-Field Electrodynamics* (Springer, Berlin, 2011).

The theory of non-radiative energy transfer between two dipoles was historically discussed in:

T. Förster, Ann. Physik **6**, 55 (1948).

42

Multiple Scattering of Electromagnetic Waves: Average Field

In this chapter, we study the propagation of electromagnetic waves in a disordered medium and consider the problem of calculating the average field. The calculation makes use of the diagrammatic perturbation theory introduced for scalar waves in Chapter 15.

42.1 Model of Disorder

We describe the disordered medium by a real dielectric function $\varepsilon(\mathbf{r}) = 1 + \delta\varepsilon(\mathbf{r})$, where $\delta\varepsilon(\mathbf{r})$ is a Gaussian random field with correlations

$$\langle \delta\varepsilon(\mathbf{r}) \rangle = 0 \,,$$
$$\langle \delta\varepsilon(\mathbf{r})\, \delta\varepsilon(\mathbf{r}') \rangle = \frac{A}{\pi^{3/2}\, \ell_c^3} \exp(-|\mathbf{r} - \mathbf{r}'|^2/\ell_c^2) \,. \tag{42.1}$$

Here $\langle \cdots \rangle$ denotes statistical averaging over the disorder. The correlation function is characterized by a correlation length ℓ_c. As in Chapter 15, the prefactor is chosen so that $\langle \delta\varepsilon(\mathbf{r})\, \delta\varepsilon(\mathbf{r}') \rangle = A\, \delta(\mathbf{r} - \mathbf{r}')$ in the limit $\ell_c \to 0$, which corresponds to the uncorrelated white-noise model. In the weak-scattering limit, we will show below that $A = 6\pi/(k_0^4\, \ell_s)$, where $k_0 = \omega/c = 2\pi/\lambda$ is the wavenumber in vacuum and ℓ_s is the scattering mean free path. Throughout this chapter, we will assume that the inequalities

$$\ell_c \ll \lambda \ll \ell_s \tag{42.2}$$

are satisfied.

42.2 Average Green's Function

The expression for the free-space Green's function $\mathbf{G}_0(\mathbf{r}, \mathbf{r}')$ is derived in Chapter 37. There we define $\mathbf{G}_0(\mathbf{k})$ as the Fourier integral

$$\mathbf{G}_0(\mathbf{r}, \mathbf{r}') = \int \mathbf{G}_0(\mathbf{k}) \exp[i\mathbf{k} \cdot (\mathbf{r} - \mathbf{r}')] \frac{d^3k}{(2\pi)^3} . \tag{42.3}$$

In Section 37.4, we have shown that

$$\mathbf{G}_0(\mathbf{k}) = [(k^2 - k_0^2)\mathbf{I} - \mathbf{k} \otimes \mathbf{k}]^{-1} , \tag{42.4}$$

where $\mathbf{I}$ is the unit tensor.

The equation satisfied by the average Green's function $\langle \mathbf{G}(\mathbf{r}, \mathbf{r}') \rangle$ is derived following the same diagrammatic approach as in Chapter 15, simply replacing the scalar Green's function $G_0(\mathbf{r}, \mathbf{r}')$ by $\mathbf{G}_0(\mathbf{r}, \mathbf{r}')$. Defining the tensor self-energy $\boldsymbol{\Sigma}(\mathbf{r}, \mathbf{r}')$ by the diagrammatic expansion in (15.2), with horizontal lines representing $\mathbf{G}_0(\mathbf{r}, \mathbf{r}')$, we obtain the Dyson equation

$$\langle \mathbf{G}(\mathbf{r}, \mathbf{r}') \rangle = \mathbf{G}_0(\mathbf{r}, \mathbf{r}') + \int \mathbf{G}_0(\mathbf{r}, \mathbf{r}_1) \boldsymbol{\Sigma}(\mathbf{r}_1, \mathbf{r}_2) \langle \mathbf{G}(\mathbf{r}_2, \mathbf{r}') \rangle d^3r_1 d^3r_2 . \tag{42.5}$$

In Fourier space, the Dyson equation becomes

$$\langle \mathbf{G}(\mathbf{k}) \rangle = \mathbf{G}_0(\mathbf{k}) + \mathbf{G}_0(\mathbf{k})\, \boldsymbol{\Sigma}(\mathbf{k})\, \langle \mathbf{G}(\mathbf{k}) \rangle , \tag{42.6}$$

where $\langle \mathbf{G}(\mathbf{k}) \rangle$ and $\boldsymbol{\Sigma}(\mathbf{k})$ are defined with Fourier integrals similar to (42.3). Solving for $\langle \mathbf{G}(\mathbf{k}) \rangle$, and making use of (42.4), we find that

$$\langle \mathbf{G}(\mathbf{k}) \rangle = [(k^2 - k_0^2)\mathbf{I} - \mathbf{k} \otimes \mathbf{k} - \boldsymbol{\Sigma}(\mathbf{k})]^{-1} , \tag{42.7}$$

where $k = |\mathbf{k}|$. We note that the above expression for the average Green's function holds for any model of disorder, provided that the random medium is statistically homogeneous.

42.3 Self-Energy and the Effective Medium

In a statistically homogeneous and isotropic medium, it is useful to separate the average Green's function and the self-energy into their transverse and longitudinal components. To proceed, we make use of the transverse and longitudinal projection operators $\mathbf{P}_\perp(\mathbf{k}) = \mathbf{I} - \hat{\mathbf{k}} \otimes \hat{\mathbf{k}}$ and $\mathbf{P}_\parallel(\mathbf{k}) = \hat{\mathbf{k}} \otimes \hat{\mathbf{k}}$ introduced in Section 37.5 and write

$$\boldsymbol{\Sigma}(\mathbf{k}) = \Sigma_\perp(\mathbf{k})\mathbf{P}_\perp(\mathbf{k}) + \Sigma_\parallel(\mathbf{k})\mathbf{P}_\parallel(\mathbf{k}) . \tag{42.8}$$

Note that both $\Sigma_\perp$ and $\Sigma_\parallel$ are scalar quantities as a consequence of statistical isotropy. From (42.7) and (42.8), we see that

$$\langle \mathbf{G}(\mathbf{k})\rangle = \frac{1}{k^2 - k_0^2 - \Sigma_\perp(k)}\mathbf{P}_\perp(\mathbf{k}) - \frac{1}{k_0^2 + \Sigma_\parallel(k)}\mathbf{P}_\parallel(\mathbf{k})\,, \tag{42.9}$$

where the first and second terms on the right-hand side define the transverse and longitudinal average Green's functions $\langle \mathbf{G}_\perp(\mathbf{k})\rangle$ and $\langle \mathbf{G}_\parallel(\mathbf{k})\rangle$. The transverse Green's function describes excitations with a dispersion relation $k^2 = k_0^2 + \Sigma_\perp(k)$. As discussed in Section 37.5 in the case of free-space, the longitudinal Green's function describes near-field excitations, governed by the dispersion relation $k_0^2 + \Sigma_\parallel(k) = 0$.

The $\mathbf{k}$-dependence of the self-energy $\boldsymbol{\Sigma}(\mathbf{k})$ causes the propagation of the average field to be non-local. In a disordered medium with short-range correlations ($\ell_c \ll \lambda$), non-locality can be neglected (as will become clear in the calculation below). Moreover, assuming that $|\Sigma(k)| \ll k_0^2$, $\langle \mathbf{G}(k)\rangle$ is peaked around k_0, we can evaluate $\Sigma(k)$ on shell (i.e., for $k = k_0$). As a result, we have $\boldsymbol{\Sigma}(\mathbf{k}) = \Sigma(k_0)\mathbf{I}$, with $\Sigma(k_0) = \Sigma_\perp(k_0) = \Sigma_\parallel(k_0)$. Defining the effective wavenumber k_{eff} by the relation

$$k_{\text{eff}}^2 = k_0^2 + \Sigma(k_0)\,, \tag{42.10}$$

we can rewrite the average Green's function as

$$\langle \mathbf{G}(\mathbf{k})\rangle = [(k^2 - k_{\text{eff}}^2)\mathbf{I} - \mathbf{k}\otimes\mathbf{k}]^{-1}\,. \tag{42.11}$$

We see that the average Green's function takes the same form as the free-space Green's function, with the effective wavenumber k_{eff} replacing the vacuum wavenumber k_0. The average Green's function propagates the average field in a homogeneous effective medium. It is sometimes convenient to use the effective dielectric function defined as

$$\varepsilon_{\text{eff}} = \frac{k_{\text{eff}}^2}{k_0^2}\,, \tag{42.12}$$

which describes the effective medium seen by the average Green's function. From (37.18), we can deduce the expression for the average Green's function in real space:

$$\langle \mathbf{G}(\mathbf{r},\mathbf{r}')\rangle = \mathrm{P}\left(\mathbf{I} + \frac{1}{k_{\text{eff}}^2}\nabla\otimes\nabla\right)\frac{\exp[i\mathrm{Re}(k_{\text{eff}})R]}{4\pi R}\exp[-\mathrm{Im}(k_{\text{eff}})R] - \frac{\mathbf{I}}{3k_{\text{eff}}^2}\delta(R)\,, \tag{42.13}$$

where $R = |\mathbf{r} - \mathbf{r}'|$. In view of the exponential decay of the average Green's function, we define the scattering mean free path

$$\ell_s = \frac{1}{2\,\mathrm{Im}(k_{\text{eff}})}\,. \tag{42.14}$$

In order to evaluate the self-energy, we assume that the medium is weakly scattering and keep only the first diagram in the expansion (15.2) of $\mathbf{\Sigma}(\mathbf{k})$ [which amounts to calculating $\mathbf{\Sigma}_1(\mathbf{k})$, as defined in Chapter 15]. The explicit expression for this diagram is

$$\mathbf{\Sigma}(\mathbf{k}) = k_0^4 \int \langle \delta\varepsilon(\mathbf{r})\delta\varepsilon(\mathbf{r}')\rangle \, \mathbf{G}_0(\mathbf{r}, \mathbf{r}') \exp(-i\mathbf{k}\cdot\mathbf{R}) \, d^3R \, , \tag{42.15}$$

where $\mathbf{R} = \mathbf{r} - \mathbf{r}'$. The free-space Green's function $\mathbf{G}_0$ given by (37.17) can be recast in a form suitable for further calculation:

$$\mathbf{G}_0(\mathbf{r}, \mathbf{r}') = k_0 \left[\beta(k_0R)(\mathbf{I} - \hat{\mathbf{R}} \otimes \hat{\mathbf{R}}) + \gamma(k_0R)\hat{\mathbf{R}} \otimes \hat{\mathbf{R}}\right] - \frac{\mathbf{I}}{3k_0^2}\,\delta(\mathbf{R}) \, , \tag{42.16}$$

where

$$\beta(x) == \frac{\exp(ix)}{4\pi x}\left(1 - \frac{1}{ix} - \frac{1}{x^2}\right) , \tag{42.17}$$

$$\gamma(x) == \frac{\exp(ix)}{2\pi x}\left(\frac{1}{ix} + \frac{1}{x^2}\right) . \tag{42.18}$$

Inserting (42.1) and (42.16) into (42.15) leads to

$$\begin{aligned}\mathbf{\Sigma}(\mathbf{k}) = &-\frac{k_0^2 A \mathbf{I}}{3\pi^{3/2}\ell_c^3} + \frac{k_0^5 A}{\pi^{3/2}\ell_c^3}\int [\beta(k_0R)(\mathbf{I} - \hat{\mathbf{R}} \otimes \hat{\mathbf{R}}) \\ &+ \gamma(k_0R)\hat{\mathbf{R}} \otimes \hat{\mathbf{R}}] \exp(-R^2/\ell_c^2) \exp(-i\mathbf{k}\cdot\mathbf{R}) \, d^3R \, . \end{aligned} \tag{42.19}$$

In order to perform the integral, we separate the radial and angular parts and make use of the relations

$$\int_{4\pi} \mathbf{I} \exp(-i\mathbf{k}\cdot\mathbf{R}) \, d\hat{\mathbf{R}} = 4\pi \, j_0(kR)\mathbf{I} \, ,$$

$$\begin{aligned}\int_{4\pi} \hat{\mathbf{R}} \otimes \hat{\mathbf{R}} \exp(-i\mathbf{k}\cdot\mathbf{R}) \, d\hat{\mathbf{R}} = {}& 4\pi \, \frac{j_1(kR)}{kR}\,\mathbf{I} \\ &+ \frac{\mathbf{k} \otimes \mathbf{k}}{k^2}\left[4\pi \, j_0(kR) - 12\pi \, \frac{j_1(kR)}{kR}\right] ,\end{aligned}$$

where $j_0(x)$ and $j_1(x)$ are spherical Bessel functions of the first kind. We calculate the integral on-shell for $k = k_0$ and also assume $\ell_c \ll \lambda$. Since the term $\exp(-R^2/\ell_c^2)$ effectively creates a cutoff at $R \simeq \ell_c$, we assume that $k_0R \ll 1$ and expand the integrand to leading order in k_0R. We obtain $\mathbf{\Sigma}(\mathbf{k}) = \Sigma(k_0)\mathbf{I}$, with

$$\Sigma(k_0) = -\frac{k_0^2 A}{3\pi^{3/2}\ell_c^3} + i\frac{k_0^5 A}{6\pi} . \tag{42.20}$$

Defining

$$k_R^2 = k_0^2 + \text{Re}\,\Sigma(k_0) \tag{42.21}$$

$$k_I^2 = \text{Im}\,\Sigma(k_0)\,, \tag{42.22}$$

so that $k_{\text{eff}}^2 = k_R^2 + ik_I^2$, we have

$$k_R^2 = k_0^2 - \frac{k_0^2 A}{3\pi^{3/2}\ell_c^3} \tag{42.23}$$

$$k_I^2 = \frac{k_0^5 A}{6\pi}\,. \tag{42.24}$$

Writing the effective wavenumber as

$$k_{\text{eff}} = k_R + \frac{i}{2\ell_s}\,, \tag{42.25}$$

we find that

$$\ell_s = \frac{6\pi}{k_0^4 A}\,. \tag{42.26}$$

Note that the second term on the right-hand side in (42.23) is specific to vector electromagnetic waves. It can be neglected provided that the condition $k_0^4 \ell_s \ell_c^3 \gg 1$ is satisfied. Finally, in the weak-scattering limit, the average Green's function becomes

$$\langle \mathbf{G}(\mathbf{r},\mathbf{r}')\rangle = \text{P}\left(\mathbf{I} + \frac{1}{k_R^2}\nabla\otimes\nabla\right)\frac{\exp(ik_R R)}{4\pi R}\exp[-R/(2\ell_s)] - \frac{\mathbf{I}}{3k_R^2}\,\delta(\mathbf{R})\,. \tag{42.27}$$

Using (37.17), the principal value can be evaluated, and we obtain

$$\begin{aligned}\langle \mathbf{G}(\mathbf{r},\mathbf{r}')\rangle = {} & \frac{\exp(ik_R R)}{4\pi R}\exp[-R/(2\ell_s)] \\ & \times\left[\mathbf{I} - \hat{\mathbf{R}}\otimes\hat{\mathbf{R}} - \left(\frac{1}{ik_R R} + \frac{1}{k_R^2 R^2}\right)\left(\mathbf{I} - 3\hat{\mathbf{R}}\otimes\hat{\mathbf{R}}\right)\right] - \frac{\mathbf{I}}{3k_R^2}\,\delta(\mathbf{R})\,.\end{aligned} \tag{42.28}$$

References and Additional Reading

These are textbooks with a comprehensive presentation of multiple scattering theory:

S.M. Rytov, Yu. A. Kravtsov and V.I. Tatarskii, *Principles of Statistical Radiophysics* (Springer-Verlag, Berlin, 1989), vol. 4.

P. Sheng, *Introduction to Wave Scattering, Localization, and Mesoscopic Phenomena* (Academic Press, San Diego, 1995).

E. Akkermans and G. Montambaux, *Mesoscopic Physics of Electrons and Photons* (Cambridge University Press, Cambridge, 2007).

A full treatment of multiple scattering of electromagnetic waves can be found in:

L. Tsang and J.A. Kong, *Scattering of Electromagnetic Waves: Advanced Topics* (Wiley, New York, 2001), chap. 4 and 5.

These are scientific papers with treatments of multiple scattering of vector fields, including calculations of the averaged field and self-energy:

M.J. Stephen and G. Cwilich, Phys. Rev. B **34**, 7564 (1986).

B.A. van Tiggelen, A. Lagendijk and A. Tip, J. Phys.: Condens. Matter **2**, 7653 (1990).

B.A. van Tiggelen, R. Maynard and T.M. Nieuwenhuizen, Phys. Rev. E **53**, 2881 (1996).

C.A. Müller and C. Miniatura, J. Phys. A: Math. Gen. **35**, 10163 (2002).

C.C. Kwong, D. Wilkowski, D. Delande and R. Pierrat, Phys. Rev. A **99**, 043806 (2019).

43

Multiple Scattering of Electromagnetic Waves: Radiative Transport

In this chapter, we study spatial correlations of the electric field in a disordered medium. We derive the Bethe–Salpeter equation for electromagnetic waves, making use of the diagrammatic perturbation theory introduced in Chapter 16. In the weak-scattering limit, we derive the radiative transport equation. Finally, we study the diffusive transport of polarized light in uncorrelated media.

43.1 Bethe–Salpeter Equation

For electromagnetic waves, correlation functions can be defined between any two components of the electric field. Denoting by $E_i(\mathbf{r})$ and $E_j(\mathbf{r})$ the i and j components of a monochromatic field at frequency ω, we consider the correlation function $\langle E_i(\mathbf{r}_1)\, E_j^*(\mathbf{r}_2)\rangle$. We begin by observing that

$$\left\langle E_i(\mathbf{r}_1)E_j^*(\mathbf{r}_2)\right\rangle = -(\mu_0\omega)^2 \int \left\langle G_{ik}(\mathbf{r}_1,\mathbf{r}_1')G_{jl}^*(\mathbf{r}_2,\mathbf{r}_2')\right\rangle j_k(\mathbf{r}_1')j_l^*(\mathbf{r}_2')\, d^3r_1' d^3r_2' , \tag{43.1}$$

which follows from (37.4). Here the current density $\mathbf{j}$ is taken to be deterministic and summation over repeated indices is implied. Moreover, we employ the following conventions regarding tensor notations. If $\boldsymbol{A}$ and $\boldsymbol{B}$ are second-rank tensors, their tensor product $\boldsymbol{A} \otimes \boldsymbol{B}$ is a fourth rank tensor with components $(\boldsymbol{A} \otimes \boldsymbol{B})_{ijkl} = A_{ij}B_{kl}$. If $\boldsymbol{C}$ and $\boldsymbol{D}$ are fourth rank tensors, their product is also fourth rank and is defined by $(\boldsymbol{C}\boldsymbol{D})_{ijkl} = C_{ijmn}D_{mnkl}$. Finally, $\boldsymbol{C}\boldsymbol{A}$ is a second rank tensor with components $(\boldsymbol{C}\boldsymbol{A})_{ij} = C_{ijkl}A_{kl}$.

We now turn to the problem of calculating the fourth rank tensor $\langle \mathbf{G} \otimes \mathbf{G}^* \rangle$ which appears in (43.1). To proceed, we make use of the diagrammatic technique introduced in Chapter 16 for scalar waves. The diagrammatic rules are the same as in Table 16.1, with the scalar Green's functions G_0 and G replaced by the electric Green's functions $\mathbf{G}_0$ and $\mathbf{G}$.

The diagrammatic expansion for $\mathbf{G} \otimes \mathbf{G}^*$ is given by

$$\mathbf{G} \otimes \mathbf{G}^* = \quad + \quad + \quad + \quad + \quad + \cdots \tag{43.2}$$

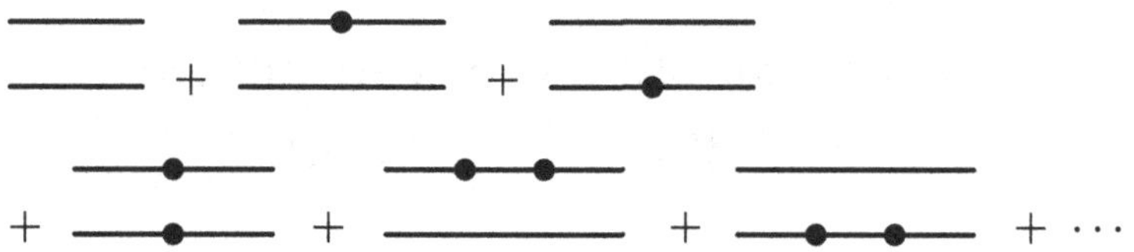

Here the upper line stands for the retarded Green's function $\mathbf{G}$, and the lower line for the advanced Green's function $\mathbf{G}^*$. Following the same steps as in Section 16.1, we first average over disorder by making use of Eq. (15.5), and obtain the following diagrammatic expansion for $\langle \mathbf{G} \otimes \mathbf{G}^* \rangle$:

$$\langle \mathbf{G} \otimes \mathbf{G}^* \rangle = \quad + \quad + \quad + \quad + \quad + \quad + \quad + \quad + \cdots \tag{43.3}$$

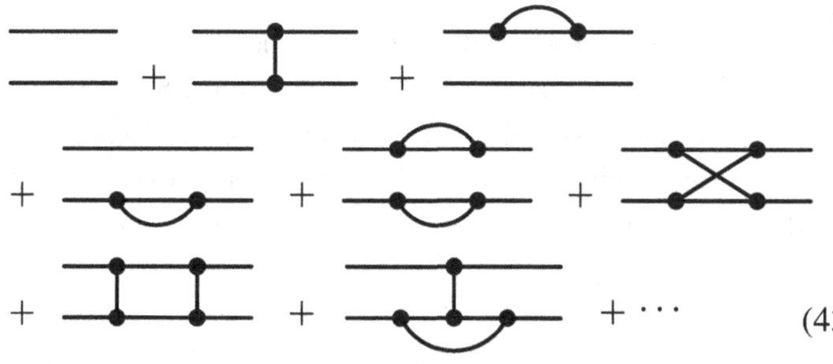

Next, we define the irreducible vertex $\mathbf{\Gamma}$, which is a fourth rank tensor, to be the sum of all connected diagrams with external legs amputated:

$$\mathbf{\Gamma} = \quad + \quad + \quad + \quad + \cdots \tag{43.4}$$

By diagrammatic calculation, we find that $\langle \mathbf{G} \otimes \mathbf{G}^* \rangle$ obeys an equation of the form

$$\begin{aligned} \left\langle \mathbf{G} \otimes \mathbf{G}^* \right\rangle &= \langle \mathbf{G} \rangle \otimes \left\langle \mathbf{G}^* \right\rangle + \langle \mathbf{G} \rangle \otimes \left\langle \mathbf{G}^* \right\rangle \mathbf{\Gamma} \, \langle \mathbf{G} \rangle \otimes \left\langle \mathbf{G}^* \right\rangle \\ &\quad + \langle \mathbf{G} \rangle \otimes \left\langle \mathbf{G}^* \right\rangle \mathbf{\Gamma} \, \langle \mathbf{G} \rangle \otimes \left\langle \mathbf{G}^* \right\rangle \mathbf{\Gamma} \, \langle \mathbf{G} \rangle \otimes \left\langle \mathbf{G}^* \right\rangle \cdots \end{aligned} \tag{43.5}$$

This equation can be summed to obtain the Bethe–Salpeter equation, which reads as

$$\left\langle \mathbf{G} \otimes \mathbf{G}^* \right\rangle = \langle \mathbf{G} \rangle \otimes \left\langle \mathbf{G}^* \right\rangle + \langle \mathbf{G} \rangle \otimes \left\langle \mathbf{G}^* \right\rangle \mathbf{\Gamma} \left\langle \mathbf{G} \otimes \mathbf{G}^* \right\rangle . \tag{43.6}$$

In terms of coordinates we have

$$\begin{aligned} &\left\langle \mathbf{G}(\mathbf{r}_1, \mathbf{r}_2) \otimes \mathbf{G}^*(\mathbf{r}_1', \mathbf{r}_2') \right\rangle = \langle \mathbf{G}(\mathbf{r}_1, \mathbf{r}_2) \rangle \otimes \left\langle \mathbf{G}^*(\mathbf{r}_1', \mathbf{r}_2') \right\rangle \\ &+ \int d^3 R_1 d^3 R_1' d^3 R_2 d^3 R_2' \, \langle \mathbf{G}(\mathbf{r}_1, \mathbf{R}_1) \rangle \otimes \left\langle \mathbf{G}^*(\mathbf{r}_1', \mathbf{R}_1') \right\rangle \mathbf{\Gamma}(\mathbf{R}_1, \mathbf{R}_2; \mathbf{R}_1', \mathbf{R}_2') \\ &\times \left\langle \mathbf{G}(\mathbf{R}_2, \mathbf{r}_2) \otimes \mathbf{G}^*(\mathbf{R}_2', \mathbf{r}_2') \right\rangle . \end{aligned} \tag{43.7}$$

The ladder approximation amounts to retaining only the first diagram in (43.4). It follows that (43.6) can be expressed as a sum of diagrams of the form

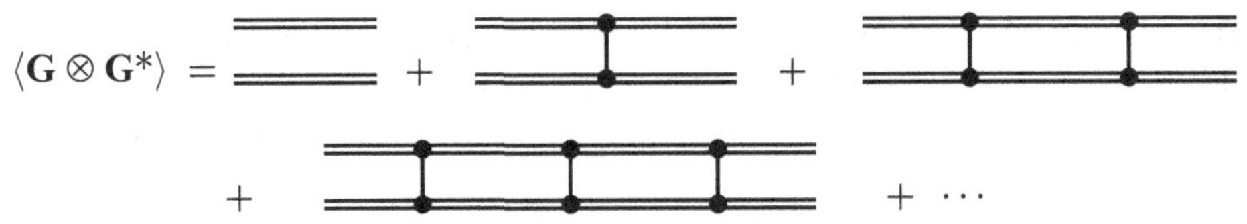

$$\langle \mathbf{G} \otimes \mathbf{G}^* \rangle = \quad + \quad + \quad + \quad + \cdots \tag{43.8}$$

As in the case of scalar waves, the higher-order diagrams in (43.4) are of order $O\left(1/(k_R \ell_s)^2\right)$, with k_R the real part of the effective wavevector defined in (42.25). Thus the ladder approximation is accurate to the same order as the weak-scattering approximation for the self-energy used in Chapter 42.

Using the model of disorder in (42.1) and the notation $C(\mathbf{r}-\mathbf{r}') = \langle \delta\varepsilon(\mathbf{r})\, \delta\varepsilon(\mathbf{r}')\rangle$, the ladder approximation is of the form

$$\boldsymbol{\Gamma}(\mathbf{R}_1, \mathbf{R}_2; \mathbf{R}_1', \mathbf{R}_2') = k_0^4\, C(\mathbf{R}_1 - \mathbf{R}_1')\delta(\mathbf{R}_1 - \mathbf{R}_2)\delta(\mathbf{R}_1' - \mathbf{R}_2')\mathbf{I} \otimes \mathbf{I}\,, \tag{43.9}$$

with $\mathbf{I}$ the unit second-rank tensor. In the ladder approximation, the Bethe–Salpeter equation in position space becomes

$$\begin{aligned}\left\langle G_{ik}(\mathbf{r}_1, \mathbf{r}_2)G_{jl}^*(\mathbf{r}_1', \mathbf{r}_2')\right\rangle &= \langle G_{ik}(\mathbf{r}_1, \mathbf{r}_2)\rangle \left\langle G_{jl}^*(\mathbf{r}_1', \mathbf{r}_2')\right\rangle + k_0^4 \int d^3r d^3r' \\ &\times \langle G_{im}(\mathbf{r}_1, \mathbf{r})\rangle \left\langle G_{jn}^*(\mathbf{r}_1', \mathbf{r}')\right\rangle C(\mathbf{r} - \mathbf{r}') \left\langle G_{mk}(\mathbf{r}, \mathbf{r}_2)G_{nl}^*(\mathbf{r}', \mathbf{r}_2')\right\rangle .\end{aligned} \tag{43.10}$$

43.2 Radiative Transport

In this section, we derive the radiative transport equation (RTE) from the Bethe–Salpeter equation in the ladder approximation. We follow the development in Sections 16.2 and 16.3 for scalar waves. We start by introducing the tensor Wigner transform $\mathbf{W}(\mathbf{r}, \mathbf{k})$, which is defined as

$$\mathbf{W}(\mathbf{r}, \mathbf{k}) = \int \frac{d^3r'}{(2\pi)^3} \exp(-i\mathbf{k} \cdot \mathbf{r}') \left\langle \mathbf{E}(\mathbf{r} + \mathbf{r}'/2) \otimes \mathbf{E}^*(\mathbf{r} - \mathbf{r}'/2)\right\rangle . \tag{43.11}$$

By Fourier inversion, we see that the coherence tensor

$$\mathcal{W}(\mathbf{r}, \mathbf{r}') = \left\langle \mathbf{E}(\mathbf{r}) \otimes \mathbf{E}^*(\mathbf{r}')\right\rangle \tag{43.12}$$

is related to the Wigner transform by

$$\mathcal{W}(\mathbf{r}, \mathbf{r}') = \int d^3k \exp(i\mathbf{k} \cdot (\mathbf{r} - \mathbf{r}'))\mathbf{W}\left(\frac{\mathbf{r} + \mathbf{r}'}{2}, \mathbf{k}\right) . \tag{43.13}$$

We immediately find that the average intensity $\langle I(\mathbf{r})\rangle = \mathrm{Tr}\, \langle \mathbf{E}(\mathbf{r}) \otimes \mathbf{E}^*(\mathbf{r})\rangle$ is related to the Wigner transform by

$$\langle I(\mathbf{r})\rangle = \int d^3k\, \mathrm{Tr}\mathbf{W}(\mathbf{r}, \mathbf{k})\,. \tag{43.14}$$

The quantity $\mathcal{W}(\mathbf{r}, \mathbf{r}')$ with $\mathbf{r} \neq \mathbf{r}'$ is a measure of the spatial coherence of the electric field, and $\mathcal{W}(\mathbf{r}, \mathbf{r})$ is a measure of the polarization.

The Wigner transform of the field is related to the Wigner transform of the source current density according to

$$\mathbf{W}(\mathbf{r}, \mathbf{k}) = \int d^3r' d^3k' \, \mathbf{\Phi}(\mathbf{r} - \mathbf{r}'; \mathbf{k}, \mathbf{k}') \mathbf{W}_0(\mathbf{r}', \mathbf{k}') , \tag{43.15}$$

which follows from (43.1) and (43.11). Here the propagator $\mathbf{\Phi}$ is the fourth rank tensor defined by

$$\begin{aligned}\mathbf{\Phi}(\mathbf{r} - \mathbf{r}'; \mathbf{k}, \mathbf{k}') = \frac{1}{(2\pi)^3} \int d^3R d^3R' \, \exp(-i\mathbf{k} \cdot \mathbf{R} + i\mathbf{k}' \cdot \mathbf{R}') \\ \times \langle \mathbf{G}(\mathbf{r} + \mathbf{R}/2, \mathbf{r}' + \mathbf{R}'/2) \otimes \mathbf{G}^*(\mathbf{r} - \mathbf{R}/2, \mathbf{r}' - \mathbf{R}'/2) \rangle\end{aligned} \tag{43.16}$$

and

$$\mathbf{W}_0(\mathbf{r}, \mathbf{k}) = -(\mu_0\omega)^2 \int \frac{d^3r'}{(2\pi)^3} \exp(-i\mathbf{k} \cdot \mathbf{r}') \mathbf{j}(\mathbf{r} + \mathbf{r}'/2) \otimes \mathbf{j}^*(\mathbf{r} - \mathbf{r}'/2) . \tag{43.17}$$

We will now use the above results to obtain the Liouville equation, which is a key step in deriving the RTE. We begin by considering the Fourier transform of $\mathbf{W}(\mathbf{r}, \mathbf{k})$, which is defined by

$$\widetilde{\mathbf{W}}(\mathbf{q}, \mathbf{k}) = \int d^3r \, \exp(-i\mathbf{q} \cdot \mathbf{r}) \, \mathbf{W}(\mathbf{r}, \mathbf{k}) . \tag{43.18}$$

Making use of (43.15), we find that

$$\widetilde{\mathbf{W}}(\mathbf{q}, \mathbf{k}) = \int \frac{d^3k'}{(2\pi)^3} \mathbf{\Phi}(\mathbf{q}; \mathbf{k}, \mathbf{k}') \, \widetilde{\mathbf{W}}_0(\mathbf{q}, \mathbf{k}') , \tag{43.19}$$

where $\mathbf{\Phi}(\mathbf{q}; \mathbf{k}, \mathbf{k}')$ is defined by

$$\mathbf{\Phi}(\mathbf{r} - \mathbf{r}'; \mathbf{k}, \mathbf{k}') = \int \frac{d^3q}{(2\pi)^3} \exp(i\mathbf{q} \cdot (\mathbf{r} - \mathbf{r}')) \mathbf{\Phi}(\mathbf{q}; \mathbf{k}, \mathbf{k}') . \tag{43.20}$$

After some calculation, it follows from (43.10) and (43.16) that $\mathbf{\Phi}(\mathbf{q}; \mathbf{k}, \mathbf{k}')$ satisfies

$$\begin{aligned}\mathbf{\Phi}(\mathbf{q}; \mathbf{k}, \mathbf{k}') = \langle \mathbf{G}(\mathbf{k} + \mathbf{q}/2) \rangle \otimes \langle \mathbf{G}^*(\mathbf{k} - \mathbf{q}/2) \rangle \Big[\delta(\mathbf{k} - \mathbf{k}') \\ + k_0^4 \int \frac{d^3k''}{(2\pi)^3} \widetilde{C}(\mathbf{k} - \mathbf{k}'') \mathbf{\Phi}(\mathbf{q}; \mathbf{k}'', \mathbf{k}') \Big] .\end{aligned} \tag{43.21}$$

This equation is equivalent to the Bethe–Salpeter equation in the ladder approximation.

To proceed further, we need an expression of the average Green's function $\langle \mathbf{G} \rangle$. Since we are interested in the transport of intensity at large scales compared to the wavelength, we only consider the transverse Green's function defined in (42.9) and write

$$\langle \mathbf{G}(\mathbf{k}) \rangle = \langle G(\mathbf{k}) \rangle \, \mathbf{P}_\perp(\mathbf{k}) \,, \tag{43.22}$$

where

$$\langle G(\mathbf{k}) \rangle = \frac{1}{k^2 - k_0^2 - \Sigma_\perp(\mathbf{k})} \,, \quad \mathbf{P}_\perp(\mathbf{k}) = \mathbf{I} - \hat{\mathbf{k}} \otimes \hat{\mathbf{k}} \,. \tag{43.23}$$

We then make use of the identities

$$\langle G(\mathbf{k}+\mathbf{q}/2) \rangle \left\langle G^*(\mathbf{k}-\mathbf{q}/2) \right\rangle = \frac{\langle G^*(\mathbf{k}-\mathbf{q}/2) \rangle - \langle G(\mathbf{k}+\mathbf{q}/2) \rangle}{\langle G(\mathbf{k}+\mathbf{q}/2) \rangle^{-1} - \langle G^*(\mathbf{k}-\mathbf{q}/2) \rangle^{-1}} \,, \tag{43.24}$$

$$|\mathbf{k}+\mathbf{q}/2|^2 - |\mathbf{k}-\mathbf{q}/2|^2 = 2\mathbf{k} \cdot \mathbf{q} \,, \tag{43.25}$$

to rewrite (43.21) as

$$\begin{aligned} &2\mathbf{k} \cdot \mathbf{q} \, \boldsymbol{\Phi}(\mathbf{q}; \mathbf{k}, \mathbf{k}') + \Delta\Sigma(\mathbf{q}, \mathbf{k}) \boldsymbol{\Phi}(\mathbf{q}; \mathbf{k}, \mathbf{k}') = \\ &\Delta G(\mathbf{q}, \mathbf{k}) \mathbf{P}_\perp(\mathbf{k}+\mathbf{q}/2) \otimes \mathbf{P}_\perp(\mathbf{k}-\mathbf{q}/2) \\ &\times \left[\delta(\mathbf{k}-\mathbf{k}') + k_0^4 \int \frac{d^3k''}{(2\pi)^3} \widetilde{C}(\mathbf{k}-\mathbf{k}'') \boldsymbol{\Phi}(\mathbf{q}; \mathbf{k}'', \mathbf{k}') \right] . \end{aligned} \tag{43.26}$$

Here we have introduced the notation

$$\Delta\Sigma(\mathbf{q}, \mathbf{k}) = \Sigma_\perp(\mathbf{k}+\mathbf{q}/2) - \Sigma_\perp^*(\mathbf{k}-\mathbf{q}/2) \,, \tag{43.27}$$

$$\Delta G(\mathbf{q}, \mathbf{k}) = \langle G(\mathbf{k}+\mathbf{q}/2) \rangle - \left\langle G^*(\mathbf{k}-\mathbf{q}/2) \right\rangle \,. \tag{43.28}$$

Finally, upon multiplying (43.26) by $\widetilde{\mathbf{W}}_0(\mathbf{q}, \mathbf{k})$ and inverting the Fourier transform (43.18), we find that the Wigner transform $\mathbf{W}$ obeys the Liouville equation

$$\begin{aligned} &\mathbf{k} \cdot \nabla_\mathbf{r} \mathbf{W}(\mathbf{r}, \mathbf{k}) + \frac{1}{2i} \int \frac{d^3q}{(2\pi)^3} \exp(i\mathbf{q} \cdot \mathbf{r}) \Delta\Sigma(\mathbf{q}, \mathbf{k}) \widetilde{\mathbf{W}}(\mathbf{q}, \mathbf{k}) = \\ &\frac{k_0^4}{2i} \int \frac{d^3q}{(2\pi)^3} \exp(i\mathbf{q} \cdot \mathbf{r}) \int \frac{d^3k'}{(2\pi)^3} \Delta G(\mathbf{q}, \mathbf{k}) \mathbf{P}_\perp(\mathbf{k}+\mathbf{q}/2) \otimes \mathbf{P}_\perp(\mathbf{k}-\mathbf{q}/2) \\ &\times \widetilde{C}(\mathbf{k}-\mathbf{k}') \widetilde{\mathbf{W}}(\mathbf{q}, \mathbf{k}') + \mathbf{S}(\mathbf{r}, \mathbf{k}) \,, \end{aligned} \tag{43.29}$$

where

$$\mathbf{S}(\mathbf{r}, \mathbf{k}) = \frac{1}{2i} \int \frac{d^3q}{(2\pi)^3} \exp(i\mathbf{q} \cdot \mathbf{r}) \Delta G(\mathbf{q}, \mathbf{k}) \mathbf{P}_\perp(\mathbf{k}+\mathbf{q}/2) \otimes \mathbf{P}_\perp(\mathbf{k}-\mathbf{q}/2) \widetilde{\mathbf{W}}_0(\mathbf{q}, \mathbf{k}) \,. \tag{43.30}$$

We now consider the large scale limit $\mathbf{q} \to 0$, also known as the Kubo limit, and replace $\Delta\Sigma(\mathbf{q}, \mathbf{k})$, $\Delta G(\mathbf{q}, \mathbf{k})$ and $\mathbf{P}_\perp(\mathbf{k} \pm \mathbf{q}/2)$ by $\Delta\Sigma(0, \mathbf{k})$, $\Delta G(0, \mathbf{k})$ and $\mathbf{P}_\perp(\mathbf{k})$, respectively, in (43.29). We thus obtain

$$\mathbf{k} \cdot \nabla_\mathbf{r} \mathbf{W}(\mathbf{r}, \mathbf{k}) + \frac{1}{2i} \Delta\Sigma(0, \mathbf{k}) \mathbf{W}(\mathbf{r}, \mathbf{k}) =$$

$$\frac{1}{2i} \Delta G(0, \mathbf{k})(\hat{\mathbf{k}}) \left[k_0^4 \int \frac{d^3 k'}{(2\pi)^3} \widetilde{C}(\mathbf{k} - \mathbf{k}') \mathbf{W}(\mathbf{r}, \mathbf{k}') + \mathbf{W}_0(\mathbf{r}, \mathbf{k}) \right], \quad (43.31)$$

where $\mathrm{M}(\hat{\mathbf{k}}) = \mathbf{P}_\perp(\mathbf{k}) \otimes \mathbf{P}_\perp(\mathbf{k})$. Next, we evaluate $\Delta G(0, \mathbf{k})$ and $\Delta\Sigma(0, \mathbf{k})$ in the weak-scattering limit. Using the on-shell approximation, we write $\Sigma_\perp(k) = \Sigma(k_0)$ and we use the fact that

$$\langle G(\mathbf{k}) \rangle = \frac{1}{k^2 - k_R^2 - i k_I^2 - i\epsilon}, \quad (43.32)$$

where k_R and k_I are defined in (42.21) and (42.22). Making use of the identities (16.41) and (16.42), we find that in the weak-scattering limit $k_R \ell_s \gg 1$, with ℓ_s the scattering mean free path, $\Delta G(0, \mathbf{k})$ is given by

$$\Delta G(0, \mathbf{k}) = \frac{i\pi}{k_R} \delta(k - k_R). \quad (43.33)$$

In the same limit, we also find that

$$\Delta\Sigma(0, \mathbf{k}) = 2i k_R / \ell_s. \quad (43.34)$$

Introducing the specific intensities $\boldsymbol{I}$ and $\boldsymbol{S}$ for the field and the source, respectively, such that

$$\delta(k - k_R) \boldsymbol{I}(\mathbf{r}, \hat{\mathbf{k}}) = k_R \mathbf{W}(\mathbf{r}, \mathbf{k}), \quad (43.35)$$

$$\delta(k - k_R) \boldsymbol{S}(\mathbf{r}, \hat{\mathbf{k}}) = \frac{\pi}{2k_R} \mathbf{M}(\hat{\mathbf{k}}) \mathbf{W}_0(\mathbf{r}, \mathbf{k}), \quad (43.36)$$

we can rewrite (43.31) in the form

$$\hat{\mathbf{k}} \cdot \nabla_\mathbf{r} \boldsymbol{I} + \mu_s \boldsymbol{I} = \frac{k_0^4}{16\pi^2} \mathbf{M}(\hat{\mathbf{k}}) \int d\hat{\mathbf{k}}' \, \widetilde{C}(k_R(\hat{\mathbf{k}} - \hat{\mathbf{k}}')) \boldsymbol{I}(\mathbf{r}, \hat{\mathbf{k}}') + \boldsymbol{S}. \quad (43.37)$$

Here we have introduced $\mu_s = 1/\ell_s$, as in Chapter 16. Defining the tensor phase function $\mathbf{\Pi}$ by

$$\mu_s \mathbf{\Pi}(\hat{\mathbf{k}}, \hat{\mathbf{k}}') = \frac{k_0^4}{16\pi^2} \mathbf{M}(\hat{\mathbf{k}}) \widetilde{C}(k_R(\hat{\mathbf{k}} - \hat{\mathbf{k}}'), \quad (43.38)$$

we obtain

$$\hat{\mathbf{k}} \cdot \nabla_\mathbf{r} \boldsymbol{I} + \mu_s \boldsymbol{I} = \mu_s \int d\hat{\mathbf{k}}' \, \mathbf{\Pi}(\hat{\mathbf{k}}, \hat{\mathbf{k}}') \, \boldsymbol{I}(\mathbf{r}, \hat{\mathbf{k}}') + \boldsymbol{S}. \quad (43.39)$$

Equation (43.39) is the RTE for electromagnetic waves in a non-absorbing medium. It differs from the scalar RTE (16.50) in two respects. The specific intensity is replaced by the tensor-valued Wigner transform $\boldsymbol{I}$, which provides information about the coherence and polarization of the field. Likewise, the scalar phase function is replaced by the tensor $\boldsymbol{\Pi}$.

43.3 Diffusion Approximation and Depolarization

Exact solutions to the RTE are known in only a small number of cases. However, under certain conditions, the solution to the RTE can be very well approximated by a solution to a corresponding diffusion equation. This so-called diffusion approximation (DA) is extensively discussed in Part IV for scalar waves. Here we derive the diffusion equation for electromagnetic waves, following the approach developed in Chapter 24. We then use the result to discuss the process of depolarization by multiple scattering.

We restrict the discussion to the case of white-noise disorder, for which the correlation function $C(\mathbf{r}) = 6\pi/(k_0^4\ell_s)\delta(\mathbf{r})$ in the weak-scattering regime (see Section 42.1). We then find that

$$\boldsymbol{\Pi}(\hat{\mathbf{k}}, \hat{\mathbf{k}}') = \frac{3}{2}\frac{\mathrm{M}(\hat{\mathbf{k}})}{4\pi}, \tag{43.40}$$

which is the tensor phase function for white-noise disorder or, equivalently, isotropic scattering.

We begin by rewriting the RTE (43.39) in the form

$$\hat{\mathbf{k}} \cdot \nabla_{\mathbf{r}}\boldsymbol{I} + \mu_s\boldsymbol{I} - L\boldsymbol{I} = \boldsymbol{S}, \tag{43.41}$$

where the operator L is defined by

$$L\boldsymbol{I}(\mathbf{r}, \hat{\mathbf{k}}) = \mu_s \int \boldsymbol{\Pi}(\hat{\mathbf{k}}, \hat{\mathbf{k}}')\boldsymbol{I}(\mathbf{r}, \hat{\mathbf{k}}')d\hat{\mathbf{k}}' . \tag{43.42}$$

The specific intensity is now decomposed as a sum of terms of the form

$$\boldsymbol{I} = \boldsymbol{I}_0 + \boldsymbol{I}_1 + \boldsymbol{I}_d . \tag{43.43}$$

The ballistic term $\boldsymbol{I}_0$ is highly singular and obeys

$$\hat{\mathbf{k}} \cdot \nabla_{\mathbf{r}}\boldsymbol{I}_0 + \mu_s\boldsymbol{I}_0 = \boldsymbol{S} . \tag{43.44}$$

The term I_1, which corresponds to single scattering of the incident field, obeys

$$\hat{\mathbf{k}} \cdot \nabla_{\mathbf{r}}\boldsymbol{I}_1 + \mu_s\boldsymbol{I}_1 = L\boldsymbol{I}_0 . \tag{43.45}$$

Finally, the diffuse intensity $\boldsymbol{I}_d$, which accounts for all orders of scattering beyond the first, is smooth and obeys

$$\hat{\mathbf{k}} \cdot \nabla_{\mathbf{r}} \boldsymbol{I}_d + \mu_s \boldsymbol{I}_d - L\boldsymbol{I}_d = \boldsymbol{Q} , \tag{43.46}$$

where $\boldsymbol{Q} = L\boldsymbol{I}_1$. Note that as for scalar waves, $\boldsymbol{Q}$ is exponentially small far boundaries (see Section 24.1).

The idea of the DA is to approximate the smooth part $\boldsymbol{I}_d$, assuming that the singular parts are treated exactly. We begin by introducing the first two angular moments of $\boldsymbol{I}_d$:

$$\boldsymbol{U}(\mathbf{r}) = \frac{1}{v_E} \int \boldsymbol{I}_d(\mathbf{r}, \hat{\mathbf{k}}) d\hat{\mathbf{k}} , \tag{43.47}$$

$$\mathbf{J}(\mathbf{r}) = \int \hat{\mathbf{k}} \otimes \boldsymbol{I}_d(\mathbf{r}, \hat{\mathbf{k}}) d\hat{\mathbf{k}} , \tag{43.48}$$

where v_E is the energy velocity in the medium. The quantities $\boldsymbol{U}$ and $\mathbf{J}$ can be considered as generalized energy density and current, respectively. It can easily be seen from (43.13) and (43.35) that $\boldsymbol{U}(\mathbf{r})$ is translational to the coherence tensor $\mathcal{W}(\mathbf{r}, \mathbf{r})$ that describes the polarization properties of the field and that Tr$\boldsymbol{U}(\mathbf{r})$ is the average energy density.

The specific intensity can be expended in its angular moments, leading to

$$\boldsymbol{I}_d = \frac{v_E}{4\pi} \boldsymbol{U} + \frac{3}{4\pi} \mathbf{J}\hat{\mathbf{k}} + \cdots . \tag{43.49}$$

We now obtain the equations obeyed by the angular moments. Upon integration of (43.46) with respect to $\hat{\mathbf{k}}$, we obtain

$$\nabla \cdot \mathbf{J} + v_E \mu_s \boldsymbol{U} = v_E \mu_s \boldsymbol{A} \boldsymbol{U} + \boldsymbol{S}_d , \tag{43.50}$$

where

$$\boldsymbol{A} = \frac{1}{4\pi} \int \frac{3}{2} \mathrm{M}(\hat{\mathbf{k}}) d\hat{\mathbf{k}} , \quad \boldsymbol{S}_d = \int \boldsymbol{Q} d\hat{\mathbf{k}} . \tag{43.51}$$

Noting that $\langle \hat{k}_i^2 \rangle_{\hat{\mathbf{k}}} = 1/3$, $\langle \hat{k}_i^2 \hat{k}_j^2 \rangle_{\hat{\mathbf{k}}} = 1/15$, where $\langle \cdots \rangle_{\hat{\mathbf{k}}}$ means an angular average over $\hat{\mathbf{k}}$, and that angular averages containing odd powers of $\hat{k}_i$ vanish, we find that

$$A_{ijkl} = \frac{1}{2} \delta_{ik} \delta_{jl} + \frac{1}{10} \left(\delta_{ij} \delta_{kl} + \delta_{ik} \delta_{jl} + \delta_{il} \delta_{jk} \right) . \tag{43.52}$$

Next, we multiply (43.46) by $\hat{\mathbf{k}}$ and integrate with respect to $\hat{\mathbf{k}}$. Making use of (43.49) and the identities (24.14), we obtain

$$\mathbf{J} = -D \nabla \boldsymbol{U} + \ell_s \boldsymbol{\mathcal{Q}} , \tag{43.53}$$

where the diffusion coefficient is $D = (v_E/3)\ell_s$, and $\boldsymbol{\mathcal{Q}} = \int \hat{\mathbf{k}} \otimes \boldsymbol{Q} d\hat{\mathbf{k}}$.

We can now obtain the diffusion equation obeyed by the generalized energy density $\boldsymbol{U}$. By substituting (43.53) into (43.50), we find that

$$-D\nabla^2 \boldsymbol{U} + v_E \mu_s \boldsymbol{U} = v_E \mu_s \boldsymbol{A}\boldsymbol{U} + \boldsymbol{S}_d - \ell_s \nabla \cdot \boldsymbol{\mathcal{Q}} . \tag{43.54}$$

Using (43.49) and (43.53), we see that the diffuse intensity is given by

$$\boldsymbol{I}_d = \frac{v_E}{4\pi} \left(\boldsymbol{U} - \ell_s \hat{\mathbf{k}} \cdot \nabla \boldsymbol{U} \right) , \tag{43.55}$$

where we have dropped the last term on the right-hand side of (43.53) that does not contribute. We also find in the same conditions that the generalized energy density $\boldsymbol{U}$ obeys the diffusion equation

$$-D\nabla^2 \boldsymbol{U} + v_E \mu_s \boldsymbol{U} = v_E \mu_s \boldsymbol{A}\boldsymbol{U} + \boldsymbol{S}_d . \tag{43.56}$$

Introducing the eigenvectors and eigenvalues of $\boldsymbol{A}$ such that $\boldsymbol{A}\boldsymbol{v}_n = \lambda_n \boldsymbol{v}_n$, we can expand $\boldsymbol{U}$ and $\boldsymbol{S}_d$ according to

$$\boldsymbol{U} = \sum_n u_n \boldsymbol{v}_n , \quad \boldsymbol{S}_d = \sum_n S_n \boldsymbol{v}_n . \tag{43.57}$$

Inserting these expansions into (43.56), we find that u_n obeys

$$-D\nabla^2 u_n + v_E (1 - \lambda_n) \mu_s \, u_n = S_n , \tag{43.58}$$

which takes the usual form of a diffusion equation, with an effective absorption coefficient $(1 - \lambda_n)\mu_s$. Once (43.58) has been solved, we can obtain the diffuse intensity from (43.55), which becomes

$$\boldsymbol{I}_d = \frac{v_E}{4\pi} \sum_n \left(u_n - \ell_s \hat{\mathbf{k}} \cdot \nabla u_n \right) \boldsymbol{v}_n . \tag{43.59}$$

In an infinite homogeneous medium, the solution to (43.58) is readily obtained. We find that

$$u_n(\mathbf{r}) = \frac{1}{D} \int G_n(\mathbf{r}, \mathbf{r}') S_n(\mathbf{r}') d^3 r' , \tag{43.60}$$

where G_n is the Green's function given by

$$G_n(\mathbf{r}, \mathbf{r}') = \frac{\exp(-k_n |\mathbf{r} - \mathbf{r}'|)}{4\pi |\mathbf{r} - \mathbf{r}'|} , \tag{43.61}$$

with

$$k_n = \sqrt{3(1 - \lambda_n)} \mu_s . \tag{43.62}$$

Physically, each $\boldsymbol{v}_n$ can be interpreted as a diffuse polarization mode, $1/k_n = \ell_s / \sqrt{3(1 - \lambda_n)}$, being the decay length associated with each mode.

The eigenvalues λ_n can be found by writing the fourth-rank tensor $\boldsymbol{A}$ as a 9×9 matrix and solving the characteristic equation. We find three eigenvalues, with degeneracy 1, 3 and 5, respectively, and given by $\lambda_1 = 1$, $\lambda_{2,3,4} = 1/2$ and $\lambda_{5,6,7,8,9} = 7/10$. We see that except for mode 1, all polarization modes exhibit a decay length on the order of the scattering mean free path ℓ_s, the exact decay length being dictated by the corresponding eigenvalue. The only mode that survives at large distances is mode 1, with $k_1 = 0$, sometimes referred to as a Goldstone mode. The corresponding eigenvector can be found to be $(1/\sqrt{3})\delta_{kl}$, which corresponds to an equal distribution of energy over the three components of the field, and therefore to unpolarized light. The diffusion approximation therefore describes the process of depolarization of the average intensity by multiple scattering.

References and Additional Reading

A treatment of multiple scattering of electromagnetic waves can be found in:

L. Tsang and J.A. Kong, *Scattering of Electromagnetic Waves: Advanced Topics* (Wiley, New York, 2001), chap. 4 and 5.

E. Akkermans and G. Montambaux, *Mesoscopic Physics of Electrons and Photons* (Cambridge University Press, Cambridge, 2007), sec. 6.6. A detailed study of the eigenvalue problem for the diffusion of polarized light can be found in sec. 6.6.2.

The following papers develop a multiple scattering transport theory for vector fields:

F.C. MacKintosh and S. John, Phys. Rev. B **37**, 1884 (1988).

B.A. van Tiggelen, A. Lagendijk and A. Tip, J. Phys.:Condens. Matter **2**, 7653 (1990).

V.D. Ozrin, Waves Random Media **2**, 141 (1992).

N. Cherroret, D. Delande and B.A. van Tiggelen, Phys. Rev. A **94**, 012702 (2016).

These papers study the transport of field correlations and polarization, and introduce the concept of polarization eigenmodes directly from the Bethe–Salpeter equation. The polarization eigenmodes exhibit not only a specific decay length but also a specific diffusion constant:

C.A. Müller and C. Miniatura, J. Phys. A: Math. Gen. **35**, 10163 (2002).

K. Vynck, R. Pierrat and R. Carminati, Phys. Rev. A **89**, 013842 (2014); Phys. Rev. A **94**, 033851 (2016).

This paper presents a derivation of the RTE for vector electromagnetic waves based on multiscale asymptotics:

L. Ryzhik, G. Papanicolaou and J.B. Keller, Wave Motion **24**, 327 (1996).

The diffusion approximation is also derived, and polarization eigenmodes for the diffusion equation are introduced. The eigenmodes share the same diffusion constant, in a similar way as in the approach followed in this chapter.

The radiative transport equation for polarized light is introduced in the following references, making use of the Stokes vectors formalism:

S. Chandrasekhar, *Radiative Transfer* (Dover, New York, 1960).
A. Peraiah, *An Introduction to Radiative Transfer* (Cambridge University Press, Cambridge, 2002), chap. 11 and 12.

The following textbooks present the basic concepts for the description of light polarization:

J.W. Goodman, *Statistical Optics* (Wiley, New York, 1985).
E. Collett, *Polarized Light, Fundamentals and Applications* (Marcel Dekker, New York, 1993).
C. Brosseau, *Fundamentals of Polarized Light, A Statistical Optics Approach* (Wiley, New York, 1998).
E. Wolf, *Introduction to the Theory of Coherence and Polarization* (Cambridge University Press, Cambridge, 2007).

44

Bulk Electromagnetic Speckle Correlations

In Chapter 30, we studied speckle correlations for scalar waves in an infinite medium. Here we consider this problem for electromagnetic waves. For Gaussian random fields, we relate the intensity correlation function to the field correlation tensor.

44.1 Intensity Correlation Function

In the statistical description of a speckle pattern, the electric field $\mathbf{E}(\mathbf{r})$ and the intensity $I(\mathbf{r}) = |\mathbf{E}(\mathbf{r})|^2$ are random variables. Spatial correlations in the speckle pattern are characterized by the normalized intensity correlation function, which is defined as

$$C(\mathbf{r}, \mathbf{r}') = \frac{\langle \delta I(\mathbf{r}) \delta I(\mathbf{r}') \rangle}{\langle I(\mathbf{r}) \rangle \langle I(\mathbf{r}') \rangle} = \frac{\langle I(\mathbf{r}) I(\mathbf{r}') \rangle}{\langle I(\mathbf{r}) \rangle \langle I(\mathbf{r}') \rangle} - 1 , \tag{44.1}$$

where $\delta I(\mathbf{r}) = I(\mathbf{r}) - \langle I(\mathbf{r}) \rangle$.

Assuming that $\mathbf{E}$ obeys Gaussian statistics, we will show that the intensity correlation function can be factorized into a sum of squares of field correlation functions. The development is similar to that followed in Section 28 for scalar waves, except that the vector nature of the field needs to be properly handled.

We begin by recalling the moment theorem, which states that if Z_1 and Z_2 are complex Gaussian random variables, then

$$\begin{aligned} \langle |Z_1|^2 \, |Z_2|^2 \rangle &= \langle Z_1 Z_1^* Z_2 Z_2^* \rangle \\ &= \langle Z_1 Z_1^* \rangle \langle Z_2 Z_2^* \rangle + \langle Z_1 Z_2^* \rangle \langle Z_1^* Z_2 \rangle \\ &= \langle |Z_1|^2 \rangle \langle |Z_2|^2 \rangle + |\langle Z_1 Z_2^* \rangle|^2 . \end{aligned} \tag{44.2}$$

Denoting by $I_i(\mathbf{r}) = |E_i(\mathbf{r})|^2$ the intensity corresponding to a vector component of the field, with $i = x, y, z$, we have

$$\langle \delta I(\mathbf{r})\delta I(\mathbf{r}')\rangle = \sum_{i,j} \langle \delta I_i(\mathbf{r})\delta I_j(\mathbf{r}')\rangle \,. \tag{44.3}$$

Applying (44.2), we obtain

$$\langle I_i(\mathbf{r}) I_j(\mathbf{r}')\rangle = \langle I_i(\mathbf{r})\rangle\langle I_j(\mathbf{r}')\rangle + |\langle E_i(\mathbf{r})E_j^*(\mathbf{r}')\rangle|^2 \,. \tag{44.4}$$

This leads to

$$\frac{\langle \delta I_i(\mathbf{r})\delta I_j(\mathbf{r}')\rangle}{\langle I_i(\mathbf{r})\rangle\langle I_j(\mathbf{r}')\rangle} = |\mu_{ij}(\mathbf{r},\mathbf{r}')|^2 \,, \tag{44.5}$$

where

$$\mu_{ij}(\mathbf{r},\mathbf{r}') = \frac{\langle E_i(\mathbf{r})E_j^*(\mathbf{r}')\rangle}{\langle I_i(\mathbf{r})\rangle^{1/2}\langle I_j(\mathbf{r}')\rangle^{1/2}} \tag{44.6}$$

is the normalized correlation function of two components of the field. From (44.3) and (44.5), we find that

$$\langle \delta I(\mathbf{r})\delta I(\mathbf{r}')\rangle = \sum_{i,j} \langle I_i(\mathbf{r})\rangle\langle I_j(\mathbf{r}')\rangle \, |\mu_{ij}(\mathbf{r},\mathbf{r}')|^2 \,. \tag{44.7}$$

This expression connects the intensity correlation function to the field correlation function for Gaussian random fields.

In Section 43.3, we have shown that at distances from the source much larger than the scattering mean free path, the field can be considered to be depolarized. For unpolarized light, $\mu_{ij}(\mathbf{r},\mathbf{r}')$ vanishes for $i \neq j$ and $\langle I_x(\mathbf{r})\rangle = \langle I_y(\mathbf{r})\rangle = \langle I_z(\mathbf{r})\rangle = (1/3)\langle I(\mathbf{r})\rangle$. Equation (44.7) can be rewritten as

$$\langle \delta I(\mathbf{r})\delta I(\mathbf{r}')\rangle = \frac{1}{9}\langle I(\mathbf{r})\rangle\langle I(\mathbf{r}')\rangle \sum_i |\mu_{ii}(\mathbf{r},\mathbf{r}')|^2 \,. \tag{44.8}$$

Finally, we find that

$$C(\mathbf{r},\mathbf{r}') = \frac{1}{9}\sum_i |\mu_{ii}(\mathbf{r},\mathbf{r}')|^2 \,, \tag{44.9}$$

which relates the intensity correlation function to the field correlation function in an unpolarized speckle.

An interesting feature of intensity fluctuations in unpolarized speckles can be deduced from (44.8). By taking $\mathbf{r} = \mathbf{r}'$, we obtain

$$\langle \delta I(\mathbf{r})^2\rangle = \frac{1}{3}\langle I(\mathbf{r})\rangle^2 \,, \tag{44.10}$$

which differs from the result obtained for scalar waves, (29.11), by a factor of $1/3$. Thus the contrast in an unpolarized three-dimensional speckle is reduced by a factor of $1/\sqrt{3}$ compared to the scalar case.

44.2 Field Correlation Function

Let us consider an infinite non-absorbing scattering medium, described by a dielectric function of the form $\epsilon(\mathbf{r}) = 1 + \delta\epsilon(\mathbf{r})$, where $\delta\epsilon$ is real-valued. The fluctuations $\delta\epsilon(\mathbf{r})$ obey Gaussian white-noise statistics with

$$\langle \delta\epsilon(\mathbf{r}) \rangle = 0 \,, \tag{44.11}$$

$$\langle \delta\epsilon(\mathbf{r})\, \delta\epsilon(\mathbf{r}') \rangle = \frac{6\pi}{k_0^4 \ell_s}\, \delta(\mathbf{r} - \mathbf{r}') \,. \tag{44.12}$$

Here ℓ_s and $k_0 = 2\pi/\lambda$ are the scattering mean free path and the wavenumber in the background medium, respectively. This form of the dielectric function corresponds to the model of disorder described by (42.1) in the limit $\ell_c \to 0$.

In the weak-scattering regime, with $\ell_s \gg \lambda$, it follows from the Bethe–Salpeter equation (43.10), that the field correlation function obeys

$$\langle E_i(\mathbf{r})\, E_j^*(\mathbf{r}') \rangle = \frac{6\pi}{\ell_s} \int \langle G_{ik}(\mathbf{r}, \mathbf{r}_1) \rangle\, \langle G_{jl}^*(\mathbf{r}', \mathbf{r}_1) \rangle \langle E_k(\mathbf{r}_1)\, E_l^*(\mathbf{r}_1) \rangle\, d^3 r_1 \,, \tag{44.13}$$

where $\langle \mathbf{G} \rangle$ is the average Green's function and summation over repeated indices is implied. Here we have assumed that the points of observation are far from the source of the field and have neglected the exponentially small contribution of the average field. We also note that $\langle E_i(\mathbf{r})\, E_j^*(\mathbf{r}') \rangle$ corresponds to the coherence tensor defined in (43.12).

Far from the source, we take the speckle to be unpolarized, consistent with the diffusion approximation derived in Section 43.3. This means that $\langle E_k(\mathbf{r}_1)\, E_l^*(\mathbf{r}_1) \rangle = (1/3) \langle |\mathbf{E}(\mathbf{r}_1)|^2 \rangle \delta_{kl}$, and we can rewrite (44.13) in the form

$$\langle E_i(\mathbf{r})\, E_j^*(\mathbf{r}') \rangle = \frac{2\pi}{\ell_s} \int \langle G_{ik}(\mathbf{r}, \mathbf{r}_1) \rangle\, \langle G_{jk}^*(\mathbf{r}', \mathbf{r}_1) \rangle\, \langle |\mathbf{E}(\mathbf{r}_1)|^2 \rangle\, d^3 r_1 \,. \tag{44.14}$$

Since the diffuse intensity $\langle |\mathbf{E}(\mathbf{r}_1)|^2 \rangle$ is slowly varying, it can be taken outside of the integral. The latter can be performed using (40.34), which in this context is given by

$$k_0^2\, \mathrm{Im}\, \varepsilon_{\mathrm{eff}} \int \langle G_{ik}(\mathbf{r}, \mathbf{r}_1) \rangle\, \langle G_{jk}^*(\mathbf{r}', \mathbf{r}_1) \rangle\, d^3 r_1 = \mathrm{Im} \langle G_{ij}(\mathbf{r}, \mathbf{r}') \rangle \,, \tag{44.15}$$

where ε_{eff} the effective dielectric function defined in Eq. (42.12) with $\text{Im}\,\varepsilon_{\text{eff}} = 1/(k_0 \ell_s)$. We thereby obtain

$$\langle E_i(\mathbf{r})\, E_j^*(\mathbf{r}') \rangle = \frac{2\pi}{k_0} \langle |\mathbf{E}(\mathbf{r})|^2 \rangle \text{Im} \langle G_{ij}(\mathbf{r}, \mathbf{r}') \rangle \,. \tag{44.16}$$

We note that the spatial dependence of the field correlation function is controlled by the imaginary part of the Green's function or, equivalently, by the cross density of states introduced in Section 38.7.

44.3 Degree of Spatial Coherence

We now focus on the calculation of the normalized trace of the coherence tensor

$$\mu(\mathbf{r}, \mathbf{r}') = \frac{\sum_i \langle E_i(\mathbf{r})\, E_i^*(\mathbf{r}') \rangle}{\sqrt{\langle |\mathbf{E}(\mathbf{r})|^2 \rangle} \sqrt{\langle |\mathbf{E}(\mathbf{r}')|^2 \rangle}} \,, \tag{44.17}$$

which can be taken as a definition of the degree of spatial coherence of the field. From (44.16), we readily obtain

$$\mu(\mathbf{r}, \mathbf{r}') = \frac{2\pi}{k_0} \text{Im}[\text{Tr} \langle \mathbf{G}(\mathbf{r}, \mathbf{r}') \rangle] \,. \tag{44.18}$$

Here we have assumed $\langle |\mathbf{E}(\mathbf{r})|^2 \rangle \simeq \langle |\mathbf{E}(\mathbf{r}')|^2 \rangle$ in the definition of $\mu(\mathbf{r}, \mathbf{r}')$ since the intensity is slowly varying.

The imaginary part of the trace of the average Green's function can be deduced from (42.28), noting that near-field terms do not contribute. Taking $k_R = k_0$ in the weak-scattering limit, we obtain

$$\mu(\mathbf{r}, \mathbf{r}') = \text{sinc}\left(\frac{2\pi R}{\lambda} \right) \exp\left(-\frac{R}{2\ell_s} \right) , \tag{44.19}$$

with $R = |\mathbf{r} - \mathbf{r}'|$.

We observe that the degree of spatial coherence varies on the scale of the wavelength λ in an uncorrelated medium (since we have assumed $\ell_s \gg \lambda$, the influence of the exponential term is negligible). The above expression has been derived for uncorrelated disorder. Nevertheless, in the presence of correlations, we expect the result to remain unchanged, provided that the condition $\ell_c \ll \lambda$ is satisfied (which we have assumed from the beginning). We conclude that the expression of the degree of spatial coherence given by (44.19) also applies to a medium with short-range correlations.

It is also interesting to note that the form of the degree of spatial coherence given by (44.19) is the same as that obtained for scalar waves in Chapter 30. In practice, it is characterized by the function $\text{sinc}(2\pi\rho/\lambda)$, leading to a spatial coherence length of $\lambda/2$. This result is independent of the internal structure of the disordered medium, thus being universal for unpolarized speckles obeying Gaussian statistics. This result for the degree of spatial coherence is also found in the case of equilibrium thermal radiation (blackbody radiation) in a homogeneous medium, which can also be seen as a speckle field resulting from thermal fluctuations in the source. A relation similar to (44.19) is obtained in this case, as a direct consequence of the fluctuation-dissipation theorem.

References and Additional Reading

The moment theorem for Gaussian random variables and the factorization of the intensity correlation function are presented in:

L. Mandel and E. Wolf, *Optical Coherence and Quantum Optics* (Cambridge University Press, Cambridge, 1995), sec. 1.6.3 and 8.4.1.

The following paper discusses the factorization of the intensity correlation function of vector field following Gaussian statistics and contains a derivation of the degree of polarization:

T. Setälä, K. Lindfors, M. Kaivola, J. Tervo and A.T. Friberg, Opt. Lett. **29**, 2587 (2004).

Bulk speckle correlations for scalar fields are calculated in:

B. Shapiro, Phys. Rev. Lett. **57**, 2168 (1986).

P. Sheng, *Introduction to Wave Scattering, Localization, and Mesoscopic Phenomena* (Academic Press, San Diego, 1995), chap. 10.

They are also studied in Chapter 30.

Polarization and spatial coherence in 3D electromagnetic speckles is treated theoretically in:

Ph. Réfrégier, V. Wasik, K. Vynck and R. Carminati, Opt. Lett. **39**, 2362 (2014).

K. Vynck, R. Pierrat and R. Carminati, Phys. Rev. A **89**, 013842 (2014).

A. Dogariu and R. Carminati, Phys. Rep. **559**, 1 (2015).

A calculation of the depolarization length in 3D electromagnetic speckles in the presence of short-range structural correlations of the medium can be found in:

K. Vynck, R. Pierrat and R. Carminati, Phys. Rev. A **94**, 033851 (2016).

Spatial correlations of blackbody radiation fields are studied in:

G.S. Agarwal, Phys. Rev. A **11**, 230 (1975).

L.D. Landau, E.M. Lifshitz and L.P. Pitaevskii, *Statistical Physics* (Pergamon Press, Oxford, 1980), 3rd ed., part 1, chap. 12, and part 2, chap. 8.

S.M. Rytov, Yu.A. Kravtsov and V.I. Tatarskii, *Principles of Statistical Radiophysics* (Springer-Verlag, Berlin, 1989), vol. 3, chap. 3.

K. Joulain, J.P. Mulet, F. Marquier, R. Carminati and J.-J. Greffet, Surf. Sci. Rep. **57**, 59 (2005).

The universality of the spatial correlation function of statistically homogeneous and isotropic electromagnetic fields was put forward in:

T. Setälä, K. Blomstedt, M. Kaivola and A.T. Friberg, Phys. Rev. E **67**, 026613 (2003).

45

Near-Field Speckle Correlations

In this chapter, we study spatial correlations of the electric field close to the surface of a disordered medium. The calculation makes use of the formalism introduced in Chapter 44 and includes the contribution from evanescent components of the field.[1]

45.1 Field Correlation Function in a Semi-Infinite Geometry

We consider the geometry in Fig. 45.1, in which a speckle pattern is produced in transmission through a slab of a disordered medium of thickness $L \gg \ell_s$, with ℓ_s the scattering mean free path. The exit surface is on average a flat interface, that we arbitrarily choose to be the plane $z = 0$. As a measure of spatial correlations, we will calculate the degree of spatial coherence in a plane at a distance z. The calculation will include the near-field regime with $z \ll \lambda$, λ being the incident

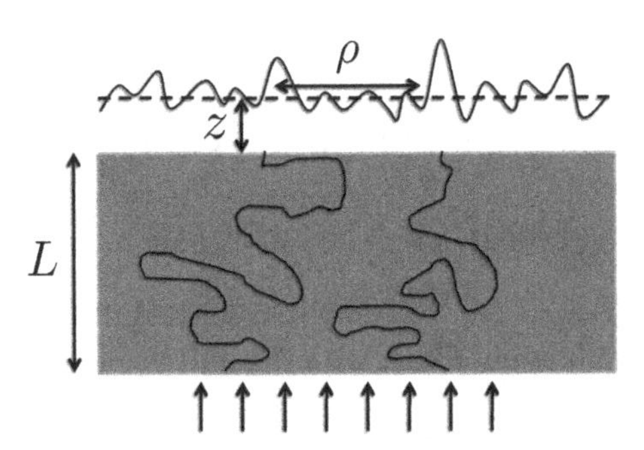

Fig. 45.1 Illustrating the geometry in which the speckle pattern is observed in a plane at a distance z from the surface of the medium.

[1] The approach is similar to that initially presented in R. Carminati, Phys. Rev. A **81**, 053804 (2010).

wavelength. The degree of spatial coherence can be defined as

$$\mu(\mathbf{r}, \mathbf{r}') = \frac{\sum_i \langle E_i(\mathbf{r})\, E_i^*(\mathbf{r}')\rangle}{\langle |\mathbf{E}(\mathbf{r})|^2\rangle}\,. \tag{45.1}$$

Here we have accounted for translational invariance along the detection plane by assuming $\langle |\mathbf{E}(\mathbf{r})|^2\rangle = \langle |\mathbf{E}(\mathbf{r}')|^2\rangle$ in the denominator.

The scattering medium is described by a dielectric function of the form $\epsilon(\mathbf{r}) = 1 + \delta\epsilon(\mathbf{r})$, which is taken to be real-valued, with fluctuations obeying Gaussian white-noise statistics governed by (42.1). We assume that the condition $\ell_c \ll \lambda \ll \ell_s \ll L$ is satisfied, where ℓ_c is the correlation length of disorder. It follows from the Bethe–Salpeter equation (43.10), in the ladder approximation, that the correlation function between two components i and j of the field obeys

$$\langle E_i(\mathbf{r})\, E_j^*(\mathbf{r}')\rangle = k_0^4 \int \langle G_{ik}(\mathbf{r}, \mathbf{r}_1)\rangle\, \langle G_{jl}^*(\mathbf{r}', \mathbf{r}_1')\rangle \\ \times \langle \delta\epsilon(\mathbf{r}_1)\, \delta\epsilon(\mathbf{r}_1')\rangle \langle E_k(\mathbf{r}_1)\, E_l^*(\mathbf{r}_1')\rangle\, d^3r_1\, d^3r_1'\,. \tag{45.2}$$

In this expression, we have neglected the exponentially small contribution of the average field. In the geometry shown in Fig. 45.1, the observation points $\mathbf{r}$ and $\mathbf{r}'$ are located in a plane at a distance z from the surface of the medium. Since the average Green's function decays on the scale of ℓ_s [with a scaling similar to that in (42.27)], the effect of the bottom interface is negligible, and it suffices to consider a semi-infinite geometry. We then require the expression for the Green's function $\langle \mathbf{G}\rangle$ in the geometry consisting of a flat interface $z = 0$ separating a vacuum half-space $z > 0$ from a homogeneous effective medium with dielectric constant ε_{eff} that fills the half-space $z < 0$. Here ε_{eff} is defined in (42.12) so that $\text{Im}\, \varepsilon_{\text{eff}} = 1/(k_0\ell_s)$. The expression for $\langle \mathbf{G}\rangle$ can be taken from Section 37.6, and written in the form of a plane-wave expansion. Here the half-space $z < 0$ is medium 1, with dielectric constant $\varepsilon_1 = \varepsilon_{\text{eff}}$, and the half-space $z > 0$ is medium 2, with dielectric constant $\varepsilon_2 = 1$. We then find that

$$\langle \mathbf{G}(\mathbf{r}, \mathbf{r}')\rangle = \frac{i}{8\pi^2} \int \frac{1}{k_z^1(q)} \left[T_s(q)\, \hat{\mathbf{s}} \otimes \hat{\mathbf{s}} + T_p(q)\, \hat{\mathbf{p}}_2 \otimes \hat{\mathbf{p}}_1\right] \\ \times \exp[i\mathbf{q} \cdot (\boldsymbol{\rho} - \boldsymbol{\rho}')]\, \exp[ik_z^2(q)z - ik_z^1(q)z']\, d^2q\,, \tag{45.3}$$

where $\boldsymbol{\rho} = (x, y)$ and $\boldsymbol{\rho}' = (x', y')$. The tensor term in the integral describes the polarization behavior at the interface, with the unit vectors

$$\hat{\mathbf{s}} = \hat{\mathbf{q}} \times \hat{\mathbf{z}} \tag{45.4}$$

$$\hat{\mathbf{p}}_j = [q\hat{\mathbf{z}} - k_z^j(q)\hat{\mathbf{q}}]/k_j\,,\quad j = 1, 2\,, \tag{45.5}$$

defining the s and p polarizations. The components of the wavevector along the z direction are $k_z^j(q) = (k_j^2 - q^2)^{1/2}$, with the conditions $\text{Re}(k_z^j(q)) > 0$ and

$\mathrm{Im}(k_z^j(q)) > 0$. In addition, $k_1 = \sqrt{\varepsilon_{\mathrm{eff}}}\, k_0$, $k_2 = k_0$, with $k_0 = 2\pi/\lambda$ being the wavenumber in vacuum. The quantities $T_s(q)$ and $T_p(q)$ are the amplitude Fresnel transmission coefficients from medium 1 to medium 2 and are given by

$$T_s(q) = 2k_z^1(q)/[k_z^1(q) + k_z^2(q)] \,, \tag{45.6}$$

$$T_p(q) = 2\sqrt{\varepsilon_{\mathrm{eff}}}\, k_z^1(q)/[\varepsilon_{\mathrm{eff}}\, k_z^2(q) + k_z^1(q)]. \tag{45.7}$$

In the region $z > 0$, the plane-wave expansion (45.3) contains both propagating and evanescent components. The latter corresponds to high spatial frequencies $q > k_0$ and imaginary values of $k_z^2(\mathbf{q})$. The evanescent waves dominate at distances $z \ll \lambda$, in the near-zone, and strongly influence the spatial correlation function at short distance from the interface.

The integration in (45.2) can be performed by assuming that the correlation tensor inside the medium coincides with that of an infinite medium, calculated far from the sources, and taken to be unpolarized according to the results in Section 43.3. This leads to $\langle E_k(\mathbf{r}_1)\, E_l^*(\mathbf{r}_1')\rangle = \langle E_k(\mathbf{r}_1)\, E_k^*(\mathbf{r}_1')\rangle\, \delta_{kl}$. Moreover, $\langle E_k(\mathbf{r}_1)\, E_l^*(\mathbf{r}_1')\rangle$ varies on length scales on the order of λ and ℓ_s, according to the calculations in Chapter 44. Since $\ell_c \ll \lambda$, we can also write $\langle \delta\epsilon(\mathbf{r}_1)\delta\epsilon(\mathbf{r}_1')\rangle \langle E_k(\mathbf{r}_1)\, E_k^*(\mathbf{r}_1')\rangle \simeq \langle \delta\epsilon(\mathbf{r}_1)\delta\epsilon(\mathbf{r}_1')\rangle \langle |\mathbf{E}|^2\rangle_{bulk}/3$, where $\langle |\mathbf{E}|^2\rangle_{bulk}$ is the diffuse intensity in the infinite medium. Substituting (45.3) into (45.2), and using the above approximation, it can be shown after some calculation that the trace of the coherence tensor takes the form

$$\sum_i \langle E_i(\mathbf{r})\, E_i^*(\mathbf{r}')\rangle = \int H(q,z)\, \exp[i\mathbf{q}\cdot(\boldsymbol{\rho}-\boldsymbol{\rho}')]\, d^2q \,, \tag{45.8}$$

which, noticing that the function $H(q,z)$ only depends on $q = |\mathbf{q}|$, yields

$$\begin{aligned} \sum_i \langle E_i(\mathbf{r})\, E_i^*(\mathbf{r}')\rangle &= \int_0^\infty dq\, q\, H(q,z) \int_0^{2\pi} d\theta\, \exp(iq\rho\cos\theta) \\ &= \int_0^\infty q\, H(q,z)\, J_0(q\rho)\, dq \,, \end{aligned} \tag{45.9}$$

where $\rho = |\boldsymbol{\rho} - \boldsymbol{\rho}'|$ is the distance between the observation points in the detection plane and J_0 is the zeroth-order Bessel function. Writing $H(q,z)$ explicitly, we obtain

$$\begin{aligned} \sum_i \langle E_i(\mathbf{r})\, E_i^*(\mathbf{r}')\rangle = \frac{\langle |\mathbf{E}|^2\rangle_{bulk}}{4\ell_s} \int_0^\infty & f(q,z)\, \exp[-(q^2 + (\mathrm{Re}\, k_z^1(q))^2)\, \ell_c^2/4] \\ & \times J_0(q\rho)\, dq \,, \end{aligned} \tag{45.10}$$

where

$$f(q,z) = \frac{q}{2\mathrm{Im}\, k_z^1(q)} \left[\frac{|T_s(q)|^2}{|k_z^1(q)|^2} + \frac{(q^2 + |k_z^2(q)|^2)(q^2 + |k_z^1(q)|^2)}{|\sqrt{\varepsilon_{\mathrm{eff}}}|^2\, k_0^4} \frac{|T_p(q)|^2}{|k_z^1(q)|^2} \right] \times \exp(-2\,\mathrm{Im}\, k_z^2(q)\, z)\,. \tag{45.11}$$

Equation (45.10) is the starting point for explicit calculations of the degree of spatial coherence. We note that the function $f(q,z)$ has two important features. It handles the transmission of polarized waves at the interface between the effective medium and the observation plane. It also describes the attenuation of high spatial frequencies, with $q > k_0$, through the exponential factor $\exp(-2\,\mathrm{Im}\, k_z^2(q)\, z)$. The effect of the evanescent components is a major difference between speckles patterns observed in the near-field or in the far-field.

45.2 Far-Field Regime

We first consider the far-field regime, with $z \gg \lambda$. We can neglect the influence of the exponential term $\exp[-(q^2 + (\mathrm{Re}\, k_z^1(q))^2)\, \ell_c^2/4]$ in (45.10), compared to that of $\exp(-2\,\mathrm{Im}\, k_z^2(q)\, z)$. The latter acts as a filter for high spatial frequencies, and reduces the range of $f(q,z)$ to $0 \le q \le k_0$, meaning that only propagating waves contribute to the far-field. We thus obtain

$$\sum_i \langle E_i(\mathbf{r})\, E_i^*(\mathbf{r}')\rangle = \frac{\langle |\mathbf{E}|^2\rangle_{bulk}}{2k_0} \int_0^{k_0} \frac{q}{\sqrt{k_0^2 - q^2}} [1 - \mathcal{R}(q)]\, J_0(q\rho)\, dq\,, \tag{45.12}$$

where $\mathcal{R}(q) = [|R_s(q)|^2 + |R_p(q)|^2]/2$ is the intensity Fresnel reflection factor averaged over polarizations. The above integral could be computed numerically without difficulty. Nevertheless, a simpler expression can be obtained by neglecting the dependence of $\mathcal{R}$ on q, which is relevant when $\mathrm{Re}(\varepsilon_{\mathrm{eff}}) \simeq 1$ (low index mismatch). Under this assumption, we can simplify the preceding expression according to

$$\sum_i \langle E_i(\mathbf{r})\, E_i^*(\mathbf{r}')\rangle = \frac{(1-\mathcal{R})}{2} \langle |\mathbf{E}|^2\rangle_{bulk}\, \mathrm{sinc}(k_0\rho)\,. \tag{45.13}$$

The prefactor $(1-\mathcal{R})/2$ is the fraction of the bulk average intensity that is transmitted through the interface. After normalization, the degree of spatial coherence in a plane at a given distance $z \gg \lambda$ from the interface is given by

$$\mu = \mathrm{sinc}(k_0\rho). \tag{45.14}$$

The degree of spatial coherence in the far-field is plotted in Fig. 45.2 (circles). The sinc function leads to a spatial coherence length $\lambda/2$, which can be taken as

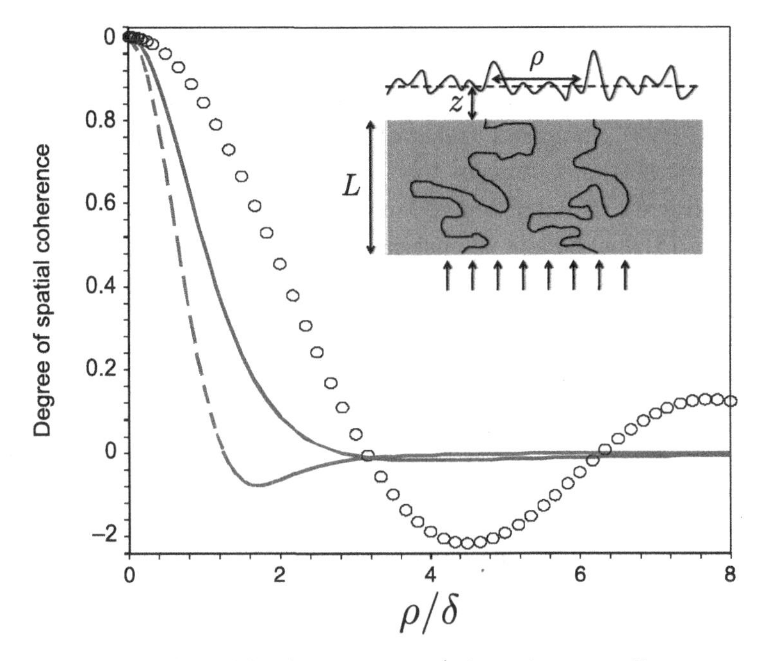

Fig. 45.2 Degree of spatial coherence $\mu(\mathbf{r}, \mathbf{r}')$ in a plane at a distance z versus ρ/δ, with ρ the distance between the observation points in the plane. Circles: far-field regime ($z \gg \lambda$), $\delta = \lambda/2\pi$. Solid line: near-field regime ($\ell_c \ll z \ll \lambda$), $\delta = z$. Dashed line: extreme near-field regime ($z \simeq \ell_c \ll \lambda$), $\delta = \ell_c$.

a measure of the average size of a speckle spot. It is interesting to note that the shape of the degree of spatial coherence is identical to that obtained for thermal equilibrium (blackbody) radiation in the far-field of a planar thermal source.

Finally, we point out that a coherence length $\lambda/2$ is found for plane-wave illumination, which corresponds in practice to an observation distance z smaller than the beam size W at the output surface of the scattering medium. In the regime $z \geq W$, the coherence length scales as $\lambda z/W$ due to diffraction.

45.3 Near-Field Regime

We now consider the near-field regime $z \ll \lambda$, for which the contribution of evanescent waves is expected to dominate. The integrand in (45.10) can be expanded to leading order in the limit $q \gg k_0$, corresponding to the quasi-static limit discussed in Chapter 38. We then find that

$$\sum_i \langle E_i(\mathbf{r})\, E_i^*(\mathbf{r}')\rangle = \frac{2\,\langle|\mathbf{E}|^2\rangle_{bulk}}{k_0^4\,\ell_s\,|\varepsilon_{\text{eff}}+1|^2} \int_0^\infty q^2 \exp(-2qz)$$

$$\times \exp(-q^2 \ell_c^2/4)\, J_0(q\rho)\, dq\,. \tag{45.15}$$

Assuming that $\ell_c \ll z \ll \lambda$, we can neglect the influence of the exponential term $\exp(-q^2 \ell_c^2/4)$, compared to that of $\exp(-2qz)$. Physically, this means that the observation distance z remains too large for the speckle to be influenced by the internal structure of the disordered medium. In this case, we can calculate the integral in (45.15) analytically, by making use of the identity

$$\int_0^\infty J_0(q\rho)\, \exp(-2qz)\, dq = \frac{1}{(4z^2 + \rho^2)^{1/2}}. \tag{45.16}$$

We obtain after normalization

$$\mu = \frac{1 - \rho^2/(8z^2)}{[1 + \rho^2/(4z^2)]^{5/2}}\,, \quad \ell_c \ll z \ll \lambda\,. \tag{45.17}$$

The degree of spatial coherence in the near-field is plotted in Fig. 45.2 (solid line). Its width, which determines the spatial coherence length, is on the order of the observation distance z. This result is a feature of the quasi-static regime, in which the spatial structure of the field is governed by geometrical length scales, as in electrostatics.

45.4 Extreme Near-Field

The decrease of the spatial coherence length with the observation distance z predicted by (45.17) saturates when $z \simeq \ell_c$. Indeed, at such distances, the internal structure of the disordered material directly influences the structure of the field. Although the multiple scattering formalism used here is, in principle, valid only when field variations remain larger than ℓ_c (this is the condition for using a local dielectric function ε_{eff}, as shown in Chapter 42), it is instructive to push the model to the limit $z \simeq \ell_c$. In this case, the exponential cut-off $\exp(-q^2 \ell_c^2/4)$ in (45.15) dominates, and the exponential term $\exp(-2qz)$ can be neglected. Making use of the result

$$\int_0^\infty q^2 \exp(-q^2 \ell_c^2/4) J_0(q\rho)\, dq = \frac{2\sqrt{\pi}}{\ell_c^3}\, M\left(\frac{3}{2}, 1, \frac{-\rho^2}{\ell_c^2}\right), \tag{45.18}$$

where M is the confluent hypergeometric function, we obtain

$$\mu = M\left(\frac{3}{2}, 1, \frac{-\rho^2}{\ell_c^2}\right), \quad z \simeq \ell_c \ll \lambda\,. \tag{45.19}$$

The degree of spatial coherence given by (45.19) is plotted in Fig. 45.2 (dashed line). In the extreme near-field regime, we find a width on the order of ℓ_c, which results from the strong correlation between the spatial variations of the near-field and of the microstructure of the medium itself. The dependence of the degree of spatial coherence on ℓ_c gives a non-universal character to electromagnetic speckle correlations in the near-field. In particular, the shape of the degree of coherence in this regime depends on the form of the correlation function $\langle \delta\epsilon(\mathbf{r})\, \delta\epsilon(\mathbf{r}')\rangle$, the result given here corresponding to the model in (42.1).

References and Additional Reading

The study of near-field speckle correlations presented in this chapter was inspired by the work initially presented in this reference:

R. Carminati, Phys. Rev. A **81**, 053804 (2010).

This paper proposed a calculation of near-field speckle correlations in diffusion theory (excluding contributions from evanescent waves):

I. Freund and D. Eliyahu, Phys. Rev. A **45**, 6133 (1992).

Theoretical studies of near-field speckles produced by scattering from a rough surface are presented in:

J.-J. Greffet and R. Carminati, Ultramicroscopy **61**, 43 (1995).
J.-J. Greffet and R. Carminati, Prog. Surf. Sci. **56**, 133 (1997).
D. Franta and I. Ohlídal, Opt. Commun. **147**, 349 (1998).
J.-J. Greffet and R. Carminati, "Speckle Pattern in the Near Field," in *Light Scattering and Nanoscale Surface Roughness*, A.A. Maradudin (ed.), Springer Series on Nanostructure Science and Technology (Springer, New York, 2007), chap. 15.

Near-field correlations of thermal radiation were calculated in:

R. Carminati and J.-J. Greffet, Phys. Rev. Lett. **82**, 1660 (1999).
C. Henkel, K. Joulain, R. Carminati and J.-J. Greffet, Opt. Commun. **186**, 57 (2000).
T. Setälä, M. Kaivola and A.T. Friberg, Phys. Rev. Lett. **88**, 123902 (2002).
C. Henkel and K. Joulain, Appl. Phys. B **84**, 61 (2006). The last paper discusses the three regimes described in this chapter in the context of thermal radiation.

The following papers report measurements of near-field speckle correlations in optics:

V. Emiliani, F. Intonti, M. Cazayous, D.S. Wiersma, M. Colocci, F. Aliev and A. Lagendijk, Phys. Rev. Lett. **90**, 250801 (2003).
A. Apostol and A. Dogariu, Phys. Rev. Lett. **91**, 093901 (2003).
A. Apostol and A. Dogariu, Opt. Lett. **29**, 235 (2004).
V. Parigi, E. Perros, G. Binard, C. Bourdillon, A. Maitre, R. Carminati, V. Krachmalnicoff and Y. De Wilde, Opt. Express **24**, 7019 (2016). In the last paper the three regimes described in this chapter have been identified experimentally.

This is a textbook with useful integrals of Bessel functions:

Handbook of Mathematical Functions, edited by M. Abramowitz and I.A. Stegun (Dover, New York, 1972).

46

Speckle Correlations Produced by a Point Source

In this chapter, we study intensity correlations and fluctuations in speckle patterns produced by an electric-dipole point source located inside a disordered medium. We analyze the role of near-field interactions between the source and its local environment.

46.1 Angular Intensity Correlation Function

We consider a non-absorbing disordered medium embedded in a sphere of radius R, as illustrated in Fig. 46.1. The medium is illuminated from the inside by a point source with electric dipole moment $\mathbf{p}$, which is located at the center of the sphere and radiates at the frequency ω. The electric field observed at a point $\mathbf{r}$ in the far-zone is of the form

$$\mathbf{E}(\mathbf{r}) = \mathbf{A}(\hat{\mathbf{r}}) \, \frac{\exp(ik_0 r)}{r} , \tag{46.1}$$

where $\mathbf{A}$ is the amplitude and $k_0 = \omega/c$ is the wavenumber in vacuum. The far-field intensity radiated per unit solid angle is given by

$$I(\hat{\mathbf{r}}) = \frac{\epsilon_0 c}{2} \, |\mathbf{A}(\hat{\mathbf{r}})|^2 . \tag{46.2}$$

The fluctuations in the speckle pattern can be characterized by the normalized angular correlation function $C(\hat{\mathbf{r}}, \hat{\mathbf{r}}')$ defined as

$$C(\hat{\mathbf{r}}, \hat{\mathbf{r}}') = \frac{\langle \delta I(\hat{\mathbf{r}}) \delta I(\hat{\mathbf{r}}') \rangle}{\langle I(\hat{\mathbf{r}}) \rangle \langle I(\hat{\mathbf{r}}') \rangle} = \frac{\langle I(\hat{\mathbf{r}}) I(\hat{\mathbf{r}}') \rangle}{\langle I(\hat{\mathbf{r}}) \rangle \langle I(\hat{\mathbf{r}}') \rangle} - 1 , \tag{46.3}$$

where $\delta I(\hat{\mathbf{r}}) = I(\hat{\mathbf{r}}) - \langle I(\hat{\mathbf{r}}) \rangle$. If the field obeys Gaussian statistics, the intensity correlation function can be written in terms of field correlation functions, as shown in Section 44.1. Here we will discuss a non-Gaussian contribution to the intensity correlation function, which is shown to be the dominant contribution for speckle patterns produced by a point source.

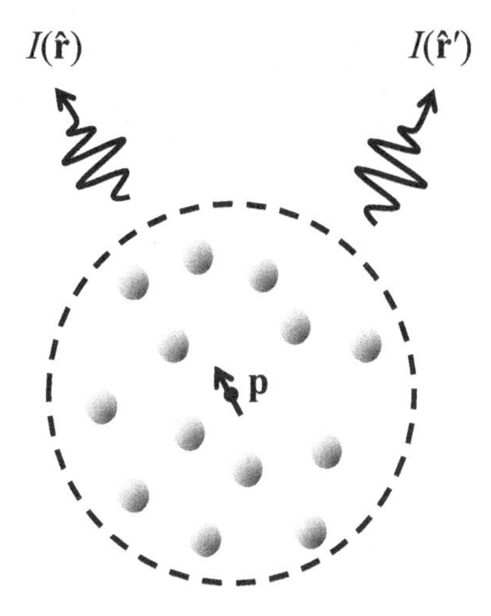

Fig. 46.1 Light emission from a point source located inside a scattering medium.

46.2 Speckle Correlations and Fluctuations of the Local Density of States

We begin by assuming that the speckle field is statistically isotropic, so that the angular intensity correlation function $C(\hat{\mathbf{r}}, \hat{\mathbf{r}}')$ is a function of $\cos\Theta = \hat{\mathbf{r}} \cdot \hat{\mathbf{r}}'$ only.[1] This condition is satisfied provided that the disordered medium is statistically isotropic and that an average over the orientations of the point dipole source is performed. Next, we define the total emitted power $P = \int I(\hat{\mathbf{r}})\, d\hat{\mathbf{r}}$. Due to statistical isotropy, we immediately find that $\langle P \rangle = 4\pi \, \langle I(\hat{\mathbf{r}}) \rangle$. We can also write the normalized variance of the total emitted power in the form

$$\frac{\langle P^2 \rangle - \langle P \rangle^2}{\langle P \rangle^2} = \frac{1}{16\pi^2} \int \int C(\hat{\mathbf{r}}, \hat{\mathbf{r}}')\, d\hat{\mathbf{r}}\, d\hat{\mathbf{r}}' \,. \tag{46.4}$$

We also introduce the expansion of the correlation function in Legendre polynomials:

$$C(\hat{\mathbf{r}}, \hat{\mathbf{r}}') = \sum_{n=0}^{\infty} C_n \, P_n(\cos\Theta). \tag{46.5}$$

Since $P_0(\cos\Theta) = 1$, the first term C_0 in the expansion corresponds to a constant term in the correlation function. Using the addition theorem for spherical harmonics, we obtain

$$\begin{aligned} P_n(\cos\Theta) = P_n(\cos\theta) P_n(\cos\theta') + 2 \sum_{m=1}^{n} \frac{(n-m)!}{(n+m)!} \\ \times P_n^m(\cos\theta) P_n^m(\cos\theta') \cos[m(\phi - \phi')] \,, \end{aligned} \tag{46.6}$$

[1] In this section, we follow the development initially presented in A. Cazé, R. Pierrat and R. Carminati, Phys. Rev. A **82**, 043823 (2010).

where (θ, ϕ) and (θ', ϕ') are the polar and azimuthal angles corresponding to the directions $\hat{\mathbf{r}}$ and $\hat{\mathbf{r}}'$, respectively, and $P_n^m(x)$ are the associated Legendre functions. Inserting this expansion into (46.4) and integrating over ϕ and ϕ', we obtain

$$\frac{\langle P^2 \rangle - \langle P \rangle^2}{\langle P \rangle^2} = \frac{1}{4} \sum_{n=0}^{\infty} C_n \left[\int_{-1}^{+1} P_n(x)\, dx \right]^2 . \tag{46.7}$$

Since $P_0(x) = 1$, we have

$$\int_{-1}^{+1} P_n(x)\, dx = \int_{-1}^{+1} P_n(x) P_0(x)\, dx = 2\, \delta_{n0} , \tag{46.8}$$

where we have made use of the orthogonality of the Legendre polynomials. This leads to the relation

$$C_0 = \frac{\langle P^2 \rangle - \langle P \rangle^2}{\langle P \rangle^2} , \tag{46.9}$$

which shows that the constant term C_0 in the angular intensity correlation function results from the fluctuations of the power P emitted by the point source.

In the absence of absorption, the power radiated outside the medium is also the power transferred from the dipole source to the field. Using (38.6), and summing over the orientations of the electric dipole $\mathbf{p}$, we find that

$$P = \frac{\mu_0\, \omega^3}{2} |\mathbf{p}|^2 \, \mathrm{Im}\, [\mathrm{Tr}\, \mathbf{G}(\mathbf{r}_s, \mathbf{r}_s)] , \tag{46.10}$$

where $\mathbf{G}$ is the Green's function of the disordered medium and $\mathbf{r}_s$ is the position of the dipole. In terms of the local density of states (LDOS) ρ defined in (38.30), we can also write

$$P = \frac{\pi\, \omega^2}{4\epsilon_0} |\mathbf{p}|^2 \, \rho(\mathbf{r}_s) . \tag{46.11}$$

Inserting (46.11) into (46.9), we finally obtain

$$C_0 = \frac{\langle \rho^2(\mathbf{r}_s) \rangle}{\langle \rho(\mathbf{r}_s) \rangle^2} - 1 = \frac{\mathrm{Var}[\rho^2(\mathbf{r}_s)]}{\langle \rho(\mathbf{r}_s) \rangle^2} . \tag{46.12}$$

This result shows that the constant term C_0 in the angular intensity correlation function equals the normalized variance of the LDOS at the source position. The derivation of this result relies only on the assumptions of energy conservation and statistical isotropy.

Equation (46.12) can be understood qualitatively in simple physical terms. For a point dipole, changes in the LDOS correspond to changes in the power transferred to the environment. In a non-absorbing medium, this power is radiated to the far-field, producing the angular speckle pattern. Therefore, LDOS fluctuations

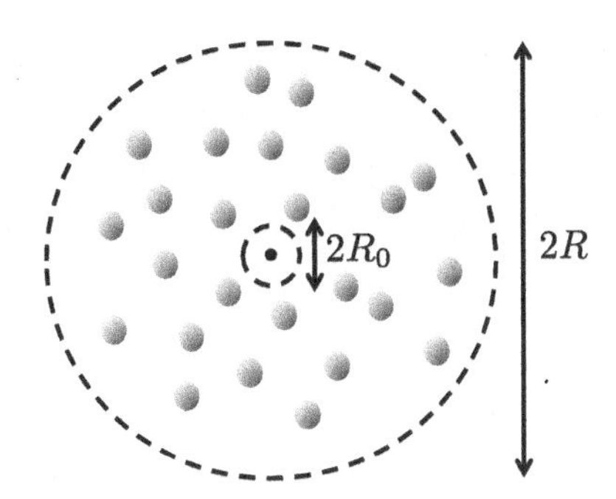

Fig. 46.2 Geometry considered for the calculation of LDOS fluctuations in the single scattering regime. The observation point is at the center of a disordered cluster of particles confined in a spherical volume with radius R and is surrounded by an exclusion volume of radius R_0.

are transformed into global fluctuations of the speckle pattern, which in turn are encoded in the form of a constant term in the angular correlation function.

46.3 Single Scattering

The C_0 contribution to the speckle correlation function is sensitive to the local environment of the source. This feature makes C_0 a non-universal speckle correlation. In this section, we compute C_0 explicitly in the single-scattering regime, and discuss its dependence on the structure of the medium.[2] The calculation also illustrates the influence of near-field interactions in the scattering of electromagnetic waves.

Defining the disordered medium more precisely, we now consider a spherical volume with radius R, made of N individual spherical scatterers with radius $a \ll \lambda$. The scatterers are randomly and independently placed within the sphere. The LDOS is calculated at the position $\mathbf{r}_s$, coinciding with the center of the volume and surrounded by a spherical exclusion volume of radius R_0, as shown in Fig. 46.2. Here R_0 is a microscopic length scale that characterizes the local environment of the source.

In the single-scattering regime, and for uncorrelated particle positions, the average value $\langle \rho \rangle_N$ and variance $\text{Var}_N(\rho)$ of the LDOS in a system with N scatterers are given by $\langle \rho \rangle_N = N \langle \rho \rangle_1$ and $\text{Var}_N(\rho) = N\text{Var}_1(\rho)$. The one-particle quantities have analytical expressions, which can be obtained using the approach of Chapter 41. A subwavelength scatterer is described by an electric polarizability $\alpha(\omega)$ given by (41.8) and (41.9). From (41.27) and (41.28), we can calculate the LDOS

[2] We summarize here the extensive study initially presented in L.S. Froufe-Pérez, R. Carminati and J.J. Sáenz, Phys. Rev. A **76**, 013835 (2007) and L.S. Froufe-Pérez and R. Carminati, Phys. Stat. Sol. (a) **205**, 1258 (2008).

$\rho = \rho_z + 2\rho_x$ summed over the possible orientations of the source and due to a single scatterer at a distance r from the observation point $\mathbf{r}_s$. We then obtain

$$\frac{\rho(r)}{\rho_0} = 1 + \frac{k_0}{2\pi}\,\mathrm{Im}\left[\alpha(\omega)\exp(2ik_0r)\left(\frac{1}{2r^2} - \frac{1}{ik_0r^3} - \frac{5}{2k_0^2r^4} + \frac{3}{ik_0^3r^5} + \frac{3}{2k_0^4r^6}\right)\right], \tag{46.13}$$

where ρ_0 is the LDOS in free space. For a spatially uniform density, the scatterers are distributed according to the probability density

$$P(r) = \frac{3\,r^2}{R^3 - R_0^3}, \tag{46.14}$$

in which the role of the exclusion volume appears explicitly. The average value of the normalized LDOS is defined as

$$\langle \rho/\rho_0 \rangle_1 = \int_{R_0}^{R} \frac{\rho(r)}{\rho_0} P(r)\,dr = \frac{3}{R^3 - R_0^3}\int_{R_0}^{R} \frac{\rho(r)}{\rho_0} r^2\,dr\,. \tag{46.15}$$

In general, the above integral cannot be calculated explicitly. The result can be greatly simplified by examining the limit of a small cluster such that $k_0R_0 \ll k_0R \ll 1$. Keeping only the leading order terms in R and R_0, and after some calculations, we find that

$$\langle \rho/\rho_0 \rangle_1 = 1 + \frac{11}{5} f\,\mathrm{Re}(\beta)\,(k_0R)^2 + \frac{3f}{(k_0R_0)^3}\,\mathrm{Im}(\beta)\,. \tag{46.16}$$

Here $f = N\,a^3/(R^3 - R_0^3)$ is the volume fraction of scatterers and $\beta = [\varepsilon - 1]/[\varepsilon + 2]$, with ε the dielectric function of the scatterers. The variance of the normalized LDOS, defined as

$$\mathrm{Var}_1(\rho/\rho_0) = \int_{R_0}^{R}\left[\frac{\rho^2(r)}{\rho_0} - \langle \rho/\rho_0 \rangle_{(1)}^2\right] P(r)\,dr\,, \tag{46.17}$$

can be calculated explicitly in the same limit. To leading order in R and R_0, we obtain

$$\begin{aligned}\mathrm{Var}_1(\rho/\rho_0) = f\,(k_0a)^3 \Bigg\{ & \frac{121}{75} k_0R\,[\mathrm{Re}(\beta)]^2 \\ & + \frac{3}{(k_0R_0)^9}\left[\mathrm{Im}(\beta) + \frac{2}{3}(k_0a)^3\,|\beta|^2\right]^2 \\ & + \frac{33}{5}\frac{\mathrm{Re}(\beta)}{(k_0R_0)^4}\left[\mathrm{Im}(\beta) + \frac{2}{3}(k_0a)^3\,|\beta|^2\right]\Bigg\}\,. \end{aligned} \tag{46.18}$$

Note that for non-absorbing scatterers, the dielectric function is real and the terms proportional to $\mathrm{Im}(\beta)$ vanish. Finally, from (46.16) and (46.18), we can deduce the expression of the C_0 contribution to the speckle correlation function

$$C_0 = \frac{\mathrm{Var}_N(\rho/\rho_0)}{\langle\rho/\rho_0\rangle_N^2} = \frac{\mathrm{Var}_1(\rho/\rho_0)}{\langle\rho/\rho_0\rangle_1^2} . \tag{46.19}$$

An interesting feature of this analysis is the appearance of terms involving the full cluster size R, due to finite-size effects, and the microscopic scale R_0. The dependence on R_0 results from near-field interactions between the source and the nearest neighbors, which gives C_0 a non-universal character. This result shows that the near-field interactions are encoded in long-range intensity correlations of the far-field speckle pattern, or equivalently in the fluctuations of the LDOS fluctuations at the source position.

Beyond the variance, it is also possible to deduce some features of the statistical distribution of the LDOS. Due to the strong dependence on r in (46.13), the LDOS fluctuations are substantially influenced by the interaction with the nearest neighbor scatterer. Expanding (46.13) to leading order in $(k_0 r)^{-1}$, in a regime in which $\rho(r)$ deviates from ρ_0, we obtain

$$\frac{\rho(r)}{\rho_0} = 1 + \frac{3}{4\pi}\frac{\mathrm{Im}[\alpha(\omega)]}{k_0^3\, r^6}. \tag{46.20}$$

From (46.14), we find that $P(r) \sim r^2$. We can now determine the scaling of the probability density of the change in the LDOS $P(\rho/\rho_0 - 1)$ using (46.20) along with a simple change of variables, which leads to

$$P(\rho/\rho_0 - 1) \sim (\rho/\rho)^{-3/2}. \tag{46.21}$$

Based on this power-law dependence, we conclude that the LDOS distribution is broad and asymmetric. We can also see that the size R_0 of the exclusion volume induces a cutoff in the distribution, given by the value of $\rho/\rho_0 - 1$ at $r = R_0$:

$$(\rho/\rho_0 - 1)_{\mathrm{cutoff}} \sim \frac{3}{4\pi}\frac{\mathrm{Im}(\alpha)}{k_0^3\, R_0^6}. \tag{46.22}$$

The existence of this cutoff guarantees the existence of a finite average value and variance of the LDOS in the single-scattering regime. Values of ρ/ρ_0 beyond this cutoff can be reached in the multiple scattering regime. These features of the LDOS distribution have been found in numerical studies and experiments.

References and Additional Reading

Gaussian and non-Gaussian speckle correlations in standard speckles generated by far-field illumination have been discussed in the following reference:

S. Feng, C. Kane, P.A. Lee and A.D. Stone, Phys. Rev. Lett. **61**, 834 (1988).

In this paper, the authors have introduced the Gaussian correlation function as the C_1 contribution, and two classes of non-Gaussian correlations denoted by C_2 and C_3 contributions.

A clear presentation of Gaussian and non-Gaussian speckle correlations can be found in:
E. Akkermans and G. Montambaux, *Mesoscopic Physics of Electrons and Photons* (Cambridge University Press, Cambridge, 2007), chap. 12.

These are the original publications on the C_0 intensity correlation due to an illumination from a point source:
B. Shapiro, Phys. Rev. Lett. **83**, 4733 (1999).
S.E. Skipetrov and R. Maynard, Phys. Rev. B **62**, 886 (2000).

The connection between C_0 correlation and LDOS fluctuations is made in the following papers, for scalar and electromagnetic waves, respectively:
B.A. van Tiggelen and S.E. Skipetrov, Phys. Rev. E **73**, 045601(R) (2006).
A. Cazé, R. Pierrat and R. Carminati, Phys. Rev. A **82**, 043823 (2010).

The following papers contain theoretical studies of LDOS statistics in scattering media:
A.D. Mirlin, Phys. Rep. **326**, 259 (2000).
L.S. Froufe-Pérez, R. Carminati and J.J. Sáenz, Phys. Rev. A **76**, 013835 (2007).
L.S. Froufe-Pérez and R. Carminati, Phys. Stat. Sol. (a) **205**, 1258 (2008).
R. Pierrat and R. Carminati, Phys. Rev. A **81**, 063802 (2010).

These papers report on measurements of LDOS fluctuations in scattering media:
R.A.L. Vallée, M. Van der Auweraer, W. Paul and K. Binder, Phys. Rev. Lett. **97**, 217801 (2006).
M.D. Birowosuto, S.E. Skipetrov, W.L. Vos and A.P. Mosk, Phys. Rev. Lett. **105**, 013904 (2010).
V. Krachmalnicoff, E. Castanié, Y. De Wilde and R. Carminati, Phys. Rev. Lett. **105**, 183901 (2010).
P.V. Ruijgrok, R. Wüest, A.A. Rebane, A. Renn and V. Sandoghdar, Opt. Express **18**, 6360 (2010).
R. Sapienza, P. Bondareff, R. Pierrat, B. Habert, R. Carminati and N.F. van Hulst, Phys. Rev. Lett. **106**, 163902 (2011).
P.D. García, S. Stobbe, I. Söllner and Peter Lodahl, Phys. Rev. Lett. **109**, 253902 (2012).

Exercises

VI.1 For vector electromagnetic waves, the Fresnel reflection and transmission factors for s polarization are defined in Chapter 36. Perform the calculation that leads to (36.19) and (36.20) based on the general expressions of the transverse component of the electric field E_y and on the boundary conditions (36.17) and (36.18).

VI.2 For the boundary condition of the previous exercise, perform the calculations that lead to (36.25) and (36.26) for p polarized electromagnetic waves, based on the transverse component of the magnetic field H_y and the boundary conditions (36.23) and (36.24).

VI.3 It is possible to define Fresnel factors for vector waves in the p polarization based on the amplitude of the electric field, instead of the y-component of the magnetic field. Explain why the transmission factor is modified and becomes

$$T'_p = \sqrt{\frac{\varepsilon_1}{\varepsilon_2}}\, T_p \,,$$

where T_p is the transmission factor defined from the y-component of the magnetic field.

VI.4 Perform the integration in (37.32) that leads to the plane-wave expansion of the electric Green function (37.37).
One approach to this problem is that developed for the calculation of the transverse delta function described in C. Cohen-Tannoudji, J. Dupont-Roc and G. Grynberg, *Photons and Atoms: Introduction to Quantum Electrodynamics* (Wiley, New York, 1989), complement A_I.

VI.5 Derive the expression for the imaginary part of the free-space Green's function in Eq. (38.7), starting from (37.18). Hint: Expand $\sin(k_0 R)/(4\pi k_0 R)$ to third order in $k_0 R$ before performing the derivatives.

VI.6 Derive (37.30). Hint: Use the series expansion of $1/(1-x)$.

VI.7 Using (37.22) and (37.25), compute the radiated power $dP/d\Omega$ for a point source with dipole moment $\mathbf{p}$ that is located at the position $\mathbf{r}_s$ in free space. The current density for the source is given by $\mathbf{j}(\mathbf{r}) = \mathbf{p}\delta(\mathbf{r}-\mathbf{r}_s)$. By performing an angular integration, calculate the total power. Compare the result to (38.8).

VI.8 Make a contour plot the electric-field intensity $|\mathbf{E}(\mathbf{r})|^2$ radiated by a monochromatic electric dipole source, using (38.1). Also plot $|\mathbf{E}(\mathbf{r})|^2$, with $\mathbf{E}(\mathbf{r})$ the electrostatic field created by an electrostatic dipole. Observe that at subwavelength distances from the location of the dipole source, the spatial structure of the field is the same. This behavior is a feature of the quasi-static regime.

VI.9 Consider a small spherical particle with radius $a \ll \lambda$, made of an absorbing material. Show that absorption dominates scattering when $a \to 0$, by comparing the scattering and absorption cross sections.

VI.10 The Lorentz reciprocity relation was derived in Chapter 40 for the case of isotropic media. In anisotropic media, the dielectric function and relative permittivity become second-rank tensors of the form $\boldsymbol{\varepsilon}$ and $\boldsymbol{\mu}$. Following the steps in Section 40.1, derive the Lorentz reciprocity theorem for anisotropic media. Show that the theorem holds provided that $\boldsymbol{\varepsilon}$ and $\boldsymbol{\mu}$ are symmetric tensors.

VI.11 Perform the perturbative calculation of the self-energy for vector electromagnetic waves given by (42.19). Do this by expanding the real and imaginary part of the integrand to leading order, assuming $k_0 R \ll 1$. You can use the results

$$\beta(x) \simeq \frac{-1}{4\pi x^3} + \frac{1}{8\pi x} + \frac{i}{6\pi} + O(x)$$

$$\gamma(x) \simeq \frac{1}{2\pi x^3} + \frac{1}{4\pi x} + \frac{i}{6\pi} + O(x)$$

when $x \to 0$, and expand $j_0(x)$ and $j_1(x)/x$ to second order in x.

VI.12 Starting from (46.15), derive the approximate expressions (46.16) and (46.18) in the limit $k_0 R_0 \ll k_0 R \ll 1$.

VI.13 Consider a semi-infinite disordered material with a correlation length $\ell_c = 50$ nm and a scattering mean free path $\ell_s = 1$ μm. The medium is illuminated by a coherent monochromatic beam with diameter $D = 50$ μm at a wavelength $\lambda = 633$ nm. Based on Chapter 45, draw qualitatively a graph showing the evolution of the speckle spot size as a function of the distance z to the surface of the medium, from the near-zone ($z \simeq 20$ nm) to the far-zone ($z \simeq 10$ μm).

Index